Precambrian Crustal Evolution of China

Springer-Verlag Berlin Heidelberg GmbH

Xingyuan Ma, Jin Bai

Precambrian Crustal Evolution of China

With 119 Figures and 17 Tables

Springer

GEOLOGICAL PUBLISHING HOUSE

Editors:

Professor Dr. Xingyuan Ma
Institute of Geology, State Seismological Bureau
100029 Beijing / China

Professor Jin Bai
Tianjin Institute of Geology and Mineral Resources
300170 Tianjin / China

Technical Editor:
Dr. Andrew C. Cadman
218 Eastfield Road
Southsea, Porthmouth PO4 9EW / UK

ISBN 978-3-662-03699-0   ISBN 978-3-662-03697-6 (eBook)
DOI 10.1007/978-3-662-03697-6

Library of Congress Cataloging-in-Publication Data
Ma, Hsing-yüan. Precambrian crustal evolution of China / Ma Xingyan, Bai Jin.
p. cm.   Includes bibliographical references and index.

1. Geology, Stratigraphic--Precambrian. 2. Earth--Crust. 3. Geology--China.
I. Pai, Chin. II. Title. QE653.M3 1998   551.7'1'0951--dc21

© Springer-Verlag Berlin Heidelberg 1998
© Geological Publishing House Beijing 1998
Originally published by Springer-Verlag Berlin Heidelberg New York in 1998.

Typesetting: Data conversion by authors
Cover: E. Kirchner, Heidelberg
SPIN:10049836   32/3020-5 4 3 2 1 0 - Printed on acid-free paper

Preface

The Precambrian of China contains a remarkably complete record of crustal evolution that spans more than three billion years (3.8 to 0.57 Ga) and represents an unusually diverse array of crustal processes. These suggest that the mode of operation of geological processes has changed progressively with time evolving in a non-uniformitarian pattern, whereby distinctive rock associations and tectonic styles were developed at a particular stage of Precambrian time.

There are three Precambrian platforms, namely the North China (Sino-Korean), Tarim, and Yangtze platforms, which serve as the nuclei of the geological framework of China. They all comprise an older Precambrian basement with varied younger Precambrian intra- and peri-platform mobile belts and platform cover. Additional Precambrian crust lies in numerous smaller inliers or median massifs within the enclosing Phanerozoic mobile belts surrounding these nuclei, among which the larger ones are the Qaidam, Dabie, Jiamusi and the Himalayan massifs or blocks, which are characterized by restricted Precambrian exposures and correspondingly widespread buried basement. The total area of exposed and buried Precambrian crust constitutes about half of the total continental area of China.

A large amount of geological work on Precambrian geology has been done since the founding of New China 43 years ago, particularly during the last decade, when China pursued a policy of opening up - and drawing upon the expertise of - the rest of the world. This policy has sped up the process and yielded significant new results in many fields of Precambrian geology. Many of these results originated from new concepts, new techniques and from increasing multidisciplinary and international cooperative research.

Several years ago Professors Cheng Yuqi and Wang Hongzen made a brief account of the Precambrian geology and geotectonic development of China in their English book *Geology of China* (Yang Zunyi et al. 1986). More recently Professor Goodwin (1991) gave an excellent synthesis on the Precambrian of the world including the Chinese (Cathaysian) Platforms. The time appears to us right to attempt a more integrated treatment of some key aspects bearing on the subject of Precambrian crustal evolution and to give an overview of crustal evolutionary events and tectonic patterns during successive epochs.

The book comprises six chapters. The introductory Chapter (1) considers the distribution of Precambrian rocks and their geological settings; Precambrian orogenies, tectonic cycles, classification scheme and the Precambrian tectonic framework of China. Chapters 2 to 5 work through the Precambrian sequence chronologically: Chapter (2) considers Archean crust; Chapter (3) the Early Proterozoic crust; Chapter (4) the Mid-Late Proterozoic crust; Chapter (5) Sinian crust. The concluding Chapter (6) remarks upon the evolution of the Precambrian continental crust in China. It is aimed to provide an up-to-date survey of the Precambrian crust, primarily concerned with the distribution, lithostratigraphy, geochronology, and petrotectonic assemblages, their structures and broad-scale deformation patterns and petrogenesis, metamorphic pattern and palaeomagnetics.

In contrast to earlier works of this subject in China, the present contribution tries to understand the evolution of the Precambrian crust on the basis of a plate tectonic framework. Globally, the trend has undoubtedly been to interpret Precambrian geology in the context of plate tectonic models. However, the application of plate tectonic processes to Precambrian crustal evolution is still a matter of considerable debate in China. In this respect we will follow Professor Goodwin's advice (1991), namely: "Correctly interpreting the Precambrian record in the light of modern plate tectonics requires a judicious blend of application, adaptation, rejection and innovation, the first two warranted in the presence of certain signatures, the last two in their absence."

Since the geology of the Precambrian embraces a time span of more than three billion years, uncertainties exist almost everywhere. Divergent or even opposing views are often heard concerning a single topic. Thanks to the commonly accepted principle in China - "let a hundred schools of thought contend" - points of argument in Precambrian studies are touched upon in this text, to provide the readers with a more complete perspective.

Given the uncertainties, it is to be hoped that this book can act as both a guide and a spur to earth scientists worldwide specializing in Precambrian geology: firstly to suggest correlations and similarities to other Precambrian terranes, and secondly in the construction of future research projects to test their hypotheses.

Note: Both Ma = 10^6 years and Ga = 10^9 years are used in the text. Chinese names are spelt out according to the Hanyu Pinyin (Chinese spelling). The boundary of China in all the attached figures and maps in the text follows those of Maps of the People's Republic of China published by the Cartographic Publishing House, Beijing in 1980.

Xingyuan Ma
Institute of Geology
State Seismological Bureau
Beijing, China, 100029

Acknowledgements

In preparing such a comprehensive book the active collaboration of a number of worldwide colleagues and experts with expertise in Chinese Precambrian geology is necessary. In this regard we would like to thank Profs. B. F. Windley and A. Kroner for their critical comments and sage advices on the early versions of the manuscript. We are also grateful to Dr. J.D.A. Piper for reading through most of the text and to Dr. James H. Monger for reviewing some chapters and correcting many errors, leading to drastic improvements in the quality of the manuscript.

We are grateful to Profs. Cheng Yuqi, Dong Shenbao, Qian Xianglin, You Zhendong and Tan Yingjia for examining the manuscript; to Profs Wang Hongzen, Shen Qihan, Sun Dazhong, Liu Dunyi, Wang Qichao and Wu Jiashan for stimulating discussions and helpful comments.

Thanks are due to You Zhendong for providing the basic material on the Qinling-Dabie mobile belt of Chapter 3; to Tan Yingjia for his contribution to the Archean geology of the Taihang Mountains in Chapter 2; to Liu Dunyi for the assessment of Precambrian radiometric dates; to Suo Shutian for some figures on the Precambrian geology of South China and to Hu Weixing and Sun Dazhong for their contribution on the Early Proterozoic crustal evolution of the Zhongtiaoshan area in Chapter 3. We also thank Hu Guowei, Guo Jinjing and wang Huichu for their data on the Early Precambrian mobile belts.

The translation of the primary manuscripts from Chinese into English was done by Wang Xichuan, Liu Linqun, Bi Lijun and Wei Chunjing.

We would like to express our particular gratitude to the authority of the Geological Publishing House for the initial motivation for writing this book. We also benefitted greatly from the facilities and professional expertise provided by them. We thank He Man and Shang Hongwei for typing and retyping the manuscripts; and thank Xu Kaiying, Wang Zhaojun and Zhong Xinbao for drawing all the figures.

Last, but by no means least, for their indispensable editorial assistance, we acknowledge our great debt to the responsible editors Bi Lijun and Wang Xichuan.

Table of Contents

1 Introduction

Ma Xingyuan
Institute of Geology, State Seismological Bureau,
Beijing, China, 100029.

1.1 Distribution of Precambrian Crust in China

The bulk of China's Precambrian crust is located in the North China, Tarim and Yangtze Platforms, which are characterised by restricted Precambrian exposures and widespread buried basement. Additional Precambrian crust lies beyond the platforms in several median massifs and numerous inliers within the enclosing and closely compressed Phanerozoic fold belts (Figs. 1.1; 1.2).

Together the three platforms are distributed across a region that from east to west measures 5000 km in length, and up to 2200 km in width. They occupy about 40% of China's territory, which approximates an area of 4 million km². Geographically they extend from the Sino-Korean border at long. 130°E in the east to the western limit of the Tarim Basin at long. 75°E, and from the southern Jilin area at lat. 43°N in the north to Yunnan at lat. 24°N in the south (Fig.1.2). Their individual dimensions are summarised below:

(a) The North China Platform is the Chinese part of the Sino-Korean Platform. Within China the extent of the platform covers most of the North China, southeastern part of northeast China, Inner Mongolian, Bohai Bay and part of the Yellow Sea regions. It is approximately 1.7 million km² in area and forms an elongated latitudinal triangle which tapers westward for 2300 km towards the Alxa area.

(b) The Tarim Platform is lozenge-shaped, being 1700 km long and a maximum of 600 km wide, it covers an area of around 710,000 km². It is situated between the Tianshan Mountains in the north and the Kunlun Mountains in the south.

(c) The Yangtze Platform is a SW-trending irregular ellipse, about 2000 km long by 300-800 km wide with an area of 1,560,000 km². It covers nearly the whole of the central-lower Yangtze River catchment area and the southern part of the Yellow Sea.

Apart from the three major Precambrian Platforms, there are Precambrian massifs and crystalline axial uplifts in the surrounding Phanerozoic fold belts (Fig. 1.2). The largest of these are the Upper Heilongjiang and Jiamusi Massifs in northeast China (the latter of which is the Chinese segment of the Bureya Massif in Russia). In addition, the roughly rhomb-shaped Songliao Massif is entirely covered by Mesozoic and Cenozoic terrestrial sediments, but aeromagnetic and other geophysical surveys have revealed the existence of Precambrian basement underneath the cover.

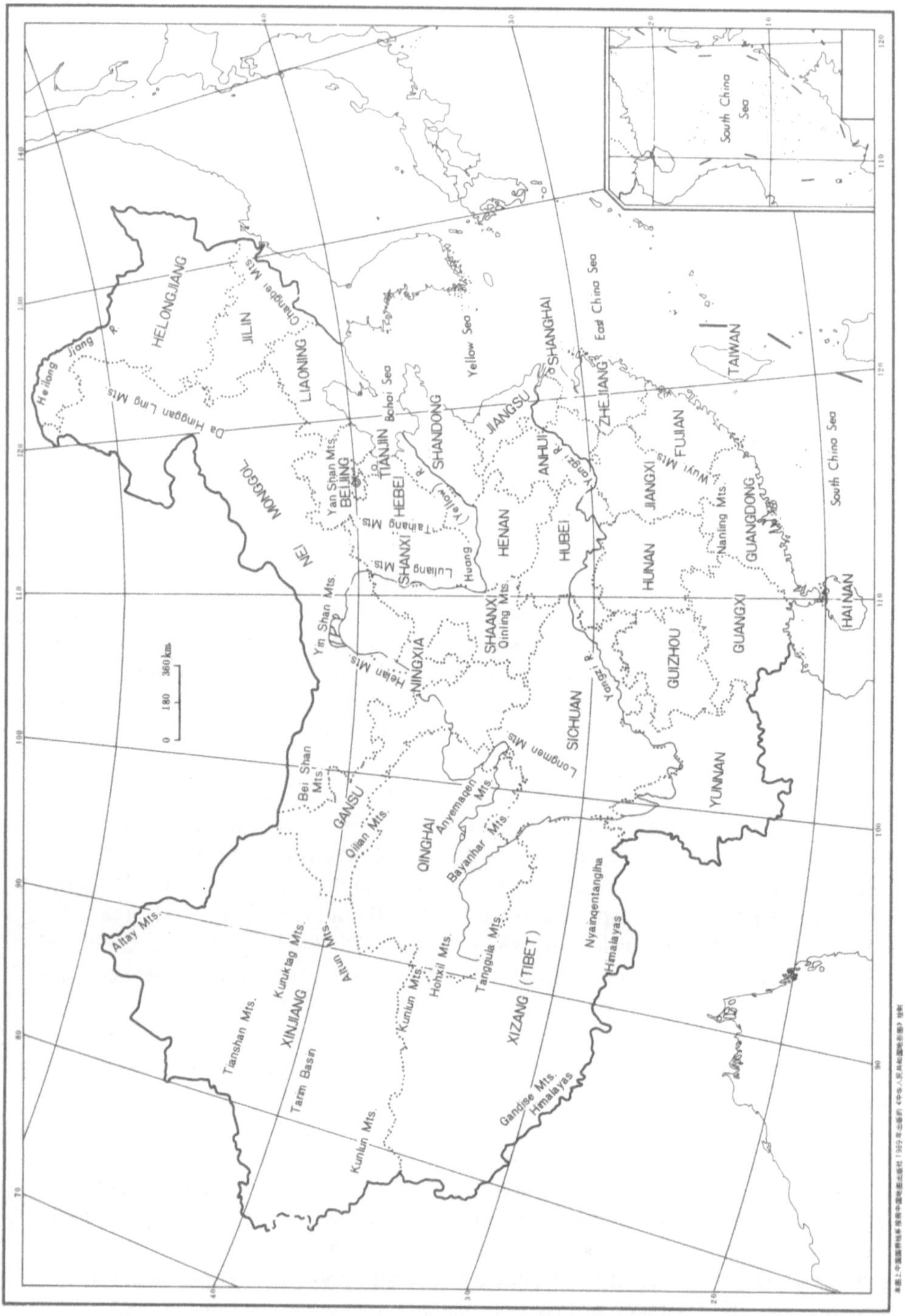

Fig. 1.1. Political provinces and geographical divisions of China

In north-western China the large Qaidam Massif is situated in the Qinghai Province and the much smaller triangle-shaped Yining Massif in the north-west of the Xinjiang Uygur Autonomous Region. There are also several long belts of Precambrian uplifts exposed along geoanticlinal axes in the Tianshan, Kunlun and Qilian orogenic belts.

In addition, Mid-Proterozoic ages have been confirmed for crystalline rocks south of the Yarlung River. These rocks form part of the Gangdise Massif, which along with the Himalayas were formerly part of Gondwanaland. Rocks of the Gangdise Massif also occur to a limited extent on the northern side of the Yarlung Zangbo River. In the Damxung region, basement rocks composed of the Middle Proterozoic Nyaiqentanglha Group are exposed and extend at least to the west of Xainza. Furthermore, the Qangtang Massif in northern Tibet has been regarded as Precambrian by Wang Hongzhen (1978), a deduction based mainly on regional stratigraphical analysis. Further east along the meridinal orogenic belts in the western Yunnan and Sichuan Provinces, there are several small Precambrian massifs, among which the Lincang Precambrian Median Massif is particularly well defined.

There are also a series of metamorphic terranes along the axis of the Qinling fold belt between the North China Platform and the Yangtze Platform in central-east China. Noted among these are the Qinling metamorphic core complex of Lower Proterozoic age, the Dabie metamorphic terrane and the Su Bei-Jiao Nan (northern Jiangsu-southern Shandong) terrane of Archean-Proterozoic age in the east.

1.2 Precambrian Tectonic Cycles and Precambrian Classification Scheme

The idea of tectonic stages is intimately connected with that of tectonic cycles. Huang Jiqing (1978) has advanced a complete subdivision of the tectonic cycles in China. He proposed a polycyclic model for the development patterns which differ from one another due to changes within the various stages of crustal evolution. Major epochs of orogeny and magmatism are commonly accepted as the key markers in Precambrian history, identified by distinctive tectonic events and radiometric age dating. In the latter respect, great advances have been made in the last decade. Apart from the work done by various scientists and academic departments, the Ministry of Geology and Mineral Resources of China organised a special Working Group (referred to as Working Group hereafter) on this subject and published a monograph entitled "The Isotopic Time Scale of China" in 1987. Since then a large number of radiometric dates of various Precambrian rocks have been published, which improved considerably the geochronological classification of Precambrian rocks.

The Archean rocks of China are highly complex and their formation involved several periods of supracrustal deposition, polyphase granitoid emplacement, prograde and retrograde metamorphism and multiphase deformation. Therefore, at this

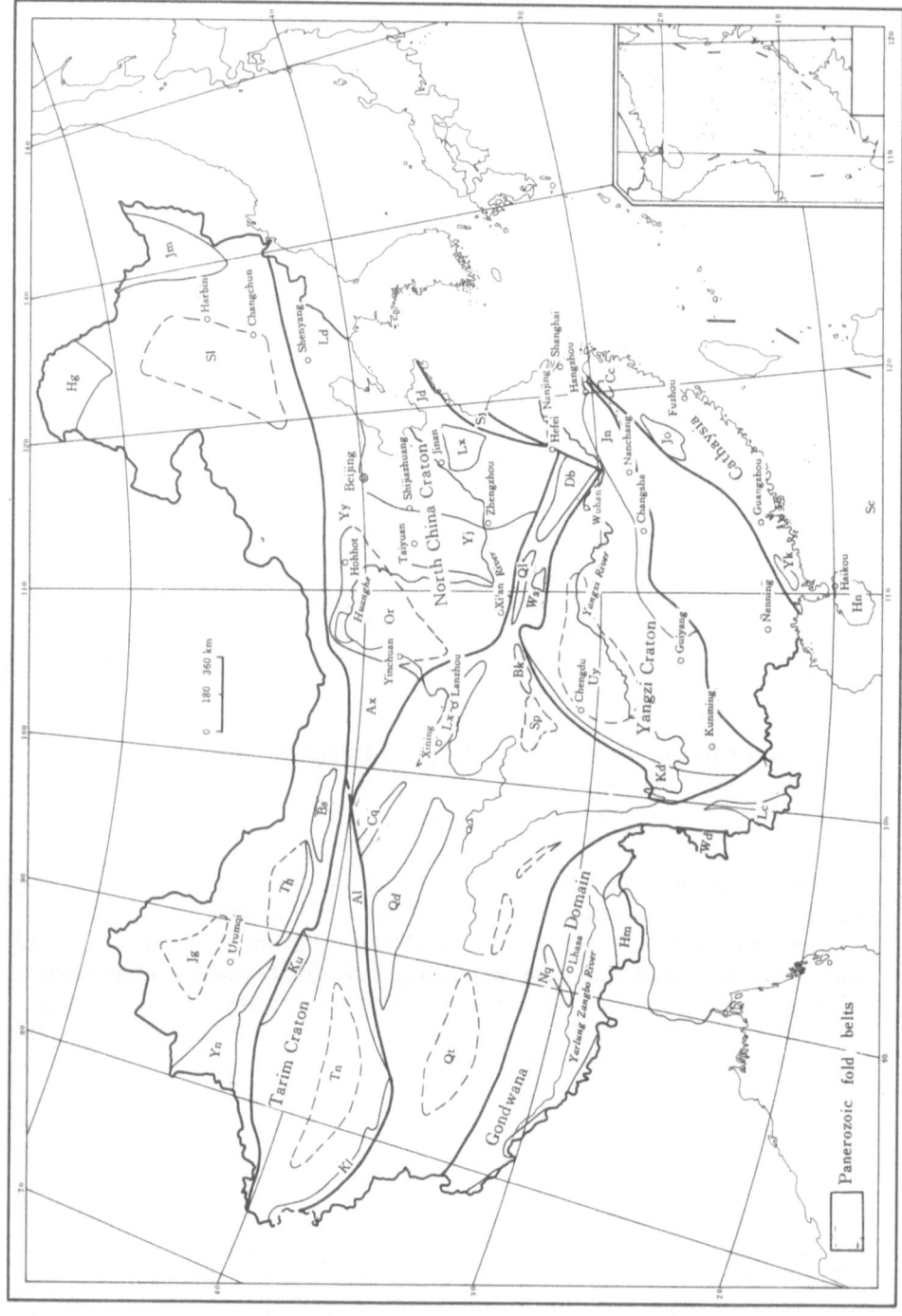

Fig 1.2. Main Precambrian geological divisions of China

- Northeast China: Hg-Northern Hinggan Uplift; Jm-Jiamusi Massif, Sl-Songliao Massif;
- North China Craton: Ax-Alxa Massif; Ld-Liaodong Uplift; Jd-Jiaodong Uplift; Yj-Yu-Jin (Henan-Shanxi) Uplift; Or-Ordos Nucleus; Yy-Yinshan-Yanshan Uplift; Lx-Luxi (western Shandong) Uplift.
- Qinling Mobile Belt: Sj-Subei-Jiaonan Terrane; Db-Dabei Massif; Ql-Qinling axial complex; Ws-Wudangshan Uplift; Bk-Biekou Uplift.
- Yangzi Craton: Uy-Upper Yangzi Nucleus; Kd-Kangdian Uplift; Jn-Jiangnan Uplift.
- Cathaysia: Cc-Chencai Uplift; Jo-Jianou Uplift; Yk-Yunkai Uplift; Hn-Hainan Uplift Sc-South China Sea Massif.
- Northwest China: Jg-Junggar Massif; Yn-Yining Massif (Central Tianshan Uplift); Bs-Beishan Uplift; Th-Turpan-Hami Massif
- Tarim Craton: Tn-South Tarim Nucleus; Ku-Kuruktag Uplift; Al-Altun Uplift; Kl-West Kunlun Uplift.

time it is only possible to make a tentative three-fold subdivision, and to set the mutual boundary between the Late and Middle Archean at 2.9-3.0 Ga, and the boundary between Middle and Early Archean at 3.5-3.6 Ga, due to the paucity of critical geochronological constraints. Only a small volume of very old crust has been detected; - in the Caozhuang area, eastern Hebei Province and near Anshan in Liaoning Province (Fig. 1.3). In the Caozhuang area detrital zircons in an Archean metaquartzite are 3.55 Ga or older in age, with about one fourth having ages between 3.8 and 3.85 Ga. Near Anshan, sheared gneiss occurs in a complex containing ca. 3.3 and 3.0 Ga granites. Some of the zircons in the mylonitized gneiss are concordant with a weighted mean $^{207}Pb/^{206}Pb$ age of 3804±5 Ma(2σ), interpreted to be the protolith age of the gneiss (Liu Dunyi et al. 1990, 1992).

In contrast, the late Archean and Proterozoic records have been much better chronologically analysed, and it is generally suggested that the greatest frequency of radiometric dates correspond to the terminal events of major crustal processes. In the platform areas four main Precambrian orogenies have been recognised, terminating approximately at 2.45-2.6, 1.7-1.85, 1.0-1.05, 0.8-0.85 Ga respectively. (See Table 1.1 for the working scheme used). The typically time-transgressive nature of orogenies (Plumb and James 1986) is clearly shown by the range in termination ages for each orogeny, which vary with location. These four terminal events delimit respectively the boundaries between the Archean and Proterozoic Aeons, the Early- to Mid-Proterozoic timescale, the Mid- to Late Proterozoic timescale, and the advent of the Sinian Era within Late Proterozoic time. In addition, the generally accepted date of 0.57 Ga is taken to mark the Precambrian-Cambrian boundary.

The 2.9-3.0 Ga Qianxi Orogeny is expressed by development of granulite facies gneisses exposed mainly in the northern part of the present North China Platform, in southern Inner Mongolia, northern Hebei, eastern Liaoning and southern Jilin Provinces. Radiometric and lithological data imply that this belt was subjected to a major tectono-thermal event. This event marks the boundary between the Middle and Late Archean with the formation of several separate cratonic nuclei in the northern and eastern part of the craton. However, it is more difficult to know where the boundary between the Archean and the Proterozoic Aeons should be drawn. Previously, the 2.5 - 2.6 Ga Fuping Orogeny was used in this context, as it involved both widespread granitoid emplacement into - and amphibolite-granulite facies metamorphism of - various lithologies, including supracrustal rocks. This orogeny also led to the process of progressive aggregation of the small cratonic nuclei and eventually built up the core of the accreted embryonic craton (Ma and Wu 1981a).

However, further study reveals that stratigraphical relationships are in conflict with the commonly accepted chronological boundary between the Archean and Proterozoic Aeons in the North China Platform. Within this region is situated the Wutai Group, which consists of a greenstone belt with associated granites. The Group is clearly unconformable with both the overlying Hutuo Group and the underlying Fuping Group (Photos 1.1, 1.2). Age dating of the Wutai Group show it to straddle the Archean-Proterozoic boundary. Among the older ages yielded by isotopic dating, four range between 2.1-2.4 Ga. Until recently, the oldest date was from the lower part of the Wutai Group, which gave an Rb-Sr whole rock isochron age of

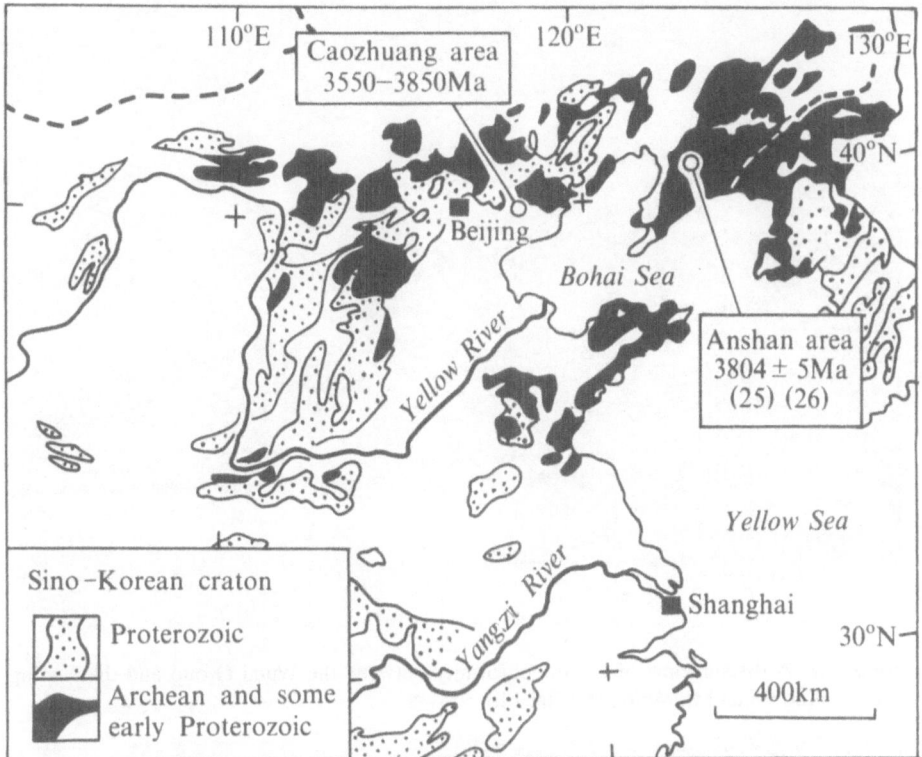

Fig. 1. 3. The oldest rocks found in China (After Liu Dunyi et al. 1992)

2522±124 Ma (Bai Jin 1986). However, recently Bai et al. (1991) published another two dates from the Wutai Group: the results showed that the zircons from the Jingangku Formation, which forms the lower stratum of the Wutai granite-green-stone belt in the Shentangbao area, have a metamorphic age of 2438±36 Ma (2σ). Furthermore, zircons from the Jingangku-Liuhuangchang garnet-gneiss, which was emplaced into the base of Wutai Group in the Jingangku area and then folded, have an age of 2607±36 Ma (2σ).

Thus, it appears that if we take the Archean-Proterozoic boundary to be 2.5 Ga, then the lower part of the Wutai Group is of Archean age (Wang Hongzhen and Qiao Xiufu 1984). However, it is to be expected that the age range of a geological unit may not necessarily comply exactly with international stratigraphic nomenclature. The Wutai Group may therefore serve as a good example that there is a diachronous transition between the Archean and Proterozoic eras. This is also supported by the fact that the underlying Fuping Group is characterised by the formation of large oval structures and gneissic domes, in contrast to the more modern style of linear mobile belts. The occurrence of banded iron formation (BIF) in the Wutai Group is clear evidence for the evolution of an oxygen rich atmosphere during Early Proterozoic time. The Proterozoic may be regarded as the main time of formation of the North China Platform: its development started in Early Proterozoic time (2.6-2.5 Ga) and

Photo 1.1. A distant view of the unconformity between the Wutai Group and the Fuping Group in the Wutai Mt. region (by Bai Jin)

Photo 1.2. The unconformity seen in Photo 1.1 runs across the man's waist in the centre of the picture. The thin-bedded biotite-schist and pebble-bearing feldspathic quartzite beds of the Wutai Group are observed to be gently dipping in vertical section

Eon	Era	Ga	North China Group/System	North China Orogeny	South China Group/System	South China Orogeny	IUGS Eon	IUGS Era	IUGS Period
Proterozoic	Late	0.57	Cambrian		Cambrian	Chengjiangian	Proterozoic	Base of Cambrian	Neoproterozoic III — 0.65 — Cryogenian — 0.85 — Tonian
		0.7	Sinian		Sinian	Jinningian		Neoproterozoic	
		0.85–0.8	Qingbaikou		Banxi	Sibaon		— 1.0 —	— 1.0 — Stenian — 1.2 — Ectasian — 1.4 — Calymmian
	Middle	1.1–1.0	Jixian		Sibao			Mesoproterozoic	
		1.4	Changcheng		...?...			— 1.6 —	— 1.6 — Statherian — 1.8 — Orosirian — 2.05 — Rhycian — 2.3 — Siderian
	Early	1.7	Guojiazhai	Zhongyuean	Dahongshan			Paleoproterozoic	
		1.8		Lüliangian					
			Hutuo	Wutaian	...?...			— 2.5 —	— 2.5 —
Archean	Late	2.4–2.3			Kangding		Archean	Neoarchean	
		2.6–2.5	Wutai	Fupingian				— 2.8 —	
	Middle	3.0–2.9	Fuping	Qianxian				Mesoarchean	
			Qianxi					— 3.2 —	
	Early	3.6–3.5	Caozhuang					Paleoarchean — 3.6 — Eoarchean	

Table 1.1. Summary tectonic and chrono-stratigraphic development of Precambrian crust in the platforms of China and the resulting Precambrian classification scheme followed in this book

ended before the Sinian Era (0.85-0.8 Ga). It may be subdivided into two stages: Early Proterozoic deposition is widely represented by the Wutai and Hutuo Groups. Subsequently, the Wutai'an Orogeny 2.3-2.4 Ga ago involved widespread granitoid emplacement in the North China Platform which effectively cratonized the basement. Additional cratonization after Hutuo sedimentation occurred during two phases of the Lüliangian (Zhontiaoan)-Zhongyuean Orogeny 1.8-1.7 Ga ago. The 1.7 Ga Zhongyuean Orogeny maybe regarded as the second phase of the main orogeny at the end of Early Proterozoic time (Ma Xingyuan et al. 1981; Wang Qichao, 1988). This completed consolidation of the North China cratonic basement with the formation of a protoplatform. Subsequently it experienced a protoplatform history, receiving a cover similar to a platform succession during the Middle Proterozoic, and a genuine cover in Late Proterozoic time. There are no widespread significant orogenies during Mid-Late Proterozoic time in North China except (1) the Xiong'er Orogeny 1.4 Ga ago (Jia Chengzao 1988), which is represented by the unconformity between the Ruyang Group and the Xiong'er Group in western Henan Province; (2) the Yinshanian Orogeny (Ma Xingyuan et al. 1989b), also termed the Seerteng Orogeny (Ma Xingyuan et al. 1989a). This orogeny resulted in the folding of the Mid-Proterozoic Chartai Group of the Chartai Aulacogen in Inner Mongolia, and the later widespread late Mid-Proterozoic Shinagan transgression. The orogeny was coeval with the Qinglong epiorogenic uplift in eastern Hebei Province. To the south along the margins of the Yangtze Craton occurs a series of Mid- to Late-Proterozoic (pre-Sinian) turbidite assemblages of mainly assorted volcanic rocks and turbidites with associated ophiolites. These are interpreted as products of island-arc and marginal sea environments which probably formed near continental margins. These were accreted to the Yangtze Protoplatform during the Sibaoan (1.0-1.1 Ga) and Jinningian Orogenies (0.85-0.8 Ga) resulting in the final consolidation of the platform prior to widespread Sinian cover.

1.3 Precambrian Tectonic Framework of China

The tectonic classification system of Precambrian rocks used in this work is based on the recognition of successive tectonic imprints on Precambrian terranes of substantial size. The crystalline basement of the Precambrian platform contains rocks of several ages which have been involved in more than one period of orogeny. The last major orogeny not only imposed the presently distinctive structural style but also tectonic stability (cratonization). Thus the assembled tectonic units of a Precambrian platform bear witness to its progressive cycle-by-cycle tectonic cratonization.

In general, each tectonic cycle preferentially affected a particular part of the platform, - commonly a linear fold belt or a block - and resulted in a mosaic of structural provinces, each with characteristic age or limited range of ages. These structural provinces have been assembled by the processes of "continental accretion". The way in which they have been accreted is denoted by prefix; i.e.- upon (epi-), within (intra-), around (peri-) and under (sub-) the pre-existing continental crust (Goodwin, 1991). The recognition of different structural provinces and tectonic elements is based largely on differences in overall structural trends and style of folding, which are products of the last radiometrically dated major orogeny to affect the rocks in a particular province.

Continental reconstructions are of fundamental importance to fully understanding the dynamic evolution of the continental crust. According to plate tectonics the continents are in general motion relative to one another, reconstructions resolve the pattern of Wilson cycles formed by the recurring fission and fusion of supercontinents. However, owing to the paucity of reliable Precambrian paleomagnetic data, little is known about the relative craton distribution in Precambrian time, although some speculations may be made according to geological syntheses. These indicate that the platforms in China were not fully assembled in their present positon until Early to Middle Mesozoic time. In the following paragraphs we will give a brief account of their geological development and tectonic framework.

Early Precambrian high-grade metamorphic terranes and granite-greenstone belts are well developed in the North China Craton. Although mainly of Archean age, some of them continue to develop in Early Proterozoic time. They are well exposed in the upturned northern and southern margins and in the meridional Shanxi Plateau and in the Liaodong and Jiaodong Peninsulas. Following the Wutain Orogeny at 2.3-2.4 Ga, a series of intracratonic troughs developed, ranging in age from 2.3 to 1.8 Ga. These are characterized by an up to 10-km-thick sequence of low grade metamorphosed clastics, dolomitic carbonates, and basic volcanic rocks, including alkalic varieties. At 1.8-1.7 Ga, the Lüliangian-Zhongyuean Orogenies completed consolidation of the North China Protoplatform. The resulting Early Precambrian crystalline basement is unconformably overlain by Mid- to Late-Proterozoic strata.

The character of Middle and Upper Proterozoic stratigraphies reflects clearly the difference in tectonic environments. The thick accumulation in the Yanshan Aulacogen continues into the northen Taihang Mountains and is partly manifested by the Middle Proterozoic flyschoid deposits, including the Dahongyu potassium-rich alkalic volcanics and Wumishan flyshoid deposits. Another type of aulacogen developed in southern Shanxi and western Henan Provinces, where transgressions reached a maximum width in the Late Changchengian as evidenced by the overlap of the Gaoyuzhuang Formation. The Upper Proterozoic Qingbaikou System is confined to the Yanshan and Jiaoliao region and forms a genuine cover sequence.

The present Tarim Platform is bounded by the South Tianshan Mts. in the north and by the West Kunlun Mts. in the south. The main part of the Tarim Platform is covered by Cenozoic deposits in the basin. Basement rocks are well exposed in the bordering highlands, notably in the Kuruktag and Kalpin regions in the north, as well as in the northern foothills of the Kunlun Mts. south of Yecheng, Hotan and Minfeng Counties and in the Altun Mountains in the southeast. Aeromagnetic surveys have revealed an ancient massif in central Tarim, the southern part of which may well represent an Archean nucleus (Fig. 1.2). In the Kuruktag region an orogeny at 2.5 Ga was responsible for the cratonization of the Tarim Platform. The Mid- and Late-Proterozoic rocks also constitute a part of the folded basement in the

Kuruktag region. Yang Zunyi et al. (1986) referred to the final consolidation of the whole platform as the Jinningian Orogeny (also locally refered to as the Tarimian Orogeny). More recently, in the Kuruktag region, a 3.3 Ga Sm-Nd isochron age of 3263±129 Ma (2σ) has been obtained from amphibolite enclaves in the grey gneiss of the Toklablak Group, which indicate the presence of rocks of Mid-Archean age in the Tarim Craton (Hu Aiqin et al. 1992). In comparison, the basement of the Qaidam Massif was mainly consolidated in the Jinningian stage, as shown by the Sinian cover sequence formed by the Guanji Group in Oulonbruk.

The Yangtze Platform began as a stable basement that was consolidated at the end of Early Proterozoic time. The basement is inferred from geological and geophysical data to occupy the present site of eastern Yunnan, central Sichuan and western Hubei Provinces. This micro-continent became the nucleus for a series of arc-trench and basin systems developed along its margin. It was initially stabilized during the Sibaoan Orogeny at about 1.05-1.0 Ga, and completely consolidated after the Jinningian Orogeny of 850-800 Ma ago. These elements together form the basement of the Yangtze Platform, with the overlying Sinian System forming a platform cover.

The Qangtang Massif of the Xizang Autonomous Rigion (Tibet) was regarded as Precambrian by Wang Hongzhen (1978). This age assignment is based mainly on the regional stratigraphic analysis, but various elements of complicated structures remained relatively uninvestigated. South of the Yarlung Zangbo River the stratigraphic sequence is fairly well established, where a Middle Proterozoic age for Precambrian crystalline rocks has been confirmed (Liu Guohui et al. 1990). Isotopic ages of 658.8 Ma, 640-660 Ma have obtained by different methods in the Nyalam Group of the crystalline basement of the High Himalayas, which confirmed that the basement was consolidated in Late Proterozoic (Wei Guanyi et al. 1989). On the northern side of the Yarlung Zangbo River, the Gangdise Massif is limited in extent. The basement rocks (the Middle Proterozoic Nyaiqentanglha Group) are exposed in the Damxung region and extend at least to the west of Xainza (Liu Guohui et al. 1990).

In pre-Sinian times the crust to the south of the Yangtze Platform was mainly oceanic, but uplifted islands that supplied detritus to nearby regions seem to have existed. The exposed Jianou Group of the northern Fujian Province and Chencai Group in western Zhejiang Province were assumed by Grabau (1924) to be an old land mass of Archean-Proterozoic age and was named Cathysia. In recent years many geologists have confirmed the existence of a Precambrian crystalline basement in the southeastern coastal region of China, including the Hainan island (Zhang Renjie et al 1990). Shui Tao et al. (1987) reported that the ages of Precambrian rocks in Cathysia range from 2000 to 900 Ma, with an unconformity between the upper and lower sequences occurring at about 1.4 Ga. The unconformity is thought to indicate an orogenic event termed the Cathysian Orogeny. Shui Tao et al. (1987) also postulated that Cathysia might be an integral part of a huge continental mass which ran along the western Pacific margin and is submerged beneath the East China Sea and northern part of the South China Sea. Ages of 1.6 Ga for the metamorphic rocks from two boreholes in the East China Sea have been reported (Liu Guangding et al. 1986), and an Rb-Sr isochron age of 1465 Ma has been obtained from migmatites in a drill hole on the Xisha Islands in the South China Sea (Zhang Renjie et al. 1990). According to these data and stratigraphic analysis of the southernmost part of Hainan Island, Ren Jishun et al. (1987) postulated the existence of a South China Sea Platform, which might be linked to the Kontum Massif of Indo-China. Recent isotopic dating by Hu Xiongjian et al. (1992) has confirmed the existence of early Proterozoic granites in the Badu Group and Mayuan Group of southwestern Zhejiang and northern Fujian Provinces. This indicates that the basement of southeast China was consolidated by 1.8 Ga.

2 Archean Crust

Bai Jin and Dai Fengyan
Tianjin Institute of Geology and Mineral Resources
Tianjin, China, 300170.

2.1 Characteristics of Early Precambrian Evolution

Since the advent of the concept of plate tectonics, heated disputes have occurred about whether and how to apply this concept in the study of Early Precambrian tectonic environments. As the Early Precambrian terranes originated in the remote geological past and have been modified by polyphase deformation and metamorphism and denudation, they show extreme complexity with respect to petrology, geochemistry and tectonics. This has made it very difficult to identify the tectonic environments of the Early Precambrian mobile belts. Hence geologists have proposed a variety of models for their tectonic environments (Fyfe 1974, 1978; Willias 1977; Glikson 1971; Goodwin and Ridler 1970; Anhaeusser 1971, 1973; Condie and Hunter 1976; Burke et al. 1976; Condie and Baragar, 1974; Windley, 1977, 1984; Tarney, 1976; Kröner 1983a; Ma Xingyuan et al. 1984). Most of the models are based on petrotectonic assemblages and their geochemical characteristics. Petrotectonic assemblages, however, can usually be given more than one interpretation (Kröner 1983a). In some cases, regional tectonic features are also taken into consideration, but it is difficult to draw any convincing conclusions owing to the lack of information of tectonic boundaries and geological setting based on field observations. Strictly speaking, even today our understanding of the history of early crustal evolution is still speculative. Nevertheless, the general tendency is for the application of plate tectonic mechanisms is being pushed further back in time.

In terms of global perspective, the early stage of Earth's history crustal differentiation occurred in permobile tectonic environments. For example, a considerable thickness of stable cratonic cover sediments began to be deposited about 3.0 Ga ago in the Kaapwaal basin of South Africa (Anhaeusser 1973; Vajner 1976). Within high grade terranes, examples of stable continental-shelf sediments are frequently found (e.g. the khondalite series). Although underplating by mantle-derived, differentiated magmas related to upwelling plumes or plate subduction might have occurred, the presence of these metasediments suggest that it would be impossible for most of the Archean crust to have formed by vertical accretion. Although the available paleomagnetic and chronological data are insufficient to prove that Archean microplates were involved in lateral crustal accretion processes or to estimate the rate of the crustal growth (Kröner 1989), some paleomagnetic data from Archean cratons

(McElhinny and Senanayake 1981; Dunlop 1981; Kröner 1983) indicate that continental plates have been moving since the Early Archean times (Kröner 1989). Moreover, the heat-generating elements K, U and Th were much more abundant in the Archean than in the present day, and it has been estimated that the global heat flow during the Archean was two to three times greater than at the present (Lambert 1981; Smith 1981; Miyashiro and Sengor 1982). Evidence for production and dispersal of substantial heat in the Earth (Bickle, 1978; Burke and Kidd 1978) suggests that the oceanic lithosphere is the main site of heat release. An additional and very striking tectonic feature is the early horizontal shear-compressive structures developed extensively in Archean terranes. The existence of these fundamental phenomena means that mechanisms analogous to plate tectonics have been operating since this very early period of Earth's crustal history. Thus, we can study the tonalite-trondhjemite-granodiorite (TTG) suite and, in some cases, regard it as a magmatic arc formation associated with plate subduction. Any difference from modern TTG suites may be considered as a clue to the different character of the crustal evolution during the Archean.

The dispute over whether uniformitarianist principles are applicable to the Archean arose following disputes over tectonic environments has now reached a critical stage. Both the uniformitarianists and the nonuniformitarianists are trying to interpret past tectonic settings with the present, better understood tectonic environments. Practice has demonstrated that the correct method of thinking is to have a thorough understanding of the theory of geological development by stages. If uniformitarianism can tolerate this theory of stage division, an event that looks like a catastrophic change can be part of a process of uniform change if such an event becomes a phenomenon that recurred cyclically. It thus follows that to apply the principle of uniformitarianism in an analytical and historical way in connection with the stages of geological development and actual geological conditions should be a basic criterion for understanding the history of Archean crustal evolution (Bai Jin 1986).

Generally, it is accepted that the Archean crust was thin, plastic and had steep thermal gradients. However, the identification of the tectonic environment of the mobile belts during the Archean must be based on intensive field geological observations and a synthetic study of the regional geophysical anomalies, petrotectonic assemblages, regional tectonic styles, paleogeothermal regime and regional geological settings. Only in this way can a number of interpretations be deduced and a relatively definite conclusion drawn.

Unlike Phanerozoic mobile belts, most of the Archean mobile belts fail to display pronounced linear features on geomorphologic or geologic maps because they have been subjected to various tectonic disturbances and later capping by sedimentary sucessions. Therefore, only by using the regional gravity data and aeromagnetic anomalies (screened for the interference caused by Phanerozoic geological events), coupled with data on regional geological settings, is it possible to indicate the existence of mobile belts and their boundaries and to explain whether they and their surrounding terranes had once belonged to a single unit.

As far as Phanerozoic records are concerned, petrotectonic assemblages, including their geochemical characteristics, can provide relatively explicit information of the tectonic environments in which they were formed. They can also show features associated with plate boundaries or special positions within a plate (Dickinson 1971; Condie 1989). Therefore, a once-popular practice was to identify the tectonic environment purely according to petrotectonic assemblages. However geological settings were very complex during the Archean, and some petrotectonic assemblages can permit several different interpretations. For example, for Archean magmatic rocks, Hargraves (1976) suggested that eruption of basaltic magma occurs in areas of lithospheric splitting and mantle upwelling and that many ultramafic volcanic rocks in greenstone belts may be considered to be indicative of the predominance of high geothermal gradients. Some geologists (e.g. Anhaeusser 1981) hold that Archean komatiite and basalt represent ancient oceanic crust, but other data indicate that these rocks tend to be erupted onto the granitic basement (Nisbet et al. 1977) in association with calc-alkaline volcanic rocks. They occur in shallow-water environments (Barley 1981) or alternatively with continental-type sediments (Archibald et al. 1981). Many greenstone belts contain not only rocks of the tholeiitic and alkali series but also rocks of the calc-alkali series. The Phanerozoic calc-alkali series mainly occurred in island arcs and on active continental margins and is considered to be genetically related to plate subduction zones (Condie 1982), while the magma of the Archean calc-alkali series may have been produced in some other way. For instance, Goodwin (1977) suggested that the downwarping of thin lithosphere with a high geothermal gradient may promote partial melting of mantle peridotite, thus bringing about successive eruptions of tholeiitic and calc-alkali magmas. Some studies even suggest that some calc-alkali rocks originated as part of the igneous rock associations within anorogenic belts in Phanerozoic times (Miyashiro and Shido 1975). Windley (1984), however, held that sedimentation, magmatism and metamorphism are bound to be controlled by similar physical-chemical rules throughout the Earth's history. During the Archean, a certain type of plate subduction could produce a large amount of calc-alkaline magma. Thus rock assemblages may allow more than one interpretation of the tectonic environment, and alone are insufficient to identify Early Precambrian tectonic environments. However, petrotectonic assemblages are still material records: no matter what disturbances may have happened during later geological processes, their content and nature are still one of the main objects of study in our identification of tectonic environments. However, in comparison with the characteristics of modern petrotectonic assemblages, we must consider more factors and then analyse them in connection with other background information to ascertain the tectonic environments in which they were formed.

The regional tectonic style is a record of the tectonic movement under a given stress state, and generally may indicate the dynamic and kinetic process of a mobile belt. Different tectonic styles may result from different tectonic environments. Therefore, in identifying paleotectonic environments, the study of regional tectonic styles may furnish very valuable evidence. For example, the close recumbent fold and its associated ductile thrust nappe structure mainly reflect horizontal shearing and compression in subduction zones and collision zones. Some shear-zones probably represent major tectonic boundaries. In contrast, the continental rift does not signify an orogenic environment. Even if it represents the initial stage of the Wilson cycle, its closure is still accomplished through subduction of the continental-margin oceanic crust on either side. This is also one of the important reasons for putting forward the model of A-subduction of ensialic crust. If folds are present, in principle they should be considered to be the result of interference from later crustal movements. Their specific styles vary with the different kinematic mechanisms of associated tectonic movements, but often the style will be marked by a fold system with steep-dipping axial planes formed in the process of lateral compression.

Study of more recent geological times indicates that accretion and subduction of oceanic plates are the main way that the heat of the Earth is dissipated (Bickle 1978). Regional metamorphism is also intimately associated with the tectonic environment and its evolution (Miyashiro 1973; Dong Shenbao et al. 1986). Therefore, the distribution pattern of paleo-isothermal planes resulting from heat dissipation is a reflection of the paleogeothermal regime resulted from the tectonic mechanism. For example, the "geothermal trough" and "geothermal ridge" are characteristic geothermal regimes of the subduction zone and mid-ocean ridge zones respectively. It is generally considered that convection in the asthenosphere and the action of related heat source lead to the pattern of geothermal distribution. A particular pattern of geothermal distribution that reflects metamorphic conditions is bound to signify a particular tectonic environment. The parallel arrangement of metamorphic minerals forms the syntectonic foliation, and the kinematic coordinates of their fabrics signify the same kinematic significance as the regional tectonic style. This is convincing evidence for the simultaneous operation of metamorphism and structural deformation. On the other hand, the spatial distribution of a metamorphic facies zone shows that the metamorphic grade is a strong reflection of the syntectonic geothermal regime at that time. Generally, the Archean terranes have all undergone polyphase metamorphism and deformation which masked their initial geothermal regimes. Therefore, a relatively correct conclusion about the paleogeothermal regime can be arrived at, only by reconstructing of the early structural features and through a study of the distribution of the paleogeotherms and metamorphic facies zones. This method may be called "geothermal structural analysis" (Bai Jin et al. 1990).

In summary, to identify the paleotectonic environments of the Archean mobile belts, it is necessary to conduct an all-round, comprehensive analysis and study of the regional geological settings, geophysical anomalies, petrotectonic assemblages, regional structural styles and paleogeothermal distribution patterns of the mobile belts. Only in this way can a relatively correct conclusion be drawn. In carrying out a comprehensive study, it is also necessary to consider the characteristics of the Archean primary crustal blocks, - such as their small scale, thin crust, high plasticity and steep geothermal gradients. These characteristics, and the severe denudation that may have taken place over geologic time, may be used to measure phenomena that appear in the past but are absent at the present time or vice-versa: for example, blueschist is mostly restricted to Phanerozoic high-pressure metamorphic zones, but it is absent in presently exposed Archean mobile belts; the metamorphic grade of the Archean mobile belts that is demonstrated now is commonly higher than that of the Phanerozoic; the Archaean mobile belts are deep, while the Phanerozoic mobile belts are shallow ones. It may be presumed that the tectonic units and their boundaries cannot be distributed linearly over a long distance owing to interference from changes in orientation and polyphase deformation. Thus Archean terranes are represented by segments with different orientations.

As stated in 2.3.6, the Late Archean might mark a stage of amalgamation of small continental nuclei. The existence of small continental nuclei marks the tectonic framework prior to Late Archean times. The greenstone belts formed since the Late Archean lie just at the peripheries of such continental nuclei. Hence the mobile belts represented by greenstone belts probably represent welding zones of these small continental nuclei.

2.2 Distribution and General Features

Within China, Archean basement rocks are mainly exposed in the North China Craton. They are composed of rocks ranging in age from about 2.5 to 3.8 Ga, which were differentiated and assembled during three periods of orogeny, at 3.8-3.5; 3.1-2.9 and 2.7-2.5 Ga respectively. Some additional exposures are found in the marginal parts of the Tarim and Yangtze Cratons. The major part of the exposed Archean crust is of Late Archean age, which amounts 85% of the total Archean crust exposures. Only small patches of Early Archean crust are found in Caozhuang area, eastern Hebei Province and near Anshan, eastern Liaoning Province. Only five cases of exposed crust of Middle Archean age are known; 1) in the Qian'an area, eastern Hebei Province; 2) within the Lower Anshan Group in northern Liaoning and southern Jilin Provinces; 3) the Tiejiashan granite, Anshan area Liaoning Provinces; 4) the Granulite rocks of Huai'an area, northern Hebei Province, and 5) the Yishui Group and Jiaodong Group of eastern Shandong Province (Fig. 2. 1.).

According to the study of Wu Jiashan et al. (1991) on metamorphic rocks and their protoliths, NE-trending craton-wide superbelts, consisting of low to medium metamorphic grade granitoid-greenstone belts occur in the central eastern part of North China. These belts extend from southern Jilin, northern Liaoning, eastern Hebei through western Shandong and cross the North China Plain southwestward into western Henan Province. Medium to high metamorphic grade metasedimentary gneiss belts occur on both sides of the arc-shaped metamorphic magmatic belt. In addition, there is a minor E-trending granitoid greenstone belt which runs east of Hohhot, Inner Mongolia to northern Shanxi Province. A variety of plate tectonic models have been used to explain these patterns.

Although there are terranes of greenstone belt affinities in the regions of Archean metamorphic rocks in China, these are not typical when compared with most of those of type localities in other countries (Condie, 1981) as greenstone belts are of higher or much higher metamorphic grade than those elsewhere. Wide occurences of amphibolitic and even granulitic facies rocks are often found in association with migmatites. This suggests that the Archean metamorphism of the cratons was of a quite uniform character over extensive areas.

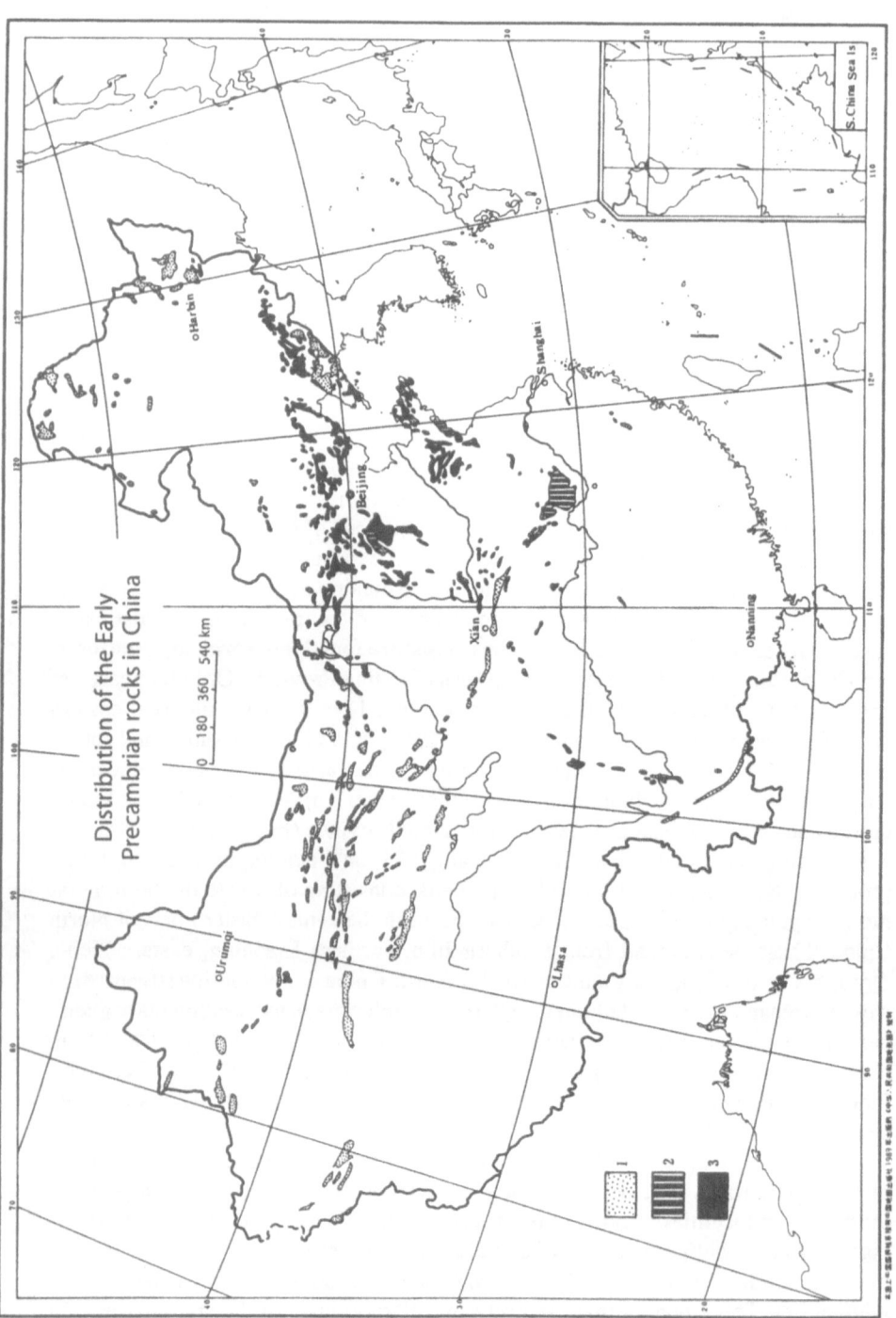

Fig. 2.1. Distribution of the Early Precambrian rocks in China: 1.Proterozoic rocks; 2.undifferentiated Archean and Proterozoic; 3. Archean rocks

2.3 Granulite-Gneiss Terrane

2.3.1 Some Classical Key Regions

Southern Liaoning-Jilin Region. The southern Liaoning-Jilin region is situated at the northern part of the North China Craton, where the Archean rocks have long been studied as a stratigraphic unit commonly referred to as the Anshan Group. The rocks are predominantly granitic gneisses metamorphosed dominantly to amphibolite facies, though granulite facies rocks occur in the northern part of the Qingyuan area and minor greenstone belt lithologies are also found. The Archean granitic rocks in this area can be grouped into three types: (1) charnockites; (2) gneissose tonalite, trondhjemite and granodiorite series rocks (TTG), and (3) potassium granites. Over 80% of the area is made up of TTG series rocks, and a biotite K-Ar age of 2900±90 Ma and a zircon concordia age of 2880±170 Ma (U-Pb) have been analysed from the tonalite samples. However, a muscovite Rb-Sr age of 2710±140 Ma and a zircon concordia age of 2730±160 Ma (U-Pb) have been obtained from the potassium granites. Thus this isotopic evidence indicates an evolutionary trend in the Archean granitic rocks from Na-rich to K-rich (Zhai Mingguo et al. 1985).

Eastern Hebei Area. The eastern Hebei area within the eastern section of Yanshan range in the central North China Craton has an area of over 25000 km^2. The metamorphic rocks in the Archean granulite-gneiss terrain of this area are mainly composed of highly deformed and metamorphosed volcano-sedimentary rocks and ancient granitic plutons of granulite and amphibolite grade. Rocks such as gneiss, granulite and various migmatites, as well as minor amphibolite and pyroxenite are common. However, the principal part of the terrane is made up of extensive tonalites, charnockites and corresponding migmatites and migmatitic gneisses. Several divisions for the Archean rocks in this area have been proposed (Sun Dazhong et al. 1984; Qian Xianglin et al. 1985; Zhang Yixia et al. 1986). Although they have different names and boundaries assigned to different lithologies, the supracrustal rocks are generally subdivided into two units: the lower unit is dominated by granulite facies rocks and is termed the Qianxi Group, or Qianxi complex. The upper unit consists mainly of amphibolite facies lithologies. This unit has been termed the Badaohe Group by Sun Dazhong et al. (1984): the Luanxian Group by Qian Xianglin et al. (1985) and the Dantazi Group by Zhang Yixia et al.(1986).

Within the lower part of the Qianxi Group the stratigraphy of granulite facies supracrustal rocks is ambiguous because of the effect of deformation, anatexis and magma intrusion. The supracrustal rocks are mainly pyroxene granulite and plagioclase-salite granulite intercalated with two-pyroxene amphibolite, salite amphibolite and thin layered pyroxene-bearing banded iron formation (BIF). These rocks occur only as isolated inclusions and blocks in vast charnockite intrusions, the ratio of the metamorphic xenoliths to charnockite intrusion being about 1:8 (Sun Dazhong 1984).

The upper part of the Qianxi Group is one of the most important iron-bearing formations in China: it is mainly composed of medium to coarse grained, amphibolite facies biotite-hypersthene-(diopside)-plagioclase gneiss intercalated with two-pyroxene amphibolite or diopside amphibolite. The amphibolites are mostly derrived from a basic volcanic protolith and are commonly associated with BIF. All these lithologies mostly occur as xenoliths in a sea of tonalites and granites, with a ratio of xenoliths to intrusion ranging from 1:8 to 1:16. An amphibolite xenolith yielded an Sm-Nd isochron age of 3470±107 Ma (whole-rock). This is one of the oldest ages reported so far in China (Jahn et al. 1987). The central part of this unit is composed mainly of medium-coarse grained amphibolites showing streaky, banded or rhythm-like structures. Because of the effect of magmatic intrusion, they are generally tranformed into streaky or banded migmatite in which the ratios of melanosome and leucosome range from 1:3 to 1:1. In addition, hornblende-plagioclase leptynite, bi-otite amphibolite and biotite-plagioclase leptynite can be found as interlayers. The upper part of this unit chiefly consists of biotite-plagioclase leptynite intercalated with minor amounts of amphibolite and BIF, locally with garnet-biotite-plagioclase leptynite and hornblende-biotite-plagioclase leptynite, usually showing a laminated or rhythmic bedding of sedimentary origin.

The amphibolite inclusions within the granulite facies metamorphic zone yield an age of 3470 Ma. According to Jahn et al. (1987), this date probably represents the age of the protolith formation. Based on some existing age data such as the Rb-Sr age of 3151 Ma (Shen Qihan et al. 1980), the U-Pb age of 3055 Ma (Sun Jiashu et al. 1982) etc, many geologists believed that the granulite facies metamorphism in east-ern Hebei Province occurred at about 3.0 Ga. However, the numerous isotopic age data around 2.5 Ga are considered to identify reworking products of the extensive later amphibolite facies metamorphism which was associated with strong deforma-tion (Sun Dazhong et al. 1984; Qian Xianglin et al. 1985; Wang Renmin et al. 1985; Zhang Yixia et al. 1986). Rb-Sr isotopic work has yielded dates of 2517 Ma, 2469 Ma and 2480 Ma (Chung F T. et al. 1979; Pidgeon 1980; Jahn and Zhang Zhongqing 1984). Similar U-Pb and Sm-Nd dating have yielded ages of 2460 Ma and 2480 Ma respectively (Pidgeon 1980; Jahn and Zhang Zhongqing 1984).

Previously it was considered that the charnockite intrusions were formed by the migmatization or early anatexis of the Qianxi complex (Sun Dazhong et al. 1984; Qian Xianglin et al. 1985; Zhang Yixia et al. 1986). However, the widely distributed charnockite yields age of 2.6 Ga (Wang Kaiyi et al. 1983). Secondly, there is a typi-cally intrusive relationship in the contact between the charnockite body and its country rock; the gneissosity becomes faint as the margin of the charnockite is approached, and inclusions of stratified rocks or even conglomerates have been identified in the body (Photo 2.1 and 2.2). This interpretation is similar to that proposed by Wang Renmin et al. (1985). Furthermore, when it is considered that the charnockite is generally formed under conditions similar to that of the granulite facies, and com-monly occurs in granulite facies areas, it is probable that a second period of granulite facies metamorphism occurred synchronously with charnockite intrusion at 2.6 Ga. Having confirmed the presence of the Early Archean rocks over 3.5 Ga in ages and noting a group of coincident ages of about 2.5 Ga by Rb-Sr, U-Pb and Sm-Nd meth-ods, Jahn and Zhang Zongqing (1984) suggest that granulite facies metamorphism might have occurred coevally to the formation of protoliths in eastern Hebei Prov-

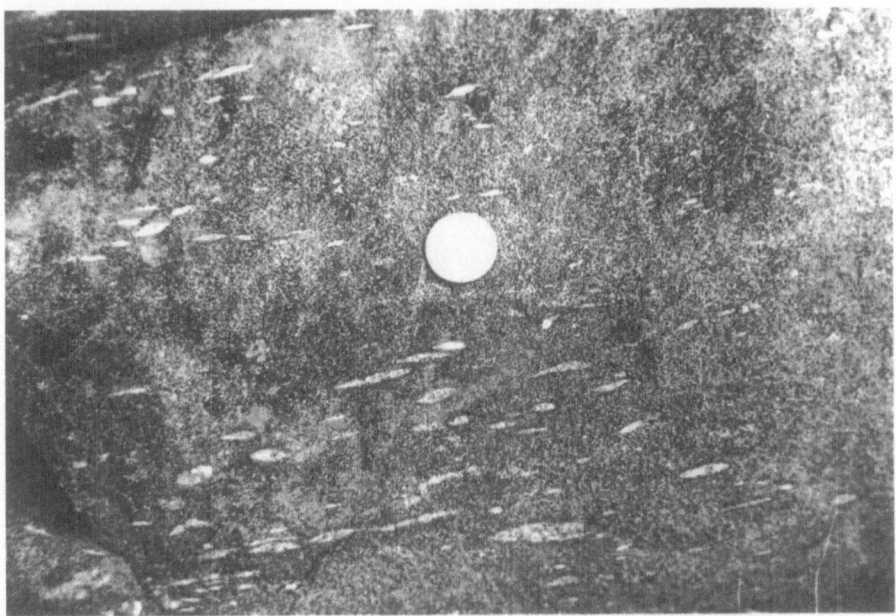

Photo. 2.1. Deformed amygdaloids run paralled to the foliation of the amphibolite occurring in the lower part of Qianxi complex, eastern Hebei, a coin 21 mm in diameter is scale marker within this photo(by Dai Fengyan)

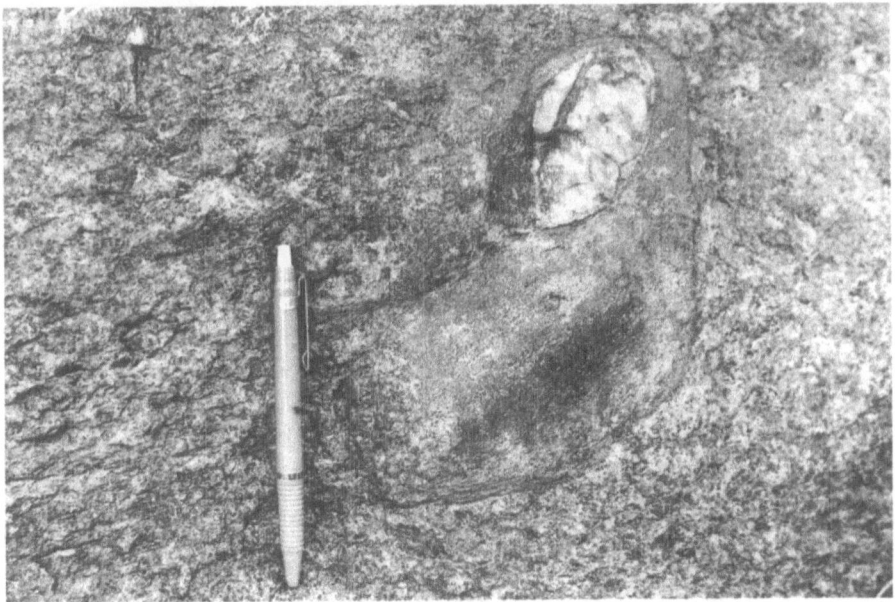

Photo. 2.2. Qian'an charnockite with an xenolith of foliated gneiss containing a deformed quartzite pebble, eastern Hebei (by Dai Fengyan)

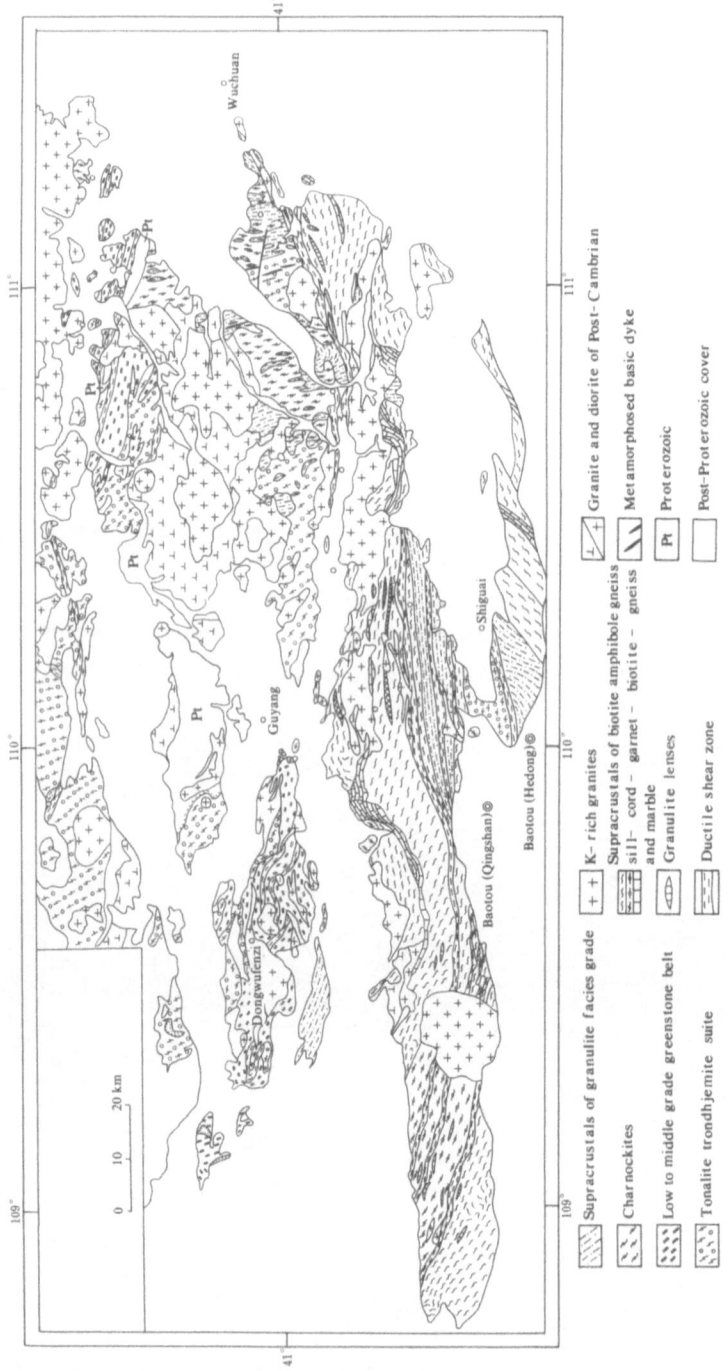

Fig. 2.2. Geological map of the western Yinshan Range showing the distribution of Archean rocks (after Jin Wei 1989).

ince, since both continued until about 2.5 Ga. However, these arguments need further study.

Yinshan Mts. Area. The Yinshan Mts area is in the western section of the northern border of North China Craton situated between coordinates 108° 40'-114° 50' E and 40°-41° 20' N. It contains outcrop of Archean complexes (see Fig 2.2), which have long been grouped into two units: a lower granulite-facies unit is termed the Jining Group, and an upper amphibolite-facies unit is termed the Wulashan Group (Shen Qihan et al. 1986; Li Shuxun and Liu Xishan 1988; Jin Wei 1989; Xu Xuechun 1989; Qian Xianglin et al. 1987). Fortunately, considerable progress has been made by recent studies of the granulite-gneiss terrane. In addition to the Late Archean greenstone belts of sub-greenschist to amphibolite facies represented by the Dongwufenzi Group, enormous meta-plutons of charnockite, tonalite-trondhjemite and potassium granites have been found. These plutons contain xenoliths of stratified rocks which have been divided into two supracrustal rock series: one is represented by granulites and the other by Al-rich gneisses.

The granulite-grade supracrustals are composed largely of granulite, pyroxenite, gneisses, garnet quartzite and magnetite quartzite which are preserved as relics within later charnockite intrusions. The structure of the stratified rocks is characterized by a N-S tectonic trend and vertically plunging isoclinal folds, which suggests that the rock series has experienced at least two phases of deformation.

The Al-rich gneissic supracrustals consist chiefly of graphite-bearing gneisses with sillimanite-garnet-biotite gneiss, garnet-cordierite-biotite-plagioclase gneiss, thick-layered diopside forsterite-phlogopite marble and minor granulite, amphibolite, magnetite quartzite, cordierite-garnet leptynite, sillimanite-garnet quartzite and thin-layered graphite (Xu Xuechun 1989; Jin Wei 1989). These lithologies have been subjected to regional metamorphism resulting in high amphibolite to granulite facies. The structural style is defined by well developed, large-scale recumbent folds and associated E-W trending ductile shear-zones. Highly deformed and metamorphosed basic dykes also intrude the gneissic supracrustals. As the structural orientation and protolith formation are distinctly different from those in granulite grade supracrustals; it is considered that there is an unconformity between the two series (Xu Xuechun 1989; Jin Wei 1989).

The greenstone belts represented by the Dongwufenzi Group exhibit bimodal features in their protolith formation and have been subjected to subgreenschist to low amphibolite facies metamorphism. Compared with the adjacent supracrustal rocks of granulite facies, temperatures and pressures of about 300°C and 0.45 GPa are indicated. No progressive metamorphism is apparent, and the structural orientation is NW-SE.

According to the structural orientations, the intrusive contacts between the plutons and the reported isotopic age data, the greenstone belts seem to be formed earlier than the supracrustal rocks represented by Al-rich gneisses (Fig 2.3; Jin Wei 1989). However, it has been proved that the lower part of the Al-rich gneissic supracrustal series contains minor meta-volcanic rocks which are quite similar protoliths to those in the granulite-facies supracrustal rock series. The high amphibolite-granulite facies metamorphism of the Al-rich gneiss series is analogous to metamorphism of the granulite-grade supracrustal rocks. This high grade amphibolite-granulite facies metamorphism has not been found to be superimposed on the greenstone belts which preserve low metamorphic grades. Therefore, it is more reasonable to regard the greenstone belts as younger than the Al-rich gneiss series.

Taishan Mt. Area of Western Shandong Province. The Archean rocks from the Taishan area of western Shandong Province were initially assigned to the Taishan complex, which includes the gneisses, crystalline schists and various intrusions in the area. However this term is currently used to refer to the strongly migmatized metasediments, various granitoids rocks and associated lithologies. Within the Taishan complex, the Taishan Group refers to the metasediments and meta-volcanics and the volcano-sedimentary formations belonging to the Archean greenstone belts present

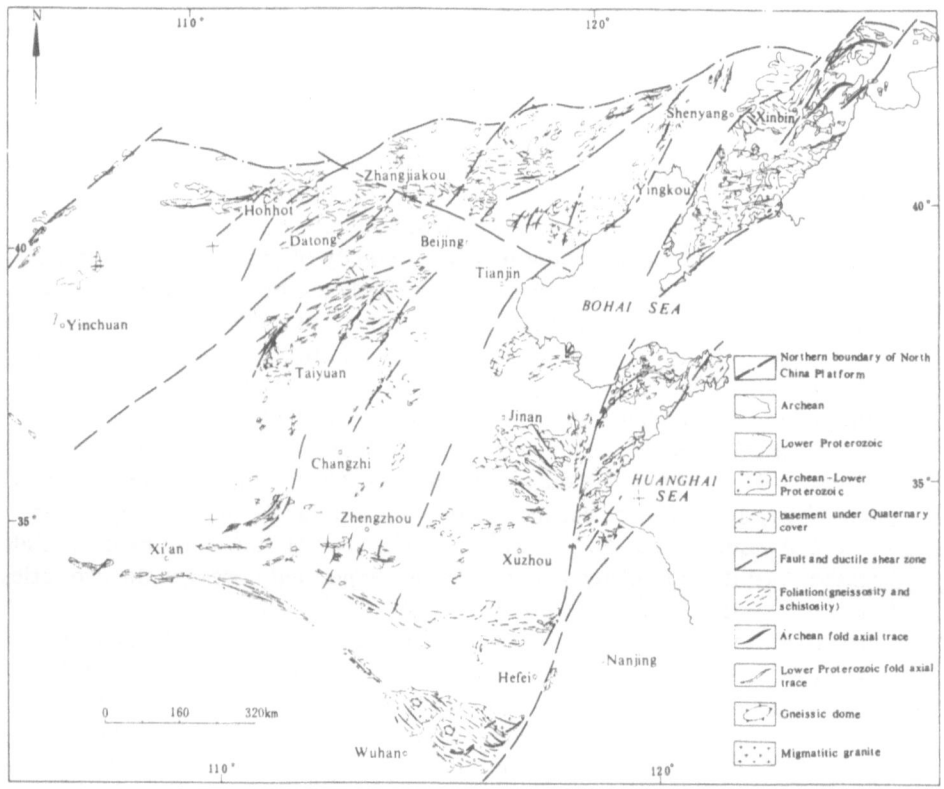

Fig. 2.3. Structural features in the crystalline basement of the North China craton.

in the Yanlingguan area. (This Group of lithologies is described separately in section 2.4.1).

The Wangfushan gneisses are the oldest supracrustal rocks in the Taishan complex and have undergone strong deformation, metamorphism and migmatization. The rocks include biotite-plagioclase gneiss and biotite-hornblende gneiss, intercalated with minor hornblendites and their Rb-Sr isochron ages range from 2690 to 2767 Ma and Sm-Nd ages from 2700 to 2820 Ma (Jahn et al. 1988).

The granodiorites and granites intruded into the basement gneisses make up nearly half of the Taishan complex. The Hushan, Aolaishan and Motianling bodies occur as a belt and are comprised of granite-series bodies with gneissose structure. The Hushan body (U-Pb zircon age ~ 2568 Ma) is intruded into the basement gneiss, and the Motianling body is intruded into both gneisses and the Hushan body. The three bodies have similar mineral compositions including quartz, plagioclase and biotite. The Hushan granite body frequently contains pegmatite dykes, and at its edges xenoliths of the basement gneisses. This intrusion was formed by the differentiation of a trondhjemitic magma produced by the partial melting of basic rocks in the deep crust, and associated with the development of metamorphism in the Taishan complex at the end of Archean time. In comparison the Aolaishan body is composed of well-lineated, medium-coarse grained, porphyroid granodiorites, formed by the partial melting of gneisses in the deep continental crust at 2450 Ma (Jahn et al. 1988). The Motianling body is lithologically massive, with rare xenoliths and faint lineation (Jahn et al. 1988). Geochemically, the three bodies show the same developing trend: the Hushan body corresponds to continental granophyre and is close to the continental trondhjemite, while the other two bodies correspond to the continental granophyre.

Diorite series rocks in the Taishan complex include meta-dioritic porphyrite, quartz diorite and some dioritic porphyrite dykes, which are mostly medium-fine grained and massive. Only the quartz diorite exhibits lineation with a biotite orientation. Compositionally, they show a calc-alkaline evolutionary trend. These intrusions are derived from the deep dioritic magma and intruded at 2595 Ma.

When the intrusive ages of the granite-series rocks in the Taishan complex are considered as a whole, it seems that the earliest intrusions were dioritic in composition. These were followed by trondhjemitic intrusions and finally granitic melts derived from the melting of continental crust.

Taihang Mts Area. The Taihang Mountains are situated on the boundary between Hebei and Shanxi Provinces, in the central part of the North China Craton. Archean metamorphic rocks are widely developed, covering 20,000 km^2. One particularly noteworthy unit is the Fuping Group, which consists of supracrustal rocks which experienced amphibolite facies metamorphism. The lower part of the Group consists of amphibolite, hornblende-plagioclase gneiss and hornblende-biotite leptynite, followed by gneiss and diopside marble in the middle of the sequence. The upper part consists of leptite and plagioclase gneiss intercalated with minor amounts of calc-silicates, marble and lenticular amphibolite (Wu Jiashan et al. 1983). As the zircon U-Pb ages of the leptites range from 2.7 to 2.9 Ga, it is suggested that sedimentation of the Group started 3.0 Ga ago (Liu Dunyi et al. 1984).

The protolith of the Fuping Group is principally composed of clastic rocks and thick carbonates, with intercalated volcanic rocks of intermediate to basic composition in the lower part of the Group. Wu Jiashan et al. (1983) considered that the sedimentary protoliths changes from argillites, semi-argillites, psammites, greywackes in the lower part of the sequence to sands, silts and marls at its top. Occassionally, blastoamygdaloidal structures can be seen in some amphibolites. However, it has been found recently that many plagioclase gneisses and leptynites in the lower-middle parts of the Fuping Group are metatonalites and trondhjemites. A suite of augen gneisses, which may be the products of the felsic volcanic-sediments which have been deformed by ductile shearing in the deep crust, are present at the top of the Fuping Group.

Huashan Mt. Area. Within this area the Taihua metamorphic complex outcrops from the central western part of Henan Province westwards to Xi' an of Shaanxi Province. The complex is separated from the adjacent Dengfeng greenstone belt by a shear zone with an E-W strike. It consists mainly of felsic gneisses (which make up over 60-70% of the whole outcrop), and sillimanite-garnet- gneiss, graphite gneiss intercalated with marble, quartzite and a banded iron formation. The protoliths of the felsic gneisses are assumed to be tonalites. An early granulite facies metamorphism occurred at 2.6-2.7 Ga. However, because of the superposition of a later amphibolite facies metamorphism, granulite facies lithologies occur only as relics in amphibolite facies rocks (Zhang Guowei et al. 1984; Cheng Yuqi et al. 1984).

Moreover, other granitic rocks in this area include Na-rich and K-rich granites. The Na-rich series consists of tonalite, trondhjemite and granodiorite which intruded as diapirs to the cores of some domes and antiforms. U-Pb dating of granodiorites within the Na-rich series yielded a zircon age of 2520±17 Ma (Wang Zejiu et al. 1987). Two types of contact relationships between this series and the Dengfeng greenstone belts are observed: intrusive and transitional. The intrusive contacts are a reflection of magmatic allochthonous granites; and the transitional contacts are the symbol of autochthonous and subautochthonous migmatic granites. The ratio of granitic rocks to greenstone belts is 4:1.

Recently, Kröner et al. (1988) obtained two zircon Pb-Pb ages from the tonalitic gneiss of Taihua complex: one is 2841±6 Ma from two light brown, euhedral zircon grains, and the other is 2806±7 Ma from a slightly dark-brown, rounded zircon grain. They considered that the former represented the age of tonalite crystallization, and the latter was the age of metamorphism from tonalite to gneiss.

2.3.2 Rock Assemblages

Granitic Gneisses. Various types of granitic gneisses form the main bodies of the Archean granulite-gneiss terranes. Most of these are derived from granitic plutons, apart from some minor amounts of supracrustal rocks subjected to strong regional metamorphism and migmatization. According to their chemical compositions, the granitic rocks can be divided into three series: (1) charnockites; (2) tonalite-trondhjemites and (3) quartz-monzonites and granites.

The charnockite series: this is widely distributed both temporally and spatially, and composed of Na-rich and K-rich rocks. In contrast to the other two series, these rocks are hypersthene-bearing, being closely associated with granulite facies and occur in high-grade metamorphic belts. Isotopic ages and regional correlations suggest that this series was formed 2.8-2.9 Ga ago and related to the Qianxi movement. Charnockites are widely distributed in the Archean high-grade metamorphic terranes from Qingyuan, Liaoning Province, westwards to Wulashan, in Inner Mongolia, on the northern border of the North China Craton. They vary distinctly from area to area in their petrography, structure and contact relationships with country rocks. Such variation may in future provide some information of their tectonic stage and origin. Their features can be described with reference to two examples: (a) the Qingyuan charnockite in northern Liaoning Province, and (b) the Qian'an charnockite in eastern Hebei Province.

(a) Qingyuan charnockite: From the charnockite margin outwards, the transitional zones (migmatite zones) and granulite zones appear in a regular order. In the transitional zone, the granulites commonly envelope both the charnockite and also relic granulite inclusions within the charnockites. The charnockites generally exhibit gneissose, granitoid and metasomatic textures. Furthermore, central (remelting) and marginal (metasomatic) facies can be identified in many charnockite bodies according to the difference in mineral and chemical compositions. The rocks in the marginal facies have more alkaline feldspar, less hypersthene (about 3%) and normal contents of K and other LIL elements, whereas the central zones have more hypersthene (about 3-10%), less alkaline feldspar, and are poor in K and other LIL elements, but relatively rich in Na. The $Na_2O:K_2O$ ratios in these two facies are 0.67 and 2.88 respectively. The rocks in the transitional facies have high REE contents and evident negative Eu anomalies with Eu:Eu* = 0.68, while those in the central facies have relatively low whole REE contents and positive Eu anomalies, with Eu:Eu* = 1.96. The studies of REE and LIL elements indicate that the Qingyuan charnockites are formed by remelting, crystalline differentiation and metasomatism (Zhai Mingguo 1984). The granulite facies metamorphism of the An'shan Group occurred at 2.8-2.9 Ga (Wu Qinsheng 1979 unpublished) which is similar to the age of Qingyuan charnockites. Consequently, it is believed that they are the products of the same tectono-thermal event.

The granulite-facies charnockites of the Yinshan Mt area intruded into the supracrustal rocks and have massive and gneissoid structures, in which the hypersthene and salite generally show idiomorphic to hypidiomorphic texture. The idiomorphic texture is of both magmatic and metamorphic origin, but the hypidiomorphic texture is of metamorphic origin only (Jin Wei 1989). The origins of these charnockites may be different from those in the Qingyuan area. Presently, although isotopic age data is not available, it has been estimated that they were formed at 2.8-2.9 Ga because the tonalites and trondhjemites which have intruded them have a zircon U-Pb age of 2.5 Ga.

(b) Qian'an charnockites: These have typical intrusive features, and contain abundant inclusions of country rocks, of which amphibolite and biotite leptynite are dominant. The charnockites have a disconcordant contact with the country rocks and poorly developed foliation, which suggest that they may be post-tectonic intrusions. They have medium-coarse grained granoblastic textures, and are mainly composed of plagioclase (45-60%), quartz (15-30%), hypersthene (5-15%), alkaline feldspar

(5-10%), biotite (2-15%), hornblende (1-3%) and some accessory minerals such as zircon, apatite and opaque phases. Most of the hypersthenes are fresh; only a small fraction of them are altered into uralite (actinolite). Samples of the Qian'an charnockites yield a whole-rock Rb-Sr isochron age of 2647±53 Ma, which may represent the intrusive age of the charnockites and an initial $^{87}Sr/^{86}Sr$ ratio of 0.70223 (Wang Kaiyi et al. 1985).

In the An-Ab-Or diagram (O'Connor 1965), the Qian'an charnockites are projected to the areas of granodiorite, quartz monzonite and tonalite, in which the K_2O content is relatively high (mean value = 3%) as compared with the Archean Na-rich tonalites. The mean values of Sr, Ba and Rb in the charnockites are 589, 794 and 91 ppm, Rb-Sr and K-Rb ratios are 0.23 and 354 respectively, and the $(Ce-Yb)_N$ ratios range from 8.9 to 83. The rocks show strongly depleted REE patterns and most have positive Eu anomalies. However, when the whole REE contents are high, negative Eu annomalies may be present in the rocks (Wang Kaiyi et al. 1985). The geochemistry indicates that charnockites may be derived from the andesitic-dacitic magma under granulite facies conditions.

As a whole, the Qian'an charnockites are post-tectonic intrusions of late Archean age, and are different from the Qingyuan charnockites which are considered as syntectonic bodies emplaced under conditions of granulite facies metamorphism. Isotopic determinations show that both the granulite facies metamorphic rocks and the charnockites in the Qian'an area were not necessarily formed in the Early Archean, but at a later stage.

Tonalitic-trondhjemitic rock series: This series is characterized by the enrichment of Na ($Na_2O>K_2O$), and is the most widespread rock in the granitic gneisses of the Archean basements of China. The rocks commonly have gneissic structures and are characterized by more plagioclase more than K-feldspar. Compositionally, the rocks vary greatly from intermediate to acidic members including tonalite, trondhjemite and granodiorite etc. (TTG series). The typical mineral assemblage is quartz + plagioclase + hornblende + biotite, and minor K-feldspar which is absent in intermediate rocks. Texturally, the rocks are medium-fine to coarse grained (granoblastic), and generally have a well-developed gneissosity. The inclusions in the rocks are complicated, but dominated by amphibolite. The geological features and origins of the Late Archean TTG series can be described by two examples: the Qinling Mts and the southern Liaoning-Jiling areas.

The white granites from the eastern section of Qinling Mts, on the southern border of North China Craton have been classified as trondhjemites (You Zhendong et al. 1986), and are commonly termed the Dengfeng trondhjemites. They usually intrude the cores of domes or anticlines in the host-rock gneiss. In most cases they have intrusive contacts with the gneisses but local concordant transitions are present in some instances, which indicates that the trondhjemites may contain subautochthonous features. In the field, some white trondhjemite dykes are found intruded into the country rock gneisses. These dykes have an initial $^{87}Sr/^{86}Sr$ ratio of 0.703±0.003, which implies that they have probably come from the upper mantle or the lower crust (Zhang Guowei et al. 1985).

The trondhjemites are dominated by plagioclase (65%), K-feldspar (10%), quartz (20%) and muscovite (10%) and have a white-grey colour, fine-medium grained granitic textures and exhibit a faint gneissosity. Plagioclase is relatively euhedral and microcline is the dominant K-feldspar. Both microcline and quartz occur as xenomorphics, and muscovite as a micro-lepidosome. In the An-Ab-Or diagram (O'Connor 1965), the Dengfeng trondhjemites project onto the trondhjemite field, and are compositionally quite different from the other types of granites in the area. The trondhjemites mostly show lower K/Na ratios (<0.5) and higher Al/(K+Na+Ca) ratios (<1.1), which show them to be chemically different from normal I-type granites. The rare earth signatures of these rocks are evidently depleted, with ΥREE contents ranging from 54.92-91.95 ppm, and HREE-LREE ratios ranging between 0.048-0.050. You Zhendong et al. (1986) considered that the Dengfeng trondhjemite intrusion was later than the early period of regional metamorphism and was an important event in the process of transition from an island arc to continental environment. Associated with this process, widespread regional metamorphism and granitization or partial melting of the continental crust occurred, resulting in formation of the subautochthonous granite bodies.

The TTG series, widely distributed in the southern Liaoning-Jiling area, is composed of a suite of granitic gneisses (Zhai Mingguo et al. 1985). Some of these occur as irregular bands around the greenstone belts. Due to strong deformation a well-developed gneissosity is present in the margins of some TTG bodies which appears to be concordant with the country rocks. However, it still can be seen that this gneissosity is disconcordant with country rock foliation in other places. Moreover, greenstone inclusions appear in the margins of some bodies which indicates that the TTG series have intruded the greenstone belt lithologies. Some of the TTG series occur as batholiths, in which xenoliths are partially assimilated and contamination features are evident along the contact with greenstone belts. In chemical composition, the TTG series exhibit an Al_2O_3 content ranging 5.36-17.0% and belonging to the high-Al type (Barker 1979). In addition, the SiO_2 content exhibits a broad range from 61.36% to 73.66%, and FeO* and MgO contents also vary greatly. The rocks have low ΥREE contents ranging from 36.4 to 142.64 ppm, with strongly depleted patterns $(La-Yb)_N$ = 13.74-21.40, and small negative or absent Eu anomalies, with Eu/Eu* ratios from 0.96 to 1.04. Based on these features, the concentrations of some trace elements such as Cr, Ni, Co and V and a Zr-Hf ratio of 38.9-67.4 within zircon mineral separates, it can be deduced that the TTG series may have a close affinity to mafic volcanics, from which they were formed by partial melting and differentiation. Furthermore, according to the differentiation index (62.95-85.71), total alkalinity $(K_2O+Na_2O=4.16-7.8\%)$, low values of FeO*/ FeO*+MgO ratio (0.67-0.83), higher CaO content (2.11-5.02%) and CaO / $(Na_2O + K_2O)$ ratio (0.26-1.08), it is suggested that the partial melting probably occurred in the compressive tectonic settings of continental margins or interiors (Petyo 1979).

Most of the TTG series which are widely distributed over the northern part of North China Protoplatform do not have clear boundaries with their country rocks. Their gneissosity is usually concordant with the regional foliation. Abundant microcline-bearing pegmatite dykes and some other types of dykes (pincipally of intermediate-basic composition) occur, and both deep source inclusions and xenoliths of the country rocks are present in the TTG bodies. The rocks belong to the Ca-rich types of calc-alkaline series with SiO_2 contents ranging from 51.01 to 75.42% (mostly less than 70%, mean value=66.82) and K/Na ratios <1. It seems that most terranes composed of this kind of granitic rocks have undergone retrogressive metamorphism or autometamorphism of different grades. In summary, this kind of granitic rock is clearly different from any granite series of the Phanerozoic Era.

The isotopic ages of the rocks are mostly 2.7-2.8 Ga, younger than that of the granulite facies metamorphism, if the latter took place 2800-2900 Ma ago. However, a group of zircon U-Pb concordia ages of 2.5 Ga have been obtained which imply that the magmatism of TTG series spanned a long time period.

Similarly, the early granitic rocks in the Taishan complex also show a tonalite-trondhjemite trend. According to Jahn et al.(1988), the eruption and intrusion of the basic and tonalitic magmas in the Taishan area occurred 2700-2750 Ma ago. They were then metamorphosed into amphibolites and grey-gneisses in the first period of metamorphism that formed the Wangfushan gneiss basement. Later, at about 2.6 Ga, LREE-enriched sub-alkaline diorites, together with related granites and trondhjemites were intruded into this basement gneisses. The Wangfushan gneisses and diorites have relatively low value of I_{Sr} (initial $^{87}Sr/^{86}Sr$ ratio=0.7006 ±4; 0.7004±2; 0.7008 ±11) and higher positive value of time-integrated ε_{Nd} (+3.1 to +4.7). This strongly verifies the conclusion that the continental accretion resulted in the generation of later magmas derived from a highly depleted mantle source. The Sr and Nd isotopic data, however, indicate that the granitic rocks may be produced by the anatexis of the basement gneisses (Jahn et al. 1988).

The granitic rocks in the Archean craton of China show a trend from early tonalite-trondhjemite to late granite, with the early intrusions rich in Na and late intrusions rich in K. This trend coincides with the process of continental accretion.

Quartz monzonite-granite series: This series is widely distributed and is characterized by K-enrichment ($K_2O>Na_2O$). Generally, the content of K-feldspar is nearly equal to or more than that of plagioclase. The rocks are typically composed of quartz, K-feldspar, plagioclase and mica, and mostly have medium-coarse grained granitic textures and gneissose structure, while the post-tectonic intrusions are mainly massive, not foliated.

The K-rich granites from the Qingyuan area are typical of the quartz-monzonite series generally, and were mostly formed 2.5-2.6 Ga ago. Analogous to calc-alkaline volcanics, they have a relatively high content of ΣREE, negative Eu-anomalies with Eu/Eu* ratio ranging from 0.85 to 0.91, and Ce-Yb$_N$ ratios from 6.03 to 5.04. The contents of some LIL elements such as K, Rb, Sr and Ba are higher than those in tonalites, with Ba-Sr ratio about 2 (Zhai Mingguo et al. 1985). This indicates that the K-rich granites may be produced by the partial melting of pre-existing calc-alkaline volcanics during their reworking and deformation.

The potassium ganites from the Yinshan Mts. area are composed of gneissose biotite- granite and -porphyritoid granite. These bodies form the lastest intrusions in the area, and are mostly distributed along the ductile shear zone in the supracrustal rock successions represented by Al-rich gneisses. A zircon U-Pb concordia age of 1.8 Ga has been obtained from one of the intrusions. In the granites, K_2O = 3.88-5.50%, K_2O/Na_2O = 1.05-1.62, SiO_2 = 69-76%. The K/Rb ratio is relatively low, ranging from 266 to 382. The Rb/Sr ratio ranges between 0.20 and 0.93, which is equivalent to the mean value of the continental crust. These facts indicate that the gneisses have a close affinity to the crustal materials of the continents (Jin Wei 1989). This kind of K-rich granites usually intrude into earlier gneisses, charnockites and tonalites and commonly contain country rock xenoliths. They are widely distributed in various high grade (granulite) to low-grade (greenschist) metamorphic belts.

Supracrustal Rocks. Supracrustal rocks, widely distributed throughout the granulite-gneiss terranes, consist of amphibolite, pyroxene granulite, and less commonly felsic gneiss and leptynite. They are the products of the metamorphism from basic (minor ultrabasic) and intermediate-acidic volcanics, and various meta-sediments. Typical metasedimentary rocks are sillimanite garnet gneiss and related rocks, magnetite quartzite (BIF), quartzite and marble.

Amphibolites and pyroxene granulites: Both these lithologies are metamorphosed from basic volcanic rocks of tholeiitic affinity to amphibolite facies and granulite facies respectively. The amphibolites are mainly composed of plagioclase (andesine) and hornblende with little or no garnet, and some accessory minerals such as magnetite and apatite. When metamorphism proceeds to granulite facies, the hornblende and andesine give way to pyroxene and labradorite or bytownite. For example, the amphibolites and pyroxene granulites in eastern Hebei Province are similar in geochemistry, contain low K_2O (mean value of K_2O ~0.92%) and TiO_2 (mean value = 1.11%), and high in Fe (Fe_2O_3 = 3.84-5.27%, FeO = 8.54-10.17%) (Sun Dazhong et al. 1984).

Jahn et al. (1987) studied the Sm-Nd isotopic geochemistry of the amphibolites from eastern Hebei Province and determined the $C_{(Nd)T}$ value to be + 2.7 which indicates that the original basaltic magma may be derived from a highly depleted mantle source. Cong Bailin (1979) considered that these kinds of rocks were geochemically similar to modern island arc tholeiites, while Sun Dazhong et al. (1984) advocated that they would be more similar to continental tholeiites because the rocks have higher contents of (FeO + Fe_2O_3), and most have K_2O>1%. Jahn (1983) thought that the sources may be quite complex because the $(La-Yb)_N$ ratio of pyroxene granulites is from 3.5 to 8.7.

The pyroxene-bearing granulites from the Yinshan area show a trend from tholeiitic series to calc-alkaline series (Shen Qihan et al. 1986; Jin Wei 1989). Their chemical compositions correspond to those of modern typical island arc calc-alkaline tholeiites, in which LIL elements such as Sr, K, Rb, Ba and Th are highly enriched. In the discrimination diagram of Pearce (1976), all the samples fall in the island arc tholeiite area. The rocks are characterized by low ΣREE content ranging from 54 to 111.3 ppm, with flat REE patterns with little enrichment in LREE (LREE-HREE = 3.7-7.2), and small negative Eu anomalies. These chemical characteristics indicate that the basic granulites are similar to the Archean TH_2 magma type (Condie 1981) and modern calc-alkaline tholeiites. Field relationships show that some amphibolites cut across their country rocks and are evidently derived from basic dyke swarms.

Gneisses and leuco-granulites: These kinds of rocks a ꞏ one of the main constituents of granulite-gneiss terranes, and are represented by mica-plagioclase gneiss, biotite (hypersthene)-plagioclase gneiss or granulite. The latter generally occurs under conditions of granulite facies. Compositionally, most of them are similar to dacite and rhyolite (Shen Qihan et al. 1986) and belong to the calc-alkaline series. The rocks have a ΣREE content ranging from 79.21 to 271.6 ppm, with LREE enrichment and HREE depletion (LREE-HREE = 9.87-13.87), which is analogous to F_{II} type Archean felsic volcanic and hypabyssal rocks (Condie 1981; Jin Wei 1989). Generally, they are associated with amphibolites in chemically bimodal successions.

Moreover, some metasedimentary biotite-plagioclase gneiss and leptynite are frequently associated in thin layers with other metasedimentary lithologies. For example, they may occur as thin layers of a few millimeters to centimeters in a banded iron formation. Their mineral assemblages are mainly plagioclase (40-70%), quartz (15-40%) biotite (10-30%) and a little garnet. In ACF and AKF diagrams, all the samples are projected to the greywacke, argillite, semi-argillite areas (Sun Dazhong et al. 1984).

Sillimanite-garnet gneisses and graphite-gneisses: Sillimanite-garnet gneisses and graphite-gneisses usually occur together and show a clear layered structure. The former contains sillimanite (10-20%), biotite (about 25%) and garnet (10%). In the northern and southern borders of the North China Craton, sillimanite gneisses, associated with quartzite and marble are commonly referred to as the Khondalite series (Shen Qihan et al. 1986; Qian Xianglin et al. 1987; Xu Xuecun 1989; and Zhang Guowei et al. 1985).

The gneisses and leptynites in the Yinshan area (on the northern border of the North China Protoplatform) vary greatly in chemical composition, but are undoubtedly metasedimentary in character. SiO_2 content ranges from 43.91 to 65.5%. These rocks are also rich in Al_2O_3 (Al_2O_3 = 14.88-22.22%, with most samples >20%), K_2O (K_2O = 1.80-5.15%, K_2O/Na_2O = 0.59-1.95) and REE (ΣREE content =162.3-529.6 ppm). Their $(La/Yb)_N$ ratio ranges from 5.41 to 42.11, with clear negative Eu anomalies and a Eu/Eu* ratio from 0.47 to 0.93 (Jin Wei 1989). They have high contents of LIL elements with Rb from 43 to 117.2 ppm, Sr from 185 to 527 ppm and Ba between 604 and 1400 ppm. Their Ce anomaly (de Ha ∂_{Ce} = -0.05 to 0.10) suggests that the original rocks might be formed in a reducing environment (Xu Xuechun 1989). The chemistry of the gneisses and leptynites suggests that their protoliths were principally of an Al-rich greywacke carbon bearing, argillite-carbonate formation, which formed on stable continental margins. In addition, there are similar metasedimentary successions composed of sillimanite augen leptynite (Photo 2. 3), marble and graphite-bearing rocks from Taihangshan area, and minor sillimanite gneiss from eastern Hebei Province.

Magnetite Quartzite (BIF): Magnetite quartzites widely occur, usually on a small-scale, in high-grade terranes. However, in eastern Liaoning and eastern Hebei Provinces (the world-famous large iron deposit area), the masses of magnetite quartzites in the granitic gneisses terrane may be up to tens of metres thick and several kilometres long. Although the magnetite quartzite-bearing successions are composed chiefly of intermediate-basic volcano-sedimentary rocks, the country rocks at the mine are intermediate-acidic greywackes (Zhang Yixia et al. 1986). Windley (1984) proposed that these successions, in contrast with the Algoma-type iron formations in greenstone belts, are composed of quartzite and argillite formed in the shallow marginal shelves of stable continental platforms. Thus they represent the proto-Superior-type iron formations formed in some stable and short-lived cratons.

Photo. 2.3. Deformed sillimanite spheres in leptite of the Fuping Group, probably the original sedimentary nodules (photo by Dai Fengyan).

Magnetite quartzites usually have laminated banded structures and (in granulite facies) are developed with gneissoid structures. The iron-bearing minerals are mainly magnetite with grain sizes of 0.1-0.3 mm in successions at amphibolite metamorphic grade, and up to 0.5 mm or 1-5 mm in successions at granulite metamorphic grade. Gangue minerals are chiefly quartz and some silicate minerals such as tremolite, cummingtonite, hypersthene and diopside, appearing in an orderly way with different metamorphic grades. Additionally, there is a 10 m thick layer of sillimanite-biotite-quartz schist metamorphosed from argillaceous siliceous rocks found in association with the magnetite quartzites from Qian'an, eastern Hebei Province.

*Quartzites:*These usually occur as layers in the Jining Group on the northern margin of the North China Craton, where feldspathic quartzite and biotite-plagioclase quartzite 160-170 m thick are associated with sillimanite garnet gneiss. In some places such as the eastern Hebei Province, quartzites occur within BIF and occasionally as thin layers. They therefore appear to have a genetic affinity with BIF, being colloid chemical sediments. The major constituent is quartz, and a small amount of feldspar may also be present. The presence of other silicate minerals, including biotite quartzite, (magnetite-bearing) amphibole quartzite, (magnetite-bearing) pyroxene quartzite and sillimanite-almandine quartzite etc. can be identified. In comparison, the garnet quartzites from Yinshan region, commonly associated with magnetite quartzites, have faintly differentiated greywacke protoliths. These rocks have lower REE contents, ranging from 59.5 to 102 ppm and no clear Eu anomalies, with Eu/Eu*ratios from 0.98 to 0.997. Their REE patterns are identical to Archean sediments elsewhere (Nance and Taylor 1976, 1977).

Marbles: Marbles are usually intercalated with other types of supracrustal rocks within far northern Hebei and Inner Mongolia (i.e.- within the northern margin of the North China Craton), the outcropped thickness is commonly tens of meters. The thickest serpentine marble and phlogopite marbles are over 1 km thick but their successions are probably duplicated by folding. In the Taihua complex from the southern margin of the North China Craton, the outcrops are over 100 m thick and 2000 m long. Marbles also outcrop extensively in the Fuping Group of the Taihangshan region, where they are over ten to tens of metres thick. The thickest serpentinized forsterite tremolite marble is 200 m thick and runs discontinuously several kilometres to tens of kilometers in length.

These kinds of rocks include calcite marble and dolomite marble, the latter being dominant. Because of the presence of many kinds of silicate minerals and complicated compositions, the protoliths of the marbles are thought to be immature and impure carbonates which probably reflect a broad and relatively stable shallow sea environment.

2.3.3 Metamorphism

All the Archean granulite-gneiss terranes in China have experienced extensive metamorphism. These were then overprinted by secondary amphibolite- (and sometimes granulite-) facies metamorphic events. Extensive isotopic age data indicate that the amphibolite facies metamorphism occurred at ca. 2.5 Ga. It has also been deduced that granulite facies metamorphism occurred at 2.8-2.9 Ga although reliable age data not available at present. Furthermore, the charnockitic intrusions emplaced in these terranes at 2.6 Ga are at granulite facies, so it is possible that a local episode of granulite facies metamorphism accompanied the more extensive amphibolite facies metamorphism. The rocks of granulite facies in the lower part of the supracrustal succession from eastern Hebei Province occur as isolated remnants and blocks of different sizes within the amphibolite facies. Many geologists (Chen Yifei, 1979;

Zhang Ruyuan et al. 1986; Dong Shenbao et al. 1987; Wang Renmin et al., 1986 Sun Dazhong et al. 1984; Qian Xianglin et al. 1985; Zhang Yixia et al. 1986) have made detailed studies of the metamorphism and its P-T (pressure-temperature) conditions in this region. According to the geothermometry and geobarometry of various mineral pairs and oxygen isotopes, it seems that the temperature of the granulite facies ranges from 750 to 900°C and that the pressure ranges from 0.7 to 1.19 GPa, while the temperature of amphibolite facies ranges between 550 and 750°C. A significant study was carried out by Qian Xianglin et al. (1985), who measured two groups of temperatures in these rocks: one group ranged in temperature between 640-758°C, averaging 710°C, which corresponds to granulite facies; the other group had a mean temperature of 510°C. The second group represents retrogressive metamorphism from granulite facies to amphibolite facies. They also suggested that the lower successions of the supracrustal sequence were firstly subjected to granulite facies metamorphism, whereas the upper successions were formed later and underwent prograde amphibolite facies metamorphism. At the same time the lower granulite facies successions underwent retrogressive metmorphism to amphibolite facies. Wang Renmin et al. (1986) pointed out that meta-equilibrium mineral assemblages and retrogressive textures, representing those of granulite facies and amphibolite facies respectively, are common and can easily be determined at outcrop scale. The cause of the retrogression, however, is still disputed. For example, Qian Xianglin et al. (1985) suggested that retrogression was due to the second period of regional amphibolite facies metamorphism. Alternatively, Zhang Yixia et al. (1986), suggested that it was the result of the anatexis of granulite facies rocks (Photo 2.4).

In the interior of the North China Craton, all of the gneiss terranes have undergone Late Archean (2.5 Ga) amphibolite facies of medium pressure and medium to high temperature regional metamorphism. The sole exception is the Fuping Group in the Taihangshan region, where small amounts of granulite facies rocks are present.

In the northern margin of the craton, including in eastern Hebei Province (see section 2.3.1), there are many granulite facies rocks exposed as variously sized blocks in widespread amphibolite facies "oceans". The entire distribution of these blocks lies in an EW-trending belt of 2000 km length. The granulite facies metamorphism belongs to the medium-high temperature and medium-pressure type with the temperature ranging from 800 to 950°C, the pressure ranged from 1.0 to 1.4 GPa and the geothermal gradient between 20-25°C/km. However, in the Qianlishan region at the western end of this belt, the temperature decreases to 700-800°C and the pressure to 0.55-0.65GPa, whereas the geothermal gradient increases to 34°C/km. Thus this section of the belt has experienced low-pressure type metamorphism (Dong Shenbao et al. 1987). Moreover, the pyroxene granulites change westwards into hornblende-pyroxene granulites; sillimanite-garnet gneisses decrease in K-feldspar and increase in plagioclase, and cordierite appears (Shen Qihan et al. 1986). These mineralogical changes indicate a decrease of pressure during metamorphism. Eastwards to the western Liaoning Province, andalusite appears; and further eastwards to towards the Qingyuan region, the eastern end of this granulite facies belt, the metamorphic temperature is from 650°C to 1000°C, the pressure between 6.6 and 10 GPa, and the geothermal gradient is 23-30°C/km, showing an increasing trend.

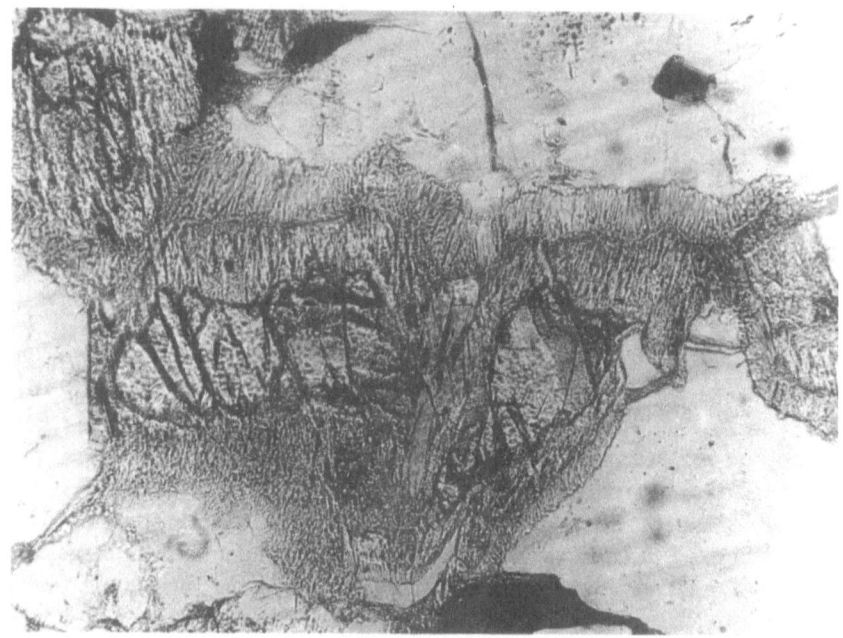

Photo 2.4. Hypersthene (high relief) retrograded to amphibole in two rims. The inner is cummingtonite, and the outer is actinolite, eastern Hebei. Plane-polarized light, X 330 (by Gan Fan)

The great granulite facies belt in the northern margin of the North China Protoplatform clearly reflects strong crustal movements. In the future more detailed work will no doubt result in more sophisticated models. For example the whole directional distribution of this granulite facies belt contrasts with the boundary between amphibolite facies and granulite facies in some specific regions, such as the N-S striking facies boundaries in eastern Hebei Province, the NE-striking boundaries in the Hengshan region and the E-W striking boundaries in Inner Mongolia. A regional analaysis of structure and protolith formation is required so that a reasonable explanation for the genesis of the granulite belt can be made in the context of crustal evolution.

2.3.4 Deformation Patterns

The wide extents of Archean granulite-gneiss terrane in China show a distinct structural pattern resulting from multiple phases of deformation. The typical deformation is a complicated interference structural pattern characterized by gneiss domes, associated refolded isoclinal folds and embryonic ductile shear zones (Fig. 2.3).

Interference fold patterns. Superimposed folding is extremely common in Archean granulite-gneiss terranes. Some cases of interference fold patterns of both large- and small-scale in such terranes have been described and analysed in some specific regions. The Archean Anshan complex in eastern Liaoning Province, the northeastern part of the North China Craton, has long been thought a simple monocline. However, detailed studies of this region (Liu Ruqi et al. 1984) reveal that there are four phases of deformation. Each phase is characterized by folds of distinct pattern and direction, and associated structural elements such as mineral lineation, mullion and boudinage. The axial planes of the first-phase folds are displayed by the gneissosity due to the differentiation of some chemical components along certain preferred mineral orientations. The folds vary from intrafolial to regional ones (Photo 2.5), and generally exhibit closed isoclinal or even recumbent fold patterns of nearly N-S direction. Locally, they are ductile nappes involved in the later large gentle folds with steep axial planes. In some places, these first-phase folds have been strongly reworked or even destroyed during or after the second-phase deformation. Consequently, they are usually enveloped in different parts of the second stage regional folds, with their axial traces varying and rotating around the hinges of the latter. The second-phase deformation appears to consist of tight to broad folds of different scales with associated foliation and lineation. The folds have east west axial traces and nearly vertical axial planes, but axial- plane foliation. Hence it is reasonable to assume that these folds are the products of flexural-slip folding and flattening. Because of the superposition of third-phase folding, these second-phase folds show

Photo 2.5. Hook-shaped folds in magnetite quartzite of the Qianxi complex. Tight recombent and even down-curling folds are regarded as D_1, whereas the open, nearly upright fold is F_2 (photo by Bai Yiliang et al. 1984).

different directions and vary, not only in the dip and dip-angle of their hinges, but also in the attitude of their axial planes. Moreover, their axial traces turn approximately around the hinges of the third-phase folds. The third-phase fold patterns range from tight to broad and from medium to large scale, with well-developed microcrenulation as the axial-plane foliation and stretching in a nearly NNW direction. The fourth-phase folds are only locally developed and are broad and upright.

The Taihua complex from the north slope of eastern Qinling Ranges, the central section of the southern margin of the North China Craton, has complicated and inhomogeneous structural patterns. Based on detailed geological mapping and structural analysis, a sequence of four phases of deformation has been established (Zhang Guowei et al. 1985). The first phase of folding (F_1) formed in the early stage of the late Archean Fuping (or Songyang) movement. This produced the regional gneissosity (S_1) which has transposed the original bedding (S_0), and also created well-developed intrafolial folds. In addition, the emplacement of mafic-ultramafic intrusions also occurred. The second-phase folding (F_2) occurred at the end of the Archean time and is characterized by the tightly folded S_1 foliaton. The F_2 event also has produced a linear fold-belt and associated structural elements such as axial-plane foliation (S_2), boudinage structure, mineral lineation and small tight folds. This event was associated with extensive plutonism. The third-phase folding (F_3) is characterized by tightly overturned folds superimposed on second-phase ones and associated with the formation of faults and mafic-felsic dyke swarms. The last phase deformation has produced broad vertical and inclined folds, kink bands and other small structures such as mullion structures. The stuctural sequence in Taihua complex mentioned is approximately similar to that in the Dengfeng complex.

The complicated interference fold pattern formed by the superposition of polyphase folding with different orientations is one of the fundamental structural patterns in Archean granulite-gneiss terranes. Although these patterns are extremely complex, it is still probable that some universal laws exist (Bai YL et al. 1984; Suo ST et al. 1982).

Most of the first-phase folds are obviously intrafolial in character and recumbent folds are generally refolded by the second-phase of tight folding. It is usually difficult to reconstruct their original orientations. The third-phase folds mostly occur as asymmetrical broad ones in which ductile strike-slip shearing zones are commonly formed at the steeply inclined or overturned limbs of the early ones, producing both horizontally and vertically complicated interference patterns.

The structural sequence mentioned indicates not only a multi-deformation history from an early ductile to late brittle deformation style, but also an evolution of the crust from an active to a stabilized state.

2.4 Gneiss Domes

Many studies indicate that various oval structures and gneiss domes of different scale and form control the fundamental structural patterns in the Archean terranes. They are mostly related to diapirism in the deep crust and in a minority of cases to the superposition of multiple folding.

In northeastern China the striking structural features of the Archean complex are characterized by the gneiss domes associated with some complicated fold belts. For example, in the Qingyuan region, the supracrustal rocks exhibit sophisticated synforms around gneiss domes. The metamorphic grades of the supracrustal rocks decrease from the dome edge outwards to the interior of the fold belts, which implies that the granitic diapirism resulted in the formation of the oval structure (Zhai Mingguo et al. 1985). However, Liu Ruqi et al. (1984) considered that the oval structures in the Archean Anshan complex of the Qingyuan region were formed by the superposition of two phases of folding. Another gneiss dome is located at Tiejiashan, 4 km to the east of Anshan, where the gneissicity formed by the recrystallization and plastic flow of previously existing tonalite-granodiorite intrusion during progressive metamorphism. Although a structure of varying attitude has been formed, it approximately constitutes a perfect regional oval fold about 25 km^2. However, this is only the exposed part of a vast Archean terrane dominated by oval folds and domes which are the consequence of structural regimes in the deep Archean crust (Windley 1984). When the geometry and axial plane features of the folds and protolith formations are synthetically considered, it is probable that the brachyfolds were formed by the vertical flow (or diapirism) associated with flexural folding.

The Archean rocks in eastern Hebei Province also show granulite-gneiss dome patterns which have been reworked by later folds (Fig. 2.4). There are at least six big gneiss domes in this area:- the Qian' an uplift is one of them. The gneissose granites of this uplift intrude into the amphibolite and granulite in the lower successions of the Qianxi Group and show a gneissose structure near the contact. The plastic flow structures in these gneissoid granites are concordant with the foliation of the supracrustal rocks. All these features indicate that the Qian'an gneiss dome is related to granite diapirism.

An antiformal oval structure with its long axis trending NE is developed in the Fuping Group in the north of the Taihang Mountains. It is 120 km long and 70 km wide (Fig. 2.5), and is formed by the superposition of plastic folding. In the north of this structure, there are a series of isoclinal folds and associated dome-basin structures. Because the axes of these folds tend to be convergent towards the east and bend towards the south, it has been considered to be a vortex structure formed by vertical shearing. In the west of the oval structure, the strike of the isoclinal folds change from W to NW. An arc-shaped belt composed of domes and oval antiforms, basins and synforms as well as recumbent folds is developed towards the east, the strike of which changes from NW to EW and then to the ENE. Detailed studies indicate that the Fuping Group has experienced three phases of deformation (Zhang Shouguang 1983). The first phase was characterized by recumbent folds which can be directly observed in the outcrops and the associated regional gneissicity. Tight and inclined folds and crenulation cleavages were formed in the second phase. Formation of gently inclined and broad brachy-antiforms and synforms accompanied regional uplift in the last phase. Eventually, the regional oval structure was formed. It is clear from field relationships that this kind of gneiss dome is entirely different from that formed by granite diapirism.

Dome structures from many high-grade Archean terranes are usually incomplete because of later deformation. For example, Zhang Guowei (1984) pointed out that the oval gneiss domes in the southern margin of North China Craton extended along the first-phase structure in an east-west direction, and because of the deformation of later tight folding, they were transformed into a long narrow fold chain with an EW direction. In addition, gneiss domes are present in the Dabie Mountain region on the southern margin of the North China Craton, the Miyun-Huairou region, north of Beijing, and the Xikang-Yunnan region of the southwestern margin of the Yangtze Craton.

2.5 Ductile Shear-Zones

Ductile shear zones, especially those associated with early low-angle or recumbent folding and parallel with the axial planes, are another important and extensive structural feature in the Archean crystalline basement. It is generally true that they are

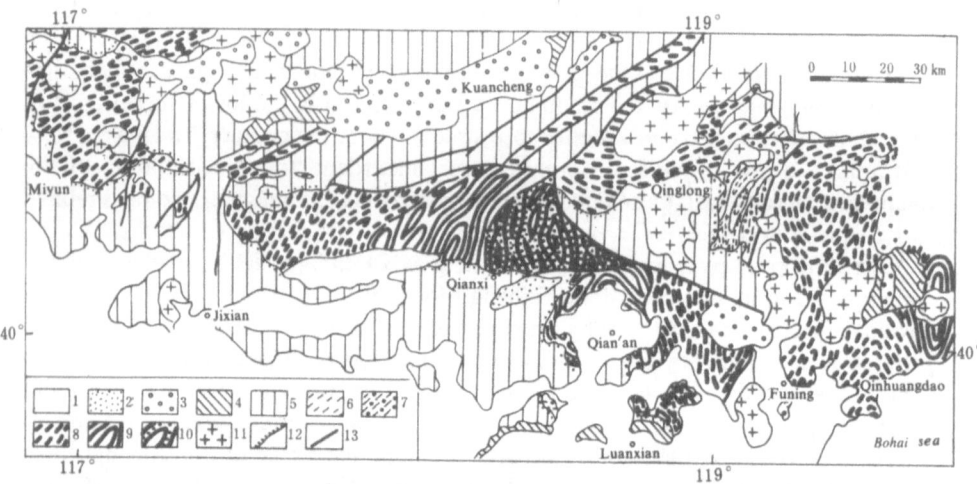

Fig. 2.4. Structural sketch map of eastern Hebei Province. (after Bai Yiliang et al. 1984). 1. Quarternary; 2. Tertiary; 3. Mesozoic; 4. Paleozoic; 5. Mid-Upper Proterozoic; 6. Lower Proterozoic: Zhuzhangzi Group; 7-10. Archean complexes; 11. granite of different ages; 12. uncomformity; 13. Fault.

only identiable on the outcrop scale by detailed studies of tectonites and the measurement of finite strain. Because of the superposition of later deformation, the ductile shear zones parallel to the fold axial planes do not, in many cases, maintain their original attitudes, but show various directions.

The regional ductile shear-zones that can be clearly seen in many of the Archean terranes occurred mostly after Late Archean times. It is well-known that ductile shear zones are mostly formed during amphibolite, or low amphibolite facies metamorphism. The mineral deformational structures in these rocks tend to be eliminated by recrystallization and only locally can a few blasto-mylonites be found. For this reason, ductile shear-zones are chiefly characterized by various mylonitized gneisses. However, when they cross Early Archean granulite facies terranes, retrograded and recrystallized mylonitic fabrics are commonly well developed. Enormous shear-flow structures at outcrop scale are preserved in many long narrow ductile shear-zones which cut the existing gneissosity. All of these features indicate high strain.

One of the most remarkable structural patterns that survives in the crystalline basement of the North China Craton was formed by Late Archean ductile shearing, is the gradual transposition of a former gneissosity by narrow ductile shears. This may result in displacement, or even rotation of the adjacent terrane. For example, a prominent ductile shear-zone with a NNE direction is named the Tancheng-Lujiang shear-zone, the central section, studied by Zhang Jiasheng (1983), located on the eastern edge of Taishan complex in western Shandong Province, is 20 km long and has apparently changed the structural trend of the Taishan complex. Detailed geological mapping and analysis of the microfabrics of deformed rocks reveals that the less strained blocks are usually separated by highly strained zones (Fig. 2.6). Deformation structures such as (1) the intrafolial and sheath folds in magnetite quartzite, biotite schist and banded felsic gneiss; (2) the plastic boudinage structures in hornblendite and magnetite quartzite, and (3) the secondary flow cleveages in felsic gneiss, are widely developed in outcrops in this area, constituting the lineation and foliation of the zone. This shear-zone transects the North China Craton from its northern to its southern margin in a NE direction.

The Late Archean ductile shear-zones in eastern Hebei and western Liaoning Provinces also extend in a NE direction. A common characteristic is that a dynamic retrograde metamorphism of amphibolite facies occurred where these shear zones cross the Early Archean granulite facies terrane. Blasto-mylonites are locally developed, and mylonitic textures and plastic-flow structures are well preserved in these retrograde zones.

As a whole, the late Archean ductile shear-zones show deformation features, high temperatures and high differential stress suggestive of deep crustal levels (Ma Xingyuan et al. 1987).

2.5.1 Geochronology

Apart from those older ages of 3.8-3.5 Ga mentioned in Chapter I.1, extensive isotopic age data ranging from 3470 to 2500 Ma has been obtained from the Archean granulite-gneiss terranes in China. New dates have been recently reported in several regions: examples include the Sm-Nd age of 3470 Ma in the metabasic volcanic rocks from eastern Hebei Province (Jahn et al. 1987), the U-Pb age of 3330 Ma in the tonalitic gneisses of eastern Liaoning Province (Jahn et al. 1987), and the Rb-Sr isochron of 3219 Ma in the metabasic volcanic rocks from Alxa, the westernmost end of the North China Craton (Yang Zhande et al. 1988). The age of 3330 Ma is probably the intrusive age of the old tonalites, but there have been different interpretations of the geological significance of the age data for the supracrustal lithologies. There is no doubt that the supracrustal rocks were deposited 3.5 Ga ago, but the nature of the underlying basement upon which they were deposited is unknown. The 3.5 Ga age therefore gives a lower age limit for the formation of the North China crust.

Although fewer ages around 2.9 Ga have so far been obtained, they show that a period of extensive mainly granulite-facies metamorphism occurred at this time. This was associated with the intrusion of mainly Na-rich granites, based on the correlation of the geological events in each granulite-gneiss terrane.

According to the zircon U-Pb age of 2880±170 Ma and biotite K-Ar age of 2900±0.9 Ma in tonalites from the Qingyuan region, Liaoning Province, it is suggested that a period of Na-rich magmatism and migmatization, together with related regional metamorphism, took place 2.8-2.9 Ga ago (Zhai Mingguo 1985). Liu Dunyi et al. (1985) obtained a zircon U-Pb age of 2.8 Ga in the feldspathic gneiss of granulite facies from the lower succession of the Fuping Group, Taihangshan region. They interpreted the felspathic gneiss to have a sedimentary protolith, and the sub-rounded brown zircons analysed may have be detrital in origin. Hence they suggested that the age of 2.8 Ga Ma may bracket the lower age limit of the Fuping Group. However, even if the gneiss has a sedimentary precursor, it is hardly conceivable that detrital zircon would remain after granulite facies metamorphism. Moreover, because the Fuping Group has experienced polyphase deformation and metamorphism, it is doubtful that the original U-Pb system in the zircon would have been maintained. Hence the sub-rounded nature of the zircon may be due to corrosion rather indicating a detrital nature. Additionally, Kröner et al. (1988) obtained two zircon $^{207}Pb/^{206}Pb$ ages of 2841±6 Ma and 2806±7 Ma in the tonalite gneiss of the Taihua complex from the southern margin of the North China Craton. Sun Jiashu et al. (1982) obtained a Pb-Pb isochron of 3055±250 Ma from the plagioclase and biotite of the Qianxi complex granulite in eastern Hebei Province. Based on the data described, it is plausible that an important tectono-thermal event characterized by strong metamorphism and migmatism occurred 2.8-2.9 Ga ago in the North China Craton, symbolizing the Qianxi movement.

Large amounts of age data between 2.5-2.6 Ga have been gained from each granulite-gneiss terrane. They obviously represent the ages of the magmatism characterized by the intrusion of early Na-rich granites and later K-rich granites, of the (predominantly amphibolite facies) regional metamorphism and of the strong secondary deformation. All these features are expressions of the Wutai movement. It was approximately between 2.5 and 2.9 Ga, or in the timespan between the Qianxi and Wutai movements, that the supracrustal rocks characterized by amphibolite facies metamorphism in the granulite-gneiss terranes were formed. These approximately correspond to the second Archean volcano-sedimentary cycle proposed by Cheng Yüqi et al. (1984).

In the high-grade terranes, ages of 1.8 Ga or younger are common and may symbolize the effects of the Lüliang movement or later geological events.

Fig. 2.5. Structural sketch map of the Archean domal structure in Fuping region (slightly modified after Zhang Shouguang 1983). K_z - Cenozoic; M_z - Mesozoic P_z - Paleozoic; P_t - Middle and Upper Proterozoic; Pt_z - P_z- Middle Proterozoic to Paleozoic; Pt_2 - Middle Protreozoic; Pt_1 h-Hutuo Group; Pt_1w - Wutai Group; g_1 - granites of Yanshan age; g_2 - granites of Fuping stage; 1. upper Fuping Group; 2. middle Fuping Group; 3. lower Fuping Group; 4. basic dyke swarms; 5. granites; 6. faults; 7. F_3 folds; 8. F_2 folds; 9. inferred F_2 folds; 10. attitude of fold; 11. transitional boundary between granite and country rocks; 12. stratigraphic boundary and unconformity.

2.5.2 Tectonic Framework

According to the rock assemblages and geochronology of some typical regions, all
the Archean granulite-gneiss terranes have experienced similar evolutionary processes of chemistry, tectonics and geochronology, though each terrane has its unique
features in rock assemblage and chemistry.

Geochronology indicates that at least two suites of supracrustal rocks are developed about 2.9 Ga and 2.5-2.6 Ga ago. These suites underwent granulite facies and
amphibolite facies metamorphism respectively. Most of the high-grade metamorphic rocks show identical structural patterns, having being subjected to poly-phase
deformation in a largely permobile tectonic regime.

The upper supracrustal rocks overlying the early granulite facies rocks were subjected to amphibolite facies metamorphism, which resulted in the retrogression of

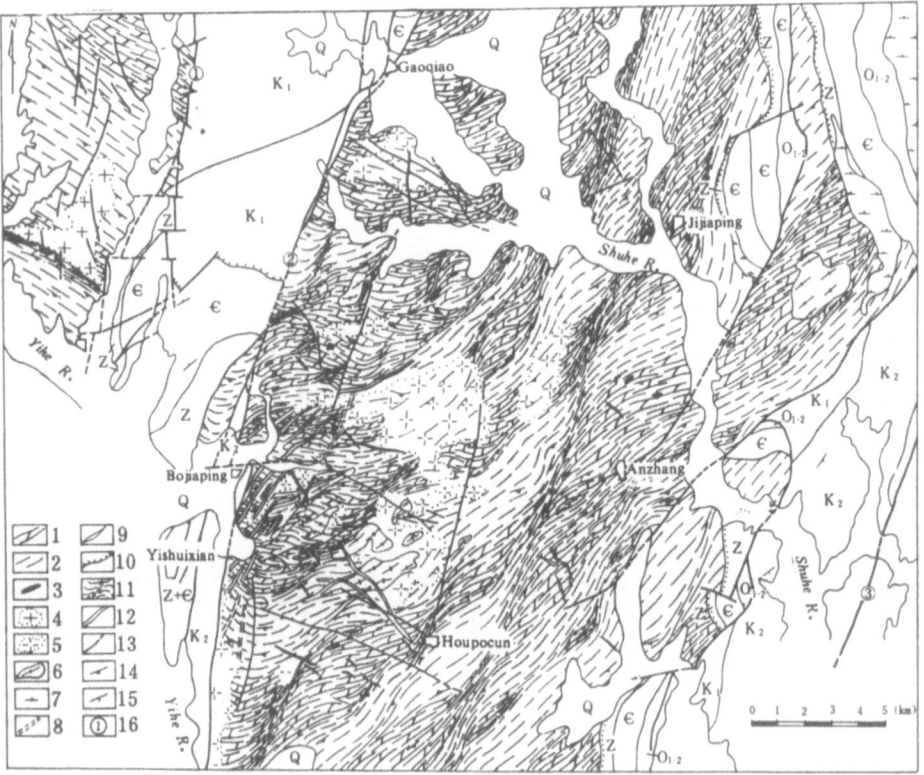

Fig. 2.6. Structural map of the central section of the palaeo-ductile Tancheng-Lujiang shear-
zone in Yishui area, Shandong Province (after Ma Xingyuan et al. 1984) 1. Hornblende
gneiss of Archean Taishan complex; 2. biotite gneiss of Taishan complex; 3. BIF; 4. granite;
5. migmatitic granite, granitic gneiss; 6. metagabbro; 7. metadiorite; 8. mylonitized quartz
porphyry vein; 9. late diabase; 10. unconformity; 11. deformation zone of ductile shear; 12.
mylonite zone; 13. fault; 14. attitude of foliation; 15. attitude of bedding; 16. number of main
faults: (1) Tangwu-Gegou fault; (2)Yishui-Tangtou fault; (3) Anqiu-Lüxian fault.

the early granulites. Obviously, before the deposition of the upper suite, its underlying crust must have been uplifted and subjected to extensive erosion. On the other hand, if the sediments represented by the "khondalite" series and some similar sediments formed in stable and shallow continental shelves have experienced amphibolite facies metamorphism, reaching temperatures of 650°C, then given an average geothermal gradient of 33 °C/km then these sedimentary rocks should have been buried to 20 km deep to reach amphibolite facies conditions. After metamorphism, these amphibolite facies rocks were uplifted and eroded prior to the deposition of further deposits. Thus the Archean crust not only experienced the process of cratonization, but also experienced polycycles of extensive uplift, erosion and subsidence during Archean times.

Metasupracrustal rocks, especially amphibolites and related rocks, occur as remnants and blocks of different sizes within granitic bodies. Locally, clear intrusive relations can be seen at the contacts between the two. However, in view of the similarity of these two rock units in geochemistry and their field contact relations, it is suggested that some of the granitoids may be the products of partial melting of the supracrustal rocks. This process also occurs in Phanerozoic orogenic belts and thus it appears that the Archean terranes and the Phanerozoic orogenic belts have experienced similar processes. Up to the present time it has generally been believed that the China Craton was formed by the end of Archean times although the formation of the earliest sialic crust is still a matter of contention.

Apart from the North China Craton represented by the typical regions described above, Archean high-grade metamorphic rocks are widely distributed over other areas of China's continental crust, such as the peripheries of the Tarim Craton, and the western and northern margins of the Yangtze Craton (Ma Xingyuan et al. 1987; Feng Benzhi et al. 1986; and Cai Xuelin et al. 1986). They were reworked by the Proterozoic orogenic belts, and many of the Archean rocks occur as small inclusions and blocks in the Proterozoic belts. Geophysical data reveal that the crystalline basements of these protoplatforms are probably composed chiefly of Archean rocks (Ma Xingyuan et al. 1984). During the mid-Proterozoic, a collisional event occurred in the northeastern margin of the Yangtze Craton. The present palaeomagnetic data also indicate that the basement of the China Plate which was formed at the end of the Archean times was not unified by the mid-Proterozoic Era. Moreover, the Early Proterozoic active continental marginal formations represented by the Qinling and Hong' an Groups were developed near the southern margin of the North China Craton. However, there is a distinct diversity between the North China and the Yangtze Craton (see 3.3). For this reason, it can be deduced that the North China, Yangtze and Tarim Cratons were separated from each other at the end of Archean time.

In the North China Craton, supracrustal rocks first appeared at about 3.5 Ga. The most prominent structure trends are nearly E-W in the Yinshan region, from NE-SW to E-W in the Fuping region of Taihang Mountains, NW-SE (see Fig 2.1) in the Taishan region of western Shandong Province, nearly N-S direction in the eastern Hebei Province, NW-SE to E-W in the Yanshan region, and NE-SW to NW-SE in the Liaoning-Jilin region. In view of the variations of structural trends and metamorphic belts, the upper rock units might not have been formed in a unified tectonic setting. Similarly, during most of the Archean there may not have been a single block of unified basement, but some isolated and different segments in the North China Craton. No matter what the compositions and origins of these segments are, they are probably a few old, small continental crustal nuclei scattered over this vast region. Due to the separation of younger platform covers of different ages, it is very difficult to determine accurately the configuration and scales of these segments, with the result that it has been impossible to establish their tectonic boundaries and original states. In other words, the basement of the North China Craton may have been built up by the accretion of small continental nuclei by the end of Archean time.

Based on the geophysical data, Liu Shoupeng (1981) has studied the deep structure of North China and believes that clear velocity layering in the crust is apparent. The depth of the PC_1 reflection wave interface corresponds to the boundary between sialic and simatic crusts and ranges from 14 to 20 km in depth. The average depth of the deep interface of magnetic bodies in the whole region is 14-16 km, which also reflects the depth of the PC_1 interface. According to the map of the depth contour line of deep interface of magnetic bodies (Fig. 2.7) and combining the analysis of the aeromagnetic anomaly and the map of shallow magnetic interface (Guan Zhining et al. 1987), the North China Craton can be divided into six uplifted blocks of simatic crust, i.e. - the Dongsheng, Linfen, Jining, Bohai, Chifeng and eastern Liaoning regions. They are bounded by the isobath lines of the deep interfaces of more strongly magnetic bodies. These sites of these six uplifted locks correlate well with the variation in structural trends between different areas. Hence the regions of geophysical anomalies might provide the evidence for small crustal nuclei which built up the North China Craton.

In conclusion, different lines of evidence suggest that older mini-continental nuclei underwent a period of amalgamation c.2.5 Ga ago (Wang Hongzhen et al. 1987). This amalgamating process might be accomplished by terrane collision which was probably associated with small-scale subduction. The ductile shear-zones, referred to as tectonic boundaries, may reflect collisional suture zones. In view of the difference of the Archean crustal thickness, geothermal gradient and crustal plasticity from those of the modern crust, it is inevitable that the scale and character of plate tectonic processes would be different from those of later times, if indeed such processes operated at all in the Archean.

2.6 Greenstone Belts

The Barberton greenstone belt in South Africa is the best studied Archean greenstone belt in the world (Viljoen and Viljoen 1969; Anhaeusser 1971). Owing to their unique role in the reconstruction of the early evolutional history of the Earth crust and their rich mineral deposits, Archean greenstone belts have attracted the special attention of geologists ever since their discovery. Successive reports of "Archean greenstone belts" in Western Australia, Canada, India, Finland and Brasil, have lead

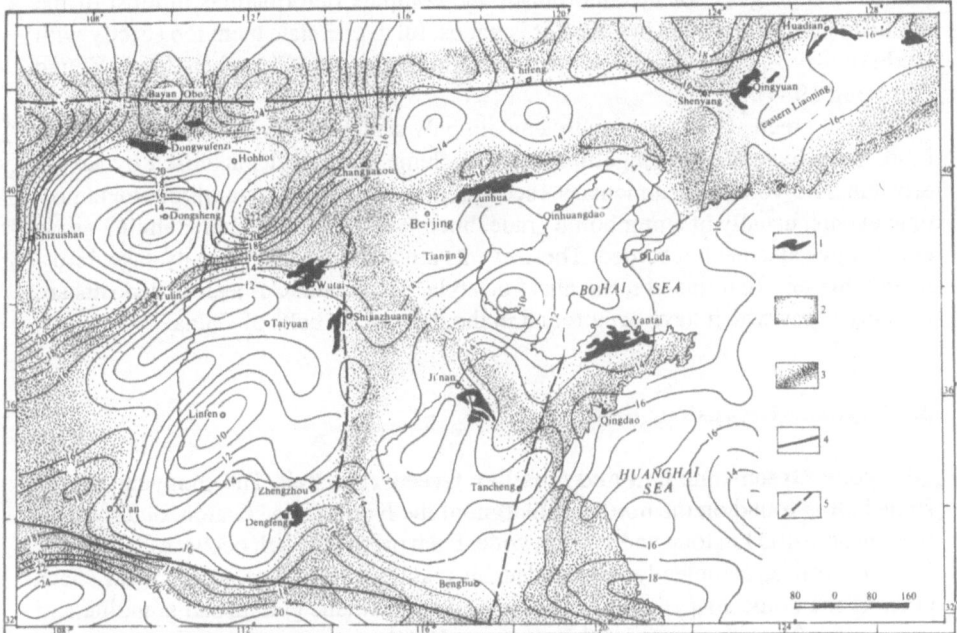

Fig. 2.7. Map showing the isobath of deep seated magnetic boundary (after Guan et al. 1987) and the schematic distribution of some exposed Archean greenstone belts in North China h=16 km, J=400.10^{2}A/m, unit for isobaths= Km: 1. Archean greenstone belt; 2. magnetic boundary upwarping; 3. magnetic boundary downwarping; 4. the boundary of North China Craton; 5. deep fault

to the acceptance of the term but also lead these terranes to be regarded with remarkable similarity worldwide (Condie 1981; Anhaeusser 1982; Windley 1984). Generally, the term greenstone belt denotes a kind of supracrustal rock largely composed of mafic volcanics, but with calc-alkaline volcanics becoming increasingly important upwards within the sequence. Ultramafic and komatiitic lavas sometimes occur in the lower part of the sequence, followed by felsic volcanics. The upper part of the sequence is dominated by wacke-dominated sedimentary rocks, which are exposed in linear or irregular shapes within a vast expanse of granitic rocks. According to this definition, Archean greenstone belts are widespread in China. Unfortunately, post-Proterozoic sedimentary cover conceals most of the greenstone belts, except for exposures around the North China Craton and on the southwest margin of the Yangtze Craton.

The highly metamorphosed greenstone belt associations in China have long been studied by Chinese geologists as common supracrustal rocks, whereas the more lightly metamorphosed ones were termed "greenschist series" rocks. However, given the relatively high grade of metamorphism and complex deformations in most of the Archean greenstone belts in China, it is not useful to correlate them too strictly with Barberton-type greenstone belts (Sun Dazhong and Wu Changhua 1982; Zhang Qiusheng 1984).

Chinese greenstone belts usually contain only scarce of olivine komatiite at their base. They have a wide range in size and volume and are characterised by severe erosion. The volcano-sedimentary succession is dominated by multiple metamorphic events, usually of amphibolite grade, but upper greenschist and granulite grade events have also been recorded. These events are associated with a polystage deformation history. In terms of mineralisation style, Au, Fe and Cu-bearing deposits are the only types which have been found in the greenstone belts of China.

2.6.1 Typical Greenstone Belts

Qingyuan Greenstone Belt and Huadian Greenstone Belt. The Qingyuan greenstone belt is found on the northeast margin of the North China Craton, outcropping over an area 400 km long and 120 km wide. Each part of the belt occurs as layered or irregular relics, completely surrounded by granitic rocks. The supracrustal rocks encircle the dome-shaped terrane made up of hypersthene granite and granulite and account for 25% of its volume (Liu Yuguan, 1982).

Since its discovery 10 years ago, a notable amount of study has been devoted to this belt (Qin Nai, 1980; Yan E et al. 1980, and unpublished work). Additional studies comparing the Qingyuan greenstone belt with classic examples world-wide has indicated a remarkable resemblance. The analysis of the Archean greenstone formation sequence of northern Liaoning Province enabled Fang Ruheng (1981) to define it as a highly metamorphosed greenstone belt which formed in a oceanic plate environment. Alternatively Li Shuguang (1986) suggested that the belt was generated in an environment similar to a modern island arc. Yang Zhensheng et al. (1988) analysed the multiple deformations of the greenstone belt in the Hongtoushan region, Northern Liaoning and concluded that it was formed at a deep crustal level.

Formerly, the supracrusal rocks in the region were grouped within the previously described Anshan complex. One supracrustal unit, the Qingyuan greenstone belt was formerly divided into four formations (Fig. 2.8). In ascending order these are termed the Jingjiagou Formation, Shipengzi Formation, Hongtoushan Formation and Taizigou Formation. The Jingjiagou Formation (also called Xiaolaihe Formation by Zhai et al. 1985) constitutes basement made up of charnockite and granulite cropping out in the core of the greenstone. In strict terms it is a high-grade relict and is therefore not a member of the Anshan Group. Furthermore, the quartzite and marble series in the upper Taizigou Formation (termed the Nantianmen Formation by Zhai et al. 1985), occurs in isolation within granitoid rocks. As it shows no direct contact with the granulite and BIF in the lower part of the Taizigou Formation, we exclude it from the greenstone belt. Therefore, we define the Qingyuan greenstone belt as the volcanic-sedimentary assemblage including the Shipengzi Formation, Hongtoushan Formation and the BIF series of Taizigou Formation.

The lower Shipengzi Formation is largely made up of amphibolite and pyroxene-bearing hornblendites with minor thin-layered biotite plagioclase gneiss and occasional ultrabasic rocks (serpentinized peridotite and cherzolite). The upper Shipengzi Formation mainly comprises amphibolite, biotite plagioclase gneiss and light-colored gneiss. The entire formation is about 1300 m thick. The protoliths to the Shipengzi Formation were ultramafic and mafic volcanic rocks: massive komatiitic rocks and tholeiite-bearing lenticular komatiitic rocks formed the lower part of the sequence and bimodal valocanics formed the upper part. The komatiitic rocks are high in MgO (>9%), low in K_2O (<1%) and with CaO / Al_2O_3 > 1. The high grades of metamorphism (Shen Baofeng et al., 1989) probably explain the absence of spinifex textures. Zhai Mingguo et al. (1985) and Yan E et al. (1986) regarded the komatiites as the first stage of an early greenstone belt.

Overlying the Shipengzi Formation is the 5000 m thick Hongtoushan Formation. The formation consists of intercalated amphibolite and biotite plagioclase gneiss. These yield massive sulphur deposits, with a little felsic gneiss bearing sillimanite and kyanite. The protoliths of this Formation are dominated by calc-alkaline rocks: tholeiites, andesites and rhyolites form the base of the Group, with interbedded tholeiites, dacites, pelitic deposits and BIF overlying them. The uppermost Taizigou Formation is 2000 m thick and is made up of a fine-grained biotite plagioclase gneiss, two-mica quartz schist, fine-grained amphibolite and minor amounts of magnetite quartzite. The evolutionary trend of the volcanics of the Shipengzi and Hongtoushan Formations are correlated by Zhai et al. (1985) with those of Belingwe, Zimbabwe and Abitibi greenstone belts.

Tectonically the Qingyuan greenstone belt makes up a synclinorium clamped between the older Xiaolaihe and Xianjinchang granitic gneiss domes. Both domes were severed by the Hunhe fault and the greenstone belt engulfed the gneiss domes in an arcuate pattern (Fig. 2.8). Yang Zhensheng (1984) suggested that the Qingyuan greenstone belt suffered from as many as four phases of structural deformation, during the three phases of the Anshan Orogeny and the first phase of the Lüliang Orogeny. The results of the first stage of deformation are small, closed isoclinal folds, combining into fold combinations of various orders. Two prominent hinge-strikes are found plunging to NNE and WSW, dipping at 20-30°. The second stage of deformation was much more severe, not only reworking the former structures but also creating the dominant structural framework. During the second stage of deformation medium to large-scale closed folds were formed with E-W trending structural lines crossing those of the first stage. Accompanying the deformation were amphibolite-granulite facies metamorphism, migmatism and deep level granite intrusion. The third stage of deformation was much weaker, creating open folds with principal hinges pointing N-S and dipping at 10-20°. Retrogressive greenschist metamorphism during this phase of deformation is indicated by biotite and chlorite on the foliation plane. This fourth stage of deformation was marked by fault activity, vein and dyke intrusion. The scale of crustal fracturing ranges from microfissures to large faults, with prevailing directions being NNE, NEE and NNW. Some basic and ultrabasic dykes and sheets are a possible by-product of this deformation. Zhai Mingguo et al. (1985) published isotopic ages spanning 2.6-2.9 Ga for the Qingyuan greenstone belt and the latest Sm-Nd age of 2.8 Ga for the amphibolite reported by Shen Baofeng (1989) supports the former age determination.

The geochemical parameters of the komatiitic rock in the Qingyuan greenstone belt classify it as basaltic komatiite, analogous to Jahn's I-type komatiite with its smooth REE pattern $(Gd/Yb)_N$ =1.08 - 1, CaO/Al_2O_3=0.81 - 0.73 and Al_2O_3 / TiO_2 =21.78 - 2.74. It also shows slightly depletion of HREE, an Eu negative anomaly and has $(Gd/Yb)_N$ =2.06, CaO/Al_2O_3=2.16 and Al_2O_3/TiO_2=12.6 values corresponding to Jahn's type II Komatiite. The tholeiites of the greenstone belt include both REE-depleted and REE-enriched types with similar HREE concentrations. According to Condie's classification (Condie, 1976), they overlap the fields of ocean ridge tholeiite (TH_1) the calc-alkaline basalt series or continental tholeiite (TH_2). Generally, the TH_1-type tholeiite makes up the Shipengzi Formation and the lower Hongtoushan Formation, whereas TH_2 type can only be found in the upper Hongtoushan Formation. Accordingly, the evolutionary mode of the preserved mafic volcanics reflects an environmental transition similar to modern island arcs, but

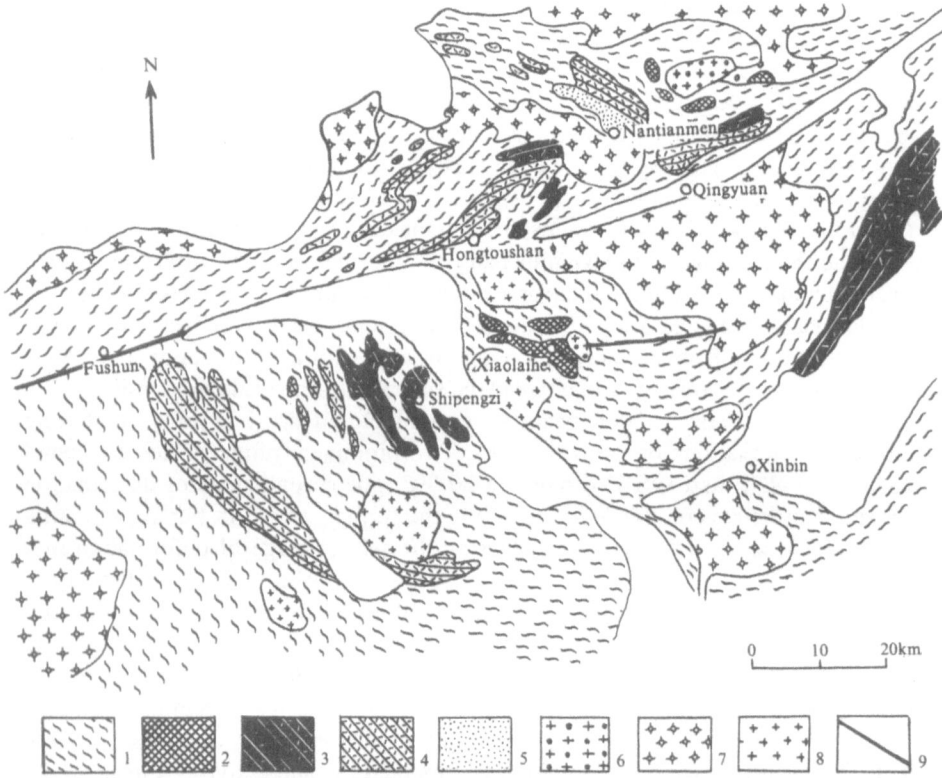

Fig. 2.8. Geological map of the Qingyuan area, eastern Liaoning Province, showing the distribution of Archean (modified after Zhai Mingguo et al. 1985): 1. Felsic gneiss and migmatite; 2. granulite (Jingjiagou Fm); 3. Shipengzi Fm.; 4. Hongtoushan Fm.; 5. taizigou Fm. (Nantianmen Fm); 6. charnockite; 7. tonalite; 8. Cenozoic granite; 9. fault

the presence of komatiites reflects a specific characteristic of Archean times. The andesite in the Qingyuan greenstone belt typically has enriched LREE, a small Eu negative anomaly and high K_2O =1.4-1.8%). Hence its chemical signature does not favour an oceanic origin. The felsic volcanics in the Shipengzi Formation are characterized by low ΣREE, strong depletion of HREE and an indistinct Eu anomaly. They fall into Condie's F_1-type (analogous to modern island arcs). Meanwhile the felsic volcanics in the Hongtoushan Formation, have a smooth REE curve and obvious Eu anomaly as well as hosting Cu, Zn mineral deposits. In contrast they belong to the F_2-type (analogous to modern continental arcs) in Condie's classification (Condie, 1976). It is suggested that the Qingyuan belt had a petrogenesis unlike most greenstone belts and was probably generated in a peculiar island arc-trench environment.

Another remarkable greenstone belt exposed in Liaoning-Jilin area is the Huadian greenstone belt. It is situated on the northeast margin of the North China Craton, stretches for 340 km E-W across southeastern Jilin. The west section of the belt has gone through severe ductile shearing (Fig. 2.9) and is surrounded by tonalite and oligoclase granite which has yielded ages of 2497 Ma and 2565 Ma (Dai Xinyi et al. 1989). This belt forms the major area of iron and gold mines in China.

Dai Xinyi et al. (1989) divided the greenstone belt into two formations. The lower formation has thick-bedded pyroxene amphibolite at the base, above it occurs stratiform plagioclase-amphibolite, lenticular amphibolite and actinolite, and further up is fine-grained amphibolite intruded by metabasic dykes and interbedded with minor amounts of biotitic-leptynitic quartzite and hornblende-magnetite quartzite. Mafic volcanics are suggested as the protolith of the lower formation, as blastophitic and blastoporphyritic textures as well as amygdaloidal structures are visibly preserved. The well-known Jiapigou gold mine is situated in the middle part of this formation. The upper formation is composed of interbedded fine-grained gneissic amphibolite, biotite leptinolite, biotite quartz schist and garnet kyanite biotite schist, with a little amphibolite and several layers of BIF. Protoliths of the Formation consisted of basic and felsic volcanics, pelitic and sandy sediments, and a large iron deposit.

The ultramafic rocks in the Huadian greenstone belt fall in to two types: one comprises metamorphosed, intrusive ultrabasic rocks derived from dunite, lherzolite, and harzburgite protoliths. Their mineralogy was substantially converted by metamorphism to serperntine, talc, tremolite and carbonates; the second type crops out as lenticular bodies conformable to the mafic volcanics in the lower level of the greenstone belt. These ultrabasic rocks were converted by metamorphism and deformation to hornblendite, actinolite, tremolite and talc-bearing rocks. The original spinifex texture was destroyed, although the komatiitic features are still recognizible by their exposure mode and chemical components. Chemically they are characterized by low TiO_2(0.35-0.95%) and K_2O (0.34-0.94%), high MgO (12.10-16.09%) and CaO / Al_2O_3= 0.87-2.05. Their character is identical with that of komatiite. The REE contents reveal a low ΣREE (8-14 times of that of chondrite), $(La-Yb)_N$ = 1.83-2.39, $(La-Sm)_N$ = 1.53-2.01 and $(Gd-Yb)_N$ = 1.32-1.49, coincident with BK_1-type basaltic komatiites.

The major oxide and REE content (Cr, Ni, Co and V) of the mafic volcanics metamorphosed into amphibolite and amphibolitic schist are similar to those of basic rocks, apart from a few samples which corresponds to high-Al basalt with higher Al_2O_3 content (16.03-18.82%). Most samples have an Al_2O_3 contents between 11.78-14.23%, equivalent to tholeiite. In comparison with present ocean tholeiite and island arc tholeiite, the mafic volcanics of the belt have distinctly higher K_2O (0.74-1.26%), Ti_2O (0.97-3.2%) and FeO*(13.23-16.08%) as well as Rb and Ba contents which resemble continental rift tholeiite. However, the REE patterns of the mafic volcanics are smooth (ΣREE = 28.95-40.76 ppm). The mafic volcanics can be divided into types based on LREE enrichment: an LREE slightly enriched type (ΣREE = 60.75-79.17 ppm), and an LREE enriched type (ΣREE = 90.46-135.33 ppm). If classified according to scheme proposed by Condie (1976), the mafic volcanics in the belt change progressively from ocean ridge tholeiite (TH_1) at the base to continental tholeiite (TH_2) in the upper part of the sequence.

The original andesite of the felsic volcanics is found in the eastern section of the Huadian greenstone belt. Metamorphism has converted the andesites into plagioclase hornblende gneiss and hornblende leptynite, chemically showing high contents of K_2O (1.93%), Zr, Ba and Sr and a lower Na_2O- K_2O ratio (1.74). It is equivalent to that of present high-K calc-alkaline andesite. The ΣREE ranges between 118.01-170.74 ppm, with higher fractionation of LREE and HREE [$(La/Yb)_N$]=7.91-12.81] and weak Eu anomaly (Eu/Eu* = 0.69-0.83). This corresponds to Archean andesite type F_{II} (i.e. continental arc type), according to the classification of Condie (1976). The biotite leptynite, biotite plagioclase gneiss and leptite, which make up the main body of the felsic volcanics, are chemically the same with rhyolite and dacite, characterized by higher K_2O (1.34-1.57%), slightly low Al_2O_3 (14.62-16.04%) and Na_2O (4.18-4.20%), with LREE/HREE = 8.85-18.18 $(La/Yb)_N$ = 14.38-38.28. The REE pattern is thus highly fractionated, with Eu/Eu*= 0.86-2.05, similar to the Archean F_1-type (i.e.- island arc derived) rhyolite and dacite (Condie 1976). The sedimentary rocks are largely tuffaceous sandstone, pelitic siltstone and siliceous-ferruginous rocks. Compared with the Shaba Formation wacke and pelitic rock in Shaberton, they have slightly higher Al_2O_3, CaO, K_2O, and a lower maturity [(Al_2O_3 /(Na_2O+K_2O) = 2.95-3.65)]. The metasediments are derived from reworked volcanic sources with only sparse input of terrigenous materials.

Dai Xinyi et al. (1989) reconstructed four stages of tectonic deformation in the Huadian greenstone belt. The first tectonic stage consisted of deep-crustal ductile deformation and metamorphism. It is only recognizible from surviving schistosity and gneissosity, rootless folds and boudinage structures. The second stage resulted in a series of isoclinal and recumbent folds as well as ductile shear-zones composed of closely packed schistosity and gneissosity, with the fold axis and shear zone extending NW-SE. A series of highly-developed inclined a-type mini folds with stretching lineations, indicates that the first two stages of deformation both took place under amphibolite facies metamorphism. The diapiric intrusion of tonalitic bodies associated with the first stage of deformation further complicated the deformationary style. During the third stage of deformation, very large scale NW-SE trending open folds were superimposed onto the recumbent folding pattern. Upright folds, a reversed S-shaped structural framework and a ductile shear-zone were also produced during this event. The orientations of the first three stages of deformation implied that they were probably created in different phases of a single Archean orogeny. The regional metamorphism associated with tonalite-tronhjemite and granodiorite intrusion was of low amphibolite facies grade, with P-T conditions of 450-650°C, 0.45-0.55 GPa. However, high amphibolite facies grade is found in the basement, with P-T conditions of 600-700°C, 0.6-0.7 GPa (Dai Xomuo et al. 1989). Lastly, the fourth stage of deformation resulted in NE-SW oriented, monotonous open folds. This tectonic event is probably of Early Proterozoic age, and was associated with a remarkable degree of retrogressive metamorphism. Finally, the intrusion of the Early Proterozoic potassium granite (K-Ar biotite age of 1754 Ma), granodiorite (U-Pb zircon age of 1754 Ma) and granodiorite (U-Pb zircon age of 1617 Ma) concluded the history of deformation and metamorphism in the region.

The following isotopic ages have been obtained from the greenstone belts and the primitive deformation-metamorphism: nine Pb-Pb zircon ages give an age range of 2479-2639 Ma for the Jiapigou greenstone belt; whereas nine further zircon Pb-Pb apparent ages range from 2444 to 2536 Ma; two whole-rock Pb-Pb isochron ages are of 2525±12 Ma and 2501±13 Ma, a Rb-Sr isochron age of 2766±266 has also been analysed. Lastly, tonalites intruded into the greenstone belt give and zircon U-Pb ages of 2497 Ma and 2565 Ma. These ages clearly suggest that the greenstone belt

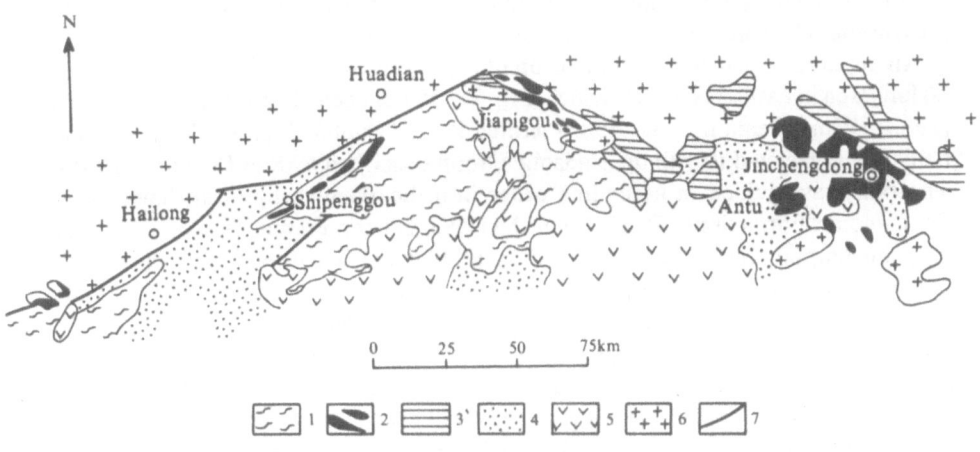

Fig. 2. 9. Geological map of the Huadian greenstone region, Jilin Province (after Dai Xinyi et al. 1989): 1. Granulite-gneiss terrane; 2. greenstone belts; 3. Proterozoic; 4. Phanerozoic; 5. Cenozoic basalt; 6. granite; 7. fault

was formed 2.7 Ga ago and that the deformation and metamorphism associated with coeval TTG suite intrusion occurred at 2.5 Ga.

The volcano-sedimentary rocks of the Huadian greenstone belt show obvious cycles, and the high alkaline components of volcanics rocks suggest a possible mixing of crustal materials. The lower part of the volcanic sequence is of bimodal type. This reflects an expanding ocean environment during the beginning of greenstone belt petrogenesis. However, the upper part of the sequence is predominantly a calc-alkaline volcanic association. The limited amount of andesite appears to be a high-K calc-alkaline type, generated in a continent marginal island-arc system with a very thick lithosphere. The sedimentary rocks of low maturity represent a transitional environment between the island-arc and active continental margin. In addition, regional geophysical data have discovered a gravity gradient zone on the south of the greenstone belt. This is interpreted as reflecting a potentially deep continental margin fault (Dai Xinyi et al. 1989).

Zunhua Greenstone Belt. At the end of the 1970s, a number of geologists suggested that many relics of supracrust rocks in the Archean terranes in Yanshan region, eastern Hebei, on the north margin of the North China Craton, had salient greenstone belt characteristics (Zhang Yixia et al. 1980; Lan Yuqi et al. 1984). Unfortunately, the relics have largely been dismembered into isolated inclusions of varing size scattered in a sea of granitoids, after undergoing high-grade regional metamorphism and complex deformation. Even though the lithology and chemistry of each relic corresponds to a specific part of a greenstone belt, it is almost impossible to reconstruct the complete stratigraphy of a primitive greenstone belt. Thus at present the relics are regarded as the components of a granulite- gneiss region.

An apparent greenstone belt made up of supracrust rocks, which covers a 130 X 20 km^2 area in eastern Hebei Province (Fig. 2. 10), has been identified by Tan Yingjia (1983). Wang Renmin et al. (1985) termed it the "Zunhua high-grade greenstone belt" and correlated it with the Badaohe Group described by Sun Dazhong (1984). The belt was divided in ascending order into the Wangchang, Wanzhangzi and Sanmendian Formations. The 2250 m thick Wangchang Formation is a column of diopside amphibolite and plagioclase diopsidite, with some ultramafic hornblendite and sparse magnetite quartzite in the lower part of the sequence. The protolith of this formation is komatiites and tholeiite, which is equivalent to the mafic-ultramafic bimodal volcanic association in the lower greenstone belt. The ultramafic rocks correspond superbly well with western Greenland komatiites dated 3.8 Ga Ma (Rivalenti 1976; Zhang Yixia et al. 1980). It is therefore likely that spinifex texture existed in these rocks prior to metamorphism (Wang Renmin et al. 1985).

The 1600 m thick Wanzhangzi Formation is dominated by amphibolite with minor interbedded biotite-plagioclase leptynite, leptite and magnetic quartzite. The protolith to this formation is basic, intermediate and acid volcanics. The Sanmendian Formation extends for 3350 m and consists of a mass of biotite plagioclase gneiss intercalated with amphibolite, hornblende-plagioclase gneiss and actinolite-magnete quartzite. This formation is regarded as the sedimentary assemblage in the upper green-stone belt, with protoliths consisting of graywacke, sandy clay silicic-ferruginous sediments and minor volcanic lava and pyroclastic rocks.

The origin of the ultramafic rocks is still questionable (Sun Dazhong et al. 1984). However, thousands of ultramafic rock masses, occur as stratiform bodies, in lenses, and in strings of various sizes along gneissic schistosity. The ultramafic units are concordant with the schistosity but their boundaries with host rock are widely talcised. This suggests that they were initially homogeneously bedded rock bodies, and their present appearance results from the various stages of tectonic deformation with stretching, boudin production, rotation and even structural emplacement. Sun Dazhong (1984) obtained a whole-rock Rb-Sr isochron age of 2552±45 Ma and a zircon U-Pb age of 2500 Ma for the tonalite intruded in to the belt, proving its Archean age.

The lower part of the mafic rocks is characterized by low K_2O-high TiO_2, slight REE fractionation [$(La/Yb)_N$= 0.75-3.29] and a ΣREE of 39-80 ppm, lower than that in the Qianxi area (70-105 ppm), The REE content is 10-14 times chondrite values and correlatable with Condie's (1976) TH_1- type; i.e. - equivalent to modern ocean-ridge tholeiites. In the SiO- <FeO>/MgO diagram, the metavolcanics demonstrate both tholeiite and calc-alkaline trends that respectively fall into modern oceanic tholei-ite and calc-alkaline volcanic areas. However, in the MgO- <FeO°> diagram they fall into the Mg-rich series fields, tholeiite series and calc-alkaline series, and re-semble the Abitibi greenstone belt in Canada (Wang Renmin et al. 1985).

To the southeast, the border between the Zunhua greenstone belt and an adjacent granulite-gneiss terrane is marked by a ductile shear-zone (Zhang Yixia 1986). Struc-tural studies in this area show two generations of folding. The first phase of folds are close, isoclinal, and overturned with axis planes dipping to the west. These struc-tures were formed under strong E-W compression. The younger folding episode superimposed large E-W trending open folds onto the first episode. However, the deformation planes of the first generation relate to a previous gneiss schistosity, with recognizible interfolial folds hinges (Sun Dazhong et al. 1984). It is suggested that this schistosity was formed by an early E-W trending recumbent fold.

The mafic rocks in the lower Zunhua greenstone belt are chemically similar to low-K oceanic tholeiite and imply an extensional tectonic setting, but the calc-alka-line series appear in the upper part of the sequence. Taking into account the regional structure analysis, it is more reasonable to conclude that the greenstone belt grew up in an evironment similar to a modern back-arc basin-island arc rather than a rift environment.

Dongwufenzi Greenstone Belt. Scattered greenstone belts have been found in a large area of granitoids in the western section of the north margin of the North China Craton. One of these is the extensive and well-studied Dongwufenzi belt (Li Shuxun et al. 1986; Jin Wei, 1989). The lower part of the belt consists of hornblende-plagio-clase gneiss, hornblendites, amphibolite and magnetite quartzites. The protoliths of this sequence were probably mafic and minor ultramafic volcanic rocks and banded

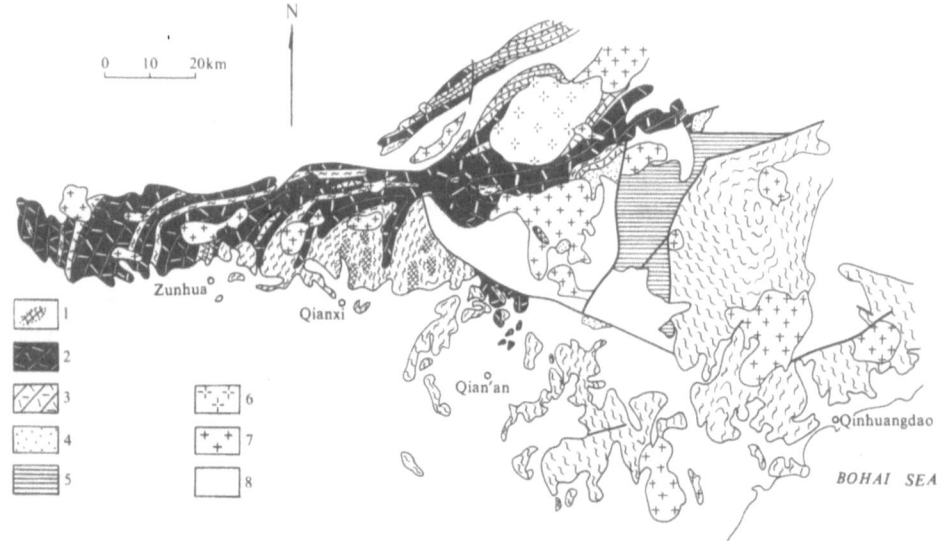

Fig. 2.10. Geological map of the Archean of the eastern Hebei Province, showing the distribution of the Zuhua high-grade greenstone belt (revised after Sun Dazhong et al. 1986):
1. Granulite-gneiss terrane with granulite relics (schematically shown in crossed areas); 2. Wangchchang Fm.; 3. Wanzhangzi Fm.; 4. Sanmendian Fm.; 5. Lower Proterozoic; 6. Proterozoic granite; 7. Mesozoic granite; 8. Mid-Proterozoic to Quaternary cover

iron formations. This is overlain by biotite-hornblende-plagioclase gneiss, leptite, amphibolite and epidote-biotite leptynite, which constitutes the middle of the sequence. Prior to metamorphism, the strata probably consisted of calc-alkaline series volcanics. The upper part of the Group consists of quartz schist, biotite leptynite, mica schist and marble. Likely protoliths were felsic pyroclastics, immature sedimentary clastics and minor impure carbonates. Owing to the low-grade of metamorphism, original structures and textures are very well preserved: vesicular, amygdaloidal and pillow structures found in the metavolcanics, palimpsest crystal and lithic fragments seen in the metapyroclastics; and rhythmic bedding is visible in the metasedimentary rocks. In general, the protolith formation of the greenstone belt appears to be an andesite-free "bimodal" volcanic association consisting of intermediate-basic and acid volcanics. The layered hornblendite in the lower part of the sequence shows a chemical similarity to typical basaltic komatiite: rich in MgO (23.5-26.8%), poor in Al_2O_3 (6-7%) and $CaO/Al_2O_3>1$. Its REE distribution corresponds to the Archean BK_1 of Condie (1981), with low ÝREE (36.46-38.12 ppm), enriched LREE, depleted HREE and a LREE/HREE ratio of 1.33-1.51. The mafic amphibolite and hornblendites have a basaltic chemical composition, but with higher K_2O and Na_2O than the average for tholeiite and calc-alkaline series. Thus high-K intraplate basalt or potassic basalts, may also form some of the protoliths of K-rich tholeiitic volcanics.

The border between Dongwufenzi Group and the neighbouring granulite-gneiss region is marked by a huge E-W striking ductile fracture zone. The greenstone belt occurs as an E-W stretching synclinorium accompanied by large ductile shear-zones with promising prospects for gold mineralization. Few detailed structural analyses have yet been made inside the greenstone belt, but the plunging vertical folds seen in the neighbouring area with axial plane parallel to schistosity strongly imply the occurrence of recumbent folds at early stages in the deformationary history. The greenstone belt has gone through a markedly zoned regional metamorphism of low greenschist to low amphibolite facies. The distribution of facies belts, with the interior of low greenschist and the two limbs of amphibolite grade, form a "thermal syncline". The temperature-pressure conditions is estimated as 400-635°C, 0.25-0.5 GPa with a geothermal gradient of 34-35°C km. Retrogressive metamorphism merely reached low greenschist facies and largely took place along the ductile shear-zones. The PT trace demonstrate a temperature rise under low pressure in the early stage of metamorphism and an increase in pressure in the later stages. A sharp and abrupt difference of 150°C and 0.45 GPa existed between the metamorphisms of the greenstone belt and of the granulite-gneiss region (Jin Wei 1989), which certainly indicates that at the time when the greenstone belt was metamorphosed, it was not in contact with the presently-adjacent high-grade region.

In terms of tectonic environment, the likely protoliths, the possible early recumbent folds, the synclinal palaeo-geothermal plane of the metamorphic facies zones and the anti-clockwise PTt trace of the Dongwufenzi belt jointly suggest a for-arc–trench tectonic environment.

Alax Greenstone Belt. Yang Zhende et al. (1988) studied the Alax Group in Alax region at the west end of the North China Craton and established it to be a greenstone belt. Vigorously developed intrabed ductile shear-zones and nappes destroyed the original sequence by pushing, splitting and piling up the segments of the greenstone belt and the remains are seen as scattered outcrops. The components of the greenstone belt emerge in new-moon crescents, stripes and lens shapes within areas of granitoids. The belt extends 300 km long by 60 km wide, and appears to be synform structure in cross-section (Fig. 2.11).

The Alax Group is divisible into five rock formations. In ascending order they are made up of ultrabasic-basic, basic volcanic, intermediate volcanic, sedimentary clastic and carbonate lithologies. This stratigraphy roughly corresponds to a greenstone sequence. The 8000 m thick lowest formation makes up the basement and principal part of the greenstone belt and occurs sporadically as inclusions in surrounding granitoid. The formation is composed of migmatized amphibolite, pyroxenite, hornblende rock, hornblende gneiss, biotite (hornblende) gneiss, marble, quartzite and magnetite quartzite. These lithologies reflect a protolith association dominated by basalt and ultramafics, with minor amounts of andesite, carbonate rock, sandstone, sandy clay and very rarely spilite. The ultramafic rocks are chemically similar to komatiites whereas the basalt is almost homologous to the present oceanic tholeiite and is chemically similar to the Hooggenoeg Formation in the Onverwacht Group of South Africa. Some high-Al and alkaline basalts also occur. The andesite is largely of the Ca-rich variety.

Mica quartz schist, metavolcanics, and biotite gneiss together with minor hornblendites make up the intermediate-acid volcanic rock formation. The protoliths to this formation were acid volcanics and pyroclastics intercalated with a little basic and intermediate volcanics. The basic volcanics in the formation were mainly composed of alkaline basalt and high-Al basalt; the intermediate rocks are largely calc-alkaline andesite and acid volcanics rhyolitic or dacitic lavas and tuffs, with a little polymineral sandstone-dominated sedimentary rocks and minor carbonates. A rhythmic sequence of basic -intermediate-acid volcanics and sedimentary rocks constitutes the above rocks. Where the intermediate volcanics are absent, the bimodal eruption cycle occurs instead.

The sedimentary-clastic rock formation in the upper part of the Alax greenstone belt contains principally of mica-quartz schist and metasandstone, with minor biotite and hornblende gneiss and marble at the top. All these rocks reveal a protolith of alternating pelitic-sandy and sandy-pelitic sediments as well as turbidite, tuffaceous clastics, tuffs and carbonates. The correlation of the Alax greenstone sequences in different areas is shown in Fig. 2.11.

The ultramafic rocks are rather well developed and are especially well preserved at the lowest level of each section. However, their preserved spinifex texture is not typical of komatiites, and may be formed from skeletal tremolite (Yang Zhende et al. 1988). Their range of major element components is as follows: SiO_2, 34.25-44.10; Al_2O_3, 2.22-8.27; Fe_2O_3, 4.65-8.77; FeO, 3.87-9.47; MgO, 23.93-35.08; CaO, 0.59-5.62; TiO_2, 0.02-0.52; Na_2O, 0.07-1.42 and K_2O, 0.02-0.97. The average composition corresponds to peridotite. Iron-rich members of the ultramafic rocks are usually interbedded with tholeiites, while the magnesian series was deformed in to mega-shear zones and huge nappes, being placed in contact with other rock masses by faults (Yang Zhende et al. 1988). These structures both reflect the tectonic relationship between the greenstone belt and the underlying rocks and represent a paleotectonic boundary.

The main sequence of the greenstone belt has been deformed into close isoclinal overturned folds, commonly accompanied with thrusts to form imbricated structures or deformed by later shearing into structural lenses. The E-W oriented fold hinges of the early stage folds, together with the regional schistosity, were reoriented by later tectonic activity and the bent fold axis and schistosity are readily seen. Figure 2.11 is a sketch map of structures of a subregion in the northeast Alax greenstone belt. It shows the relics of primitive gneissic schistosity, lineation structures and granitoids as well as basic-ultrabasic rocks. The originally E-W striking fold axes, themselves superimposed on the early primitive dome structures, are seen bending ENE due to the superimposition of NE-trending fold structures. N-S trending fold structures are also superimposed on the E-W trending folds.

Accompanying the multiple-stage deformation, several phases of metamorphism have affected the greenstone belt. In the early phase of metamorphism, E-W trending granulite to amphibolite facies zones were formed, followed by a low greenschist to greenschist phase. The occurrence of ductile shear-zones and nappe structures broke the rock mass into tectonic fragments which have piled up and the metamorphosed belt was split by intermittent outcrops of granulite facies and low greenschist facies. The paleo-geothermal regime reflected by the protoliths, metamorphic facies, and structures in the basement of the greenstone belt suggest a transitional tectonic environment from trench to island arc.

Yang Zhende et al.(1988) reported a whole-rock Rb-Sr isochron age of 3218 Ma for the amphibolite in the basic volcanic rock formation. Owing to poor homogeneity and isochroneity of the samples, the geologic significance of the age is somewhat uncertain. Nevertheless, in general the age is taken as that of the metamorphism.

Luxi and Jiaodong Greenstone Belts. On both sides of the Shandong section of the Tan-Lu fault, situated in the mid-eastern North China Platform there occur two greenstone belts named the Luxi (Western Shandong) and Jiaodong (Eastern Shandong) greenstone belts respectively.

The Luxi greenstone belt outcrops around the Taishan Mt and has a stratotype section in Yanlingguan, 50 km southeast of the Taishan Mt, where it occurs as a synclinorium surrounded by a sea of granitoid. An intrusion in the core of the

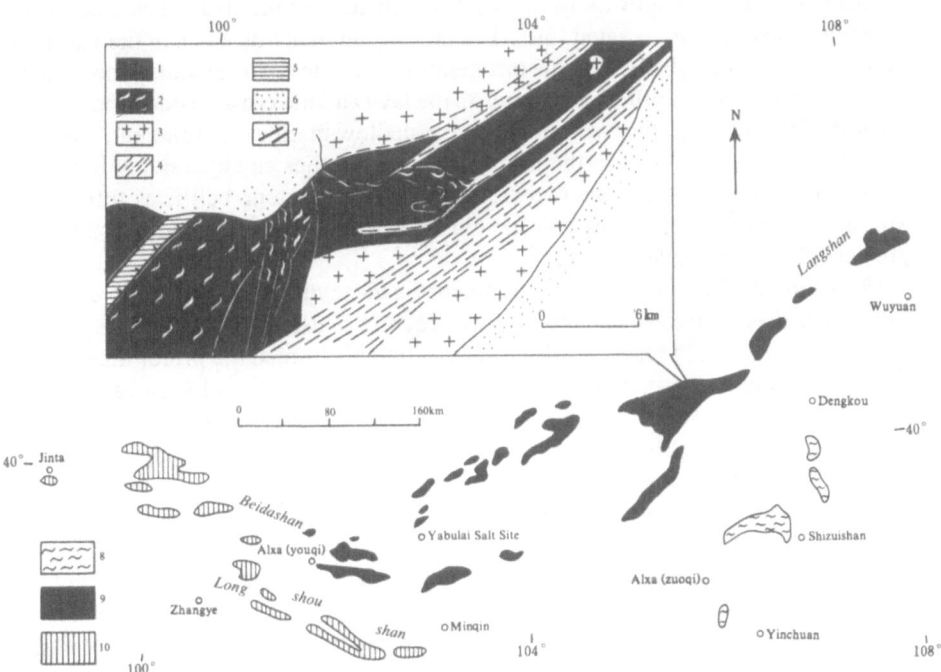

Fig. 2.11. Map of the Alax region, showing the distribution of the greenstone belt (modified after Yang Zhende et al. 1988): 1. Greenstone belt; 2. intrafolial ductile shear zone within greenstone belt; 3. granitic rocks; 4. ductile shear zone; 5. Mesozoic; 6. desert; 7. faults; 8. Archean granulite-gneiss terrane; 9. Alxa Group; 10. Proterozoic Longshou Group

synclinorium cuts the belt into two halves: the Yanlingguan section in the east and Liuhang in the west (Fig. 2.12).

The lithologies of the Luxi greenstone belt comprise the Taishan Group and is divided into two rock formations; i. e. the lower Yanlingguan Formation and the upper Shancaoyu Formation. The 1400 m Yanlingguan Formation is mainly composed of hornblendites, comprising several layers of massive amphibolite, hornblende leptynite, biotite leptynite, gravel-bearing biotite-hornblende leptynite and some siliceous-ferruginous rock, with a presumed protolith of a set of basic volcanic-sedimentary rocks. According to their mode of eruption, the volcanics can be divided into upper and lower subformations. Both of them have ultramafic rocks such as tremolite-actinolite schist, talc tremolitite and serpentinite in the basal levels. Overlying levels are dominated by mafic amphibolite in which numerous amygdaloidal and pillow structures are preserved. Between the two subformations is situated gravel-bearing hornblende-biotite leptynite and sericite quartz schists with occasional banded graphite. These units represent a quiet period of volcanic activity. Ultramafic komatiite makes up a large part of the lower subformation and is comparable with those well-known in other countries. Meanwhile the upper subformation which takes up the middle part of the greenstone belt reflects another eruption cycle. Bimodal volcanics are poorly developed in Yanlingguan section, but outcrop in the Liuhang section 10 km away to the west, dominated by ultrabasic-basic and acid volcanics as well as associated tuffs. The lack of bimodal volcanics in the east part of the belt may be explained by the difference in tectonic environments. The numerous amygdaloidal and pillow structures in the lava and the cross-bedded remnant in the tuff of the Yanlingguan section indicate a shallow marine environment, whereas the lack of such structures in the Liuhang section implies an abyssal environment. Although the gap between the two sections of the greenstone belt is as little as 10 km, the steep dip of the beds and the closed isoclinal folds reveal a substantially greater original distance .

Overlying the Yanlingguan Formation, the Shancaoyu Formation has several sections comprised of two-mica leptynite, mica quartz schist and hornblende leptynite, which suggests that a series of silty wacke sediments formed the protoliths to these rocks. This suggestion is strengthened by the occurrence of staurolite and almandine bearing biotite leptynite, almandine staurolite biotite quartz schist and andalusite biotite leptynite. The Shancaoyu Formation therefore makes up the upper sedimentary member of the greenstone belt.

The ultramafic volcanics of the Luxi greenstone belt have a total thickness of 380 m and have been metamorphosed into schistose or massive units largely made up of tremolite, actinolite, chlorite and talc. They principally occur in the lower part of the two subformations of the Yanlingguan Formation at the base of each basic volcanic cycle. Having been metamorphosed, the ultramafic rocks contain largely lepidoblastic and nematoblastic textures. Skeletal olivine or pyroxene spinifex textures are pseudomorphed by actinolite and tremolite porphyroblasts. The matrix consists of scaly chlorite in talc-chlorite-actinolite schist and talc-bearing chloritic tremolitite (Dong Yijie and Xu Huifen 1989). Chemically these rocks comprise SiO_2 (43-50%), FeO^*(7.3-11.5), MgO (13.7-26.4), Al_2O_3 (<10%), TiO_2(0.1-1.03% mostly <0.5) and

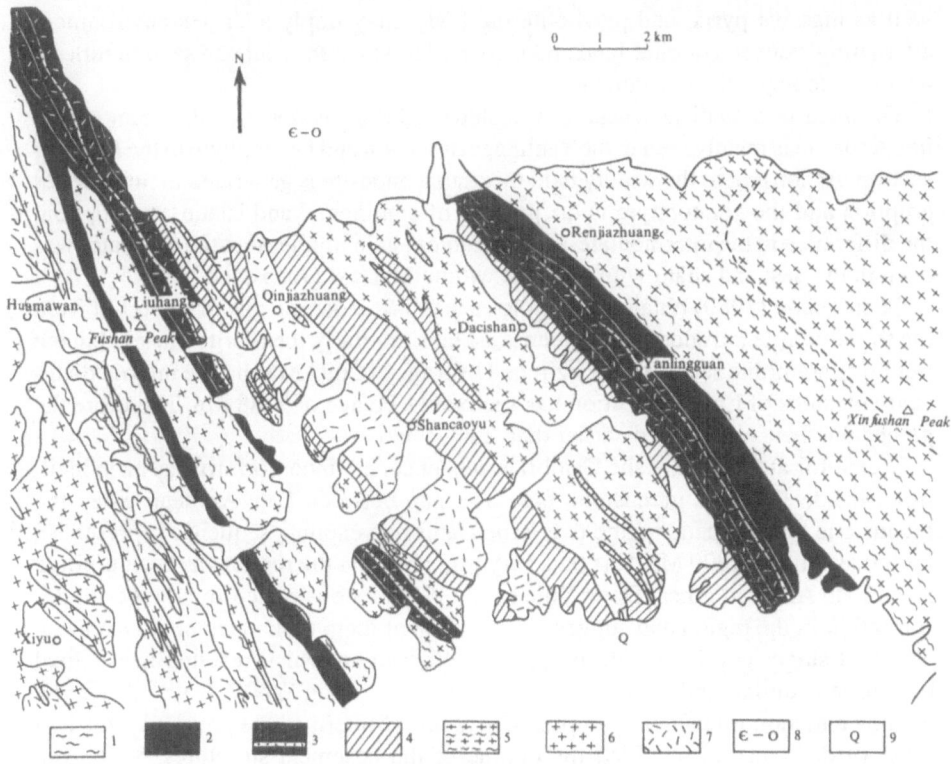

0 1 2 km

N

ε-o

oRenjiazhuang

Huamawan   Liuhang   Qinjiazhuang

Fushan Peak   Dacishan

Yanlingguan

Shancaoyu   Xin fushan Peak

Xiyuo   Q

1 2 3 4 5 6 7 ε-o 8 Q 9

Fig. 2.12. Geological map of the Archean granite-greenstone terrane in western Shandong Province (after Dong Yijie et al. 1989): 1. Biotite-plagioclase-gneiss; 2. lower Yanlingguan Fm.; 3. upper Yanlingguan Fm. with pebble-bearing biotite-leptite; 4. Shancaoyu Fm.; 5. trondhjemite; 6. granite; 7.quartz-diorite; 8. Cambrian-Ordovician; 9. Quaternary

K_2O (0.03-0.78%), with CaO/Al_2O3 spanning 0.07-1.59 (mostly <1), Cr (>300 ppm) and Ni (>100 ppm). According to the typical composition of komatiite defined by Brooks and Hart (1974), the rocks characterized by high-Mg, low-K and Ti and the spinifex texture are classified as komatiite and are divided into olivine komatiite and basaltic komatiite. This is the only spinifex texture bearing komatiite ever reported in China.

The mafic amphibolites, with percentage chemical components of SiO_2 (47-53%), Al_2O_3 (11-16%), Fe_2O_3(1-7%), FeO (5-11%), MgO(4-10%), CaO(3-11%), Na_2O(1-5%) and K_2O(0.1-0.5%), belong to the tholeiite and calc-alkaline series and largely fall in to the abyssal tholeiite area on Miyashiro's SiO_2-Fe*O-MgO and FeO*-FeO*/MgO diagrams (Miyashiro 1975), with a few points plotting in the transitional area between island-arc and abyssal tholeiites and within the island-arc tholeiite area, rarely in alkaline basalt area. The pillow and amygdaloidal structures in the Yanlingguan section indicate an environment transitional between abyssal plains to island-arc. Furthermore, the thin-layered graphite mica schist, graphite phyllite, as well as massive pyrite and pyrrhotite ore body, may imply a facies environment alternating between marine-land and lagoons. However the Liuhang section reflects a bathyal to abyssal environment.

The meta-intermediate to acid lavas include andesite keratophyre, dacite and rhyolite. Andesite is mainly seen in the Yanlingguan section and keratophyre in the Liuhang section. According to the widely held thesis that andesite is generated in underwater eruption and the keratophyre is the product of continental and island-arc eruption, the difference in the intermediate-acid volcanics again implies that the sea water was deep in the east and shallow in the west within the Luxi greenstone belt.

Shen Qihan (1980) obtained a whole-rock Rb-Sr isochron age of 2586 Ma from the trondhjemite intruded in the greenstone belt, similarly the diorite intruded in the tremolite actinolite schist yields a zircon U-Pb age of 2699 Ma and the intermediate-acid volcanics of Liuhang section has a zircon U-Pb age of 2788 Ma. Therefore the greenstone belt was formed earlier than 2.8 Ga ago.

Except for at the base of the Cambrian sequence, no unconformity has been identified inside the Luxi greenstone belt. The gap between Late Archean when the greenstone belt formed and the overlying Cambrian capping sequence represents a period as long as 2000 Ma. It is thus very probable that the region has been uplifted since Late Archean times. Dong Yijie et al. (1989) reconstructed four stages of deformation in the region and suggested that three of them took place in the Archean. The first stage was dominated by plastic deformation which resulted in isoclinal overturned folding and an axial plane schistosity striking WNW. It was accompanied by a metamorphism into amphibolite facies of intermediate pressure. This first stage of deformation moulded the outline of the basement structures. Strike-slip ductile deformation prevailed in the second stage, with the production of b-type folds resulting from sinistral strike-slip shearing. This ductile belt principally occurred in the Liuhang greenstone belt and the Fushan trondhjemite, and produced mylonite and stretched quartz. A third stage superimposed another ductile deformation upon the ductile sheared rock, resulting in a series of small smooth folds. Since no axial plane schistosity was formed, this stage had little impact on the structural framework.

Retrogressive greenschist facies metamorphism is closely related to the ductile shear-zone and fault belt, and provided paths for the mineralization of gold. Meanwhile the thermal metamorphic minerals andalusite and staurolite were possibly the products of later intrusion at the end of the Late Archean and Early Proterozoic times because some Early Proterozoic-aged intrusive bodies and dykes are also found in the region. After deposition of the cover sequence, the final stage of deformation occurred. This was a largely brittle deformation and resulted in a set of fractures which cut the Archean Luxi region into a series of faulted blocks, uplifts and downwarps.

Another greenstone belt in Shandong Province is the Jiaodong greenstone belt, situated on the Jiaodong peninsula and separated from the Luxi greenstone belt only by the Tan-Lu fault zone. Despite their proximity, remarkable diversity exists in the geological settings of the two belts (Fig. 2.13). For example, the Luxi greenstone belt has no Proterozoic cover and is capped directly by the Paleozoic, with the virtual absence of Mesozoic stratigraphy. In contrast, the Jiaodong belt has a complete cover sequence of Proterozoic deposits but lacks the entire Paleaeozoic sequence, with Mesozoic blanketed directly onto Proterozoic strata. Thus it is clear that a mutually compensational uplift and downwarp have been generated by the Tan-Lu fault on each side of the fault since Proterozoic time. As a significant tectonic boundary the Tan-Lu fault could have been generated as early as the Archean times. At that time its effect was strong enough to create the differences between the two greenstone belts although they still have some features in common.

The Jiaodong greenstone belt was studied initially by Deng Youhua et al. (1980), and Wang Konghai (1983) investigated it further. There are still diversified opinions on this sequence (Yang Shiwang 1986). However, the basic volcanic-sedimentary rock assemblage of the Jiaodong greenstone belt is similar to that of the typical Archean greenstone belts. The belt is also host to remarkably abundant gold deposits. This has lead to the belt receiving a lot of attention and it becoming known as the "golden peninsula".

The Jiaodong Group is classically divided into three formations. In ascending order these are the Pengkuang, Minshan and Fuyang Formations. The Pengkuang Formation is taken as the basic volcanic series in the lower part of the greenstone belt, and it is principally made up of amphibolite, hornblende leptynite and biotite leptite, with minor pyroxene amphibolite and hornblende actinolitite. These reflect a protolith of basic volcanics and tuff intercalated with sparse quantities of ultrabasic rock, intermediate-acid volcanics and clastics. The Minshan Formation is predominantly a mass of biotite hornblende leptynite, leptite, biotite and kyanite sillimanite biotite gneiss, with a little tremolite diopsidite, marble, feldspar quartzite and serpentite. The protolith is presumed to be basic-intermediate volcanics and tuff with minor ultramafic rocks, terrigenous clastics and chemical sediments. Two-pyroxene granulite lenses are probably derived from metamorphosed noritic gabbros intruded into the sediments. The Fuyang Formation forms the upper sedimentary sequence which contains garnet-cordierite-sillimanite-biotite gneiss, biotite leptinite, biotite schist, leptite and some amphibolite and tremolite marble. The protoliths to this formation were terrigenous clastics and minor carbonate rocks. Dacitic volcanics and tuffs with minor basic volcanic lenses occur in the lower part of the sequence.

The ultrabasic rocks within the Minshan and Pengkuang Formations include lenticular and thinly-bedded pyroxene amphibolite-actinolite-tremolite-serpentite, chlorite-hornblende-schist, talc schist, hornblende biotite rock and peridotite. Their chemical composition corresponds predominantly to basaltic komatiites, with lesser occurrences of olivine and pyroxene komatiites. The thin-bedded peridotite or sheet ultrabasic rocks are the main part of the lower greenstone belt formation, although the amphibolite-granulite facies metamorphism imposed onto it may have virtually destroyed the spinifex texture. Wang Konghai (1983) envisaged that the flow structure made up of sheeted olivine pseudomorphs in the actinolite tremolite reflected an eruptive origin. Geochemically, the basic volcanics belong to the calc-alkaline series, island arc tholeiites. They are characterized chemically by low-K, low-Ti and

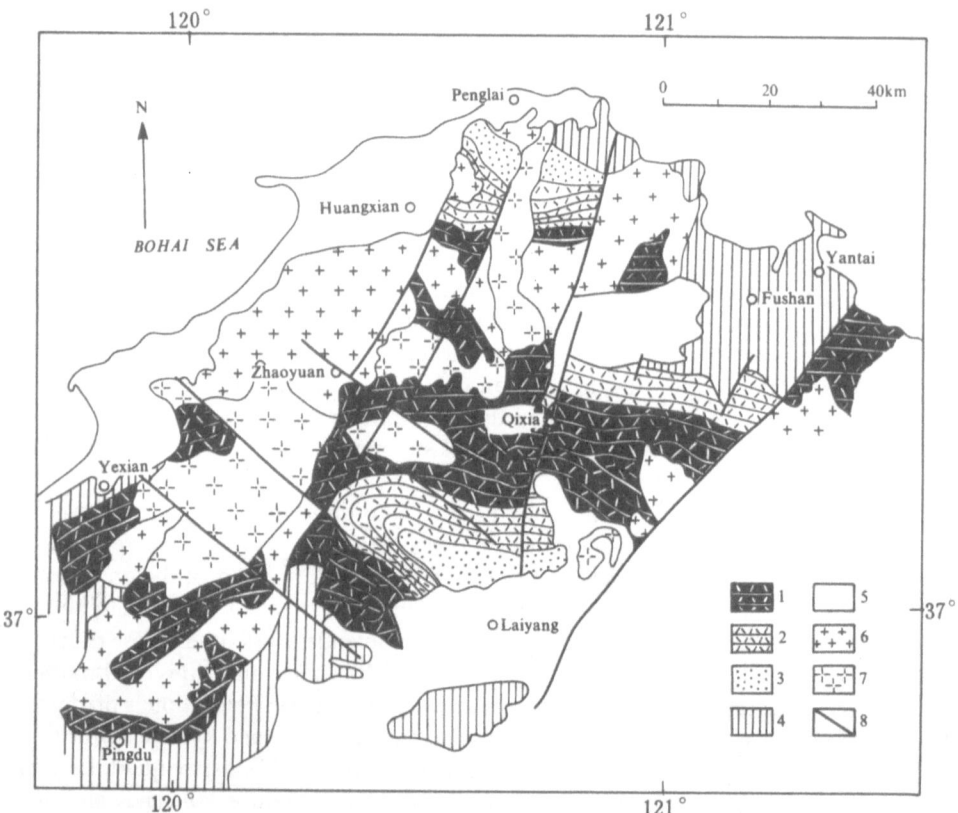

Fig. 2.13. Geological map of the Archean greenstone belt in eastern Shandong Province (after Wang Konghai 1983): 1. Pengkuang Fm.; 2. Minshan Fm.; 3. Fuyang Fm.; 4. Early Proterozoic metamorphosed series; 5. Late Proterozoic to Quartenary cover; 6. Proterozoic granite; 7. Mesozoic granite; 8. fault

high-Mg. These accordingly indicate that the greenstone belt was developed in an environment like present island arcs.

The prominent structural form of the Jiaodong greenstone belt is an anticlinorium with its hinge plunging to the east and in which a steeply-dipping gneiss schistosity belt of 600-2000 m wide (with usual angle of dip > 75°) developed, together with some local, close-fold belts a few hundred meters across. Well-developed fractures occur within the Jiaodong Group, and late stage brittle fractures formed a network of faults. Late stage faulting in particular resulted in the accumulation and exhumation of an economically huge amount of gold in the region. The rocks have undergone regional metamorphism ranging from low amphibolite to granulite facies and the PT determination of the amphibolite-granulite facies shows T=720-810 °C, P = 0.5 GPa, with a geothermal gradient as high as 41-46 °C/km.

The greenstone belt yields a number of age values spanning 2510-2628 Ma, with the maximum of 2858 Ma. Furthermore a tonalite-trondhjemite intruded in the belt has a zircon U-Pb age of 2415-2663 Ma, so the Jiaodong greenstone belt is considered to have been formed around 2.6 Ga ago and the first metamorphism took place at around 2.5 Ga.

Wutai Greenstone Belt. The Wutai Mountains in the central North China Craton expose a greenstone belt termed the Wutai greenstone belt. The sequence which makes up the belt is named the Wutai Group. The Wutai region ranks as one of the best studied Precambrian regions in China, commencing with the investigation by Richthofen (1882). He named the rock mass the 'Wutai greenschist system', which is adopted by later workers such as Wills (1904), Wang Yuelun (1953) and Ma Xingyuan et al. (1957). The study on it in terms of greenstone belt did not begin until the early 1980s (Lin Feng and Zhen Yunqing 1981 unpublished; Bai Jin et al., 1986). Multiple fold structures are superimposed on the Wutai greenstone belt. As a result, it outcrops in a "Z" shape (Fig. 2.14), assuming a synclinorium form 100 km long and 40 km wide, with long axis striking ENE. It appears to be the largest single greenstone belt in China.

The Wutai Group was deposited upon a sialic basement composed of the Fuping Group and is divided into three Subgroups-in ascending order the Shizui, Taihuai and Gaofan Subgroups (Bai Jin et al. 1986). The Subgroups are divided by two unconformities (Li Shuxun et al. 1986; Bai Jin et al. 1982, 1986). The 4000 m thick Shizui Subgroup is metamorphosed to amphibolite facies. Its base is marked by the quartzite, biotite schist and tremolite marble of the Banyukou Formation at the base. This reflects a protolith formed in a continental shelf environment, made up of terrigenous clastics, clay and carbonate rocks. On top of this formation and separated by a ductile shear zone lies a set of stratified and lenticular serpentinite, talc chlorite tremolite schist, amphibolite, biotite leptynite with interbedded cyanite schist and magnetite quartzite; which represents a protolith of ultrabasic, basic to intermediate-acid volcanics and pyroclastics. The 2000 m thick Taihuai Subgroup is in greenschist facies and consists of chlorite albite schist, chlorite schist, sericite albite schist and sericite schist. The banded Algoma-type iron-formation within the Group appears to be gold-mineralized and is the most significant iron-bearing bed in this region. At the bottom of the formation is a discontinuous layer of metaconglomerate base. The protolith of the Subgroup is basic (Photo 2.6) to intermediate-acid lavas and pyroclastics with minor argillaeous and arenaceous sediments. The amount of arenaceous sediments and acid lavas increases upwards in the sequence. The 2000 m thick Gaofan Subgroup of the upper Wutai Group occurs in subgreenschist facies and is largely composed of quartzite, metasiltstone and phyllite. The Group retains a distinct sedimentary rhythm, suggesting a protolith made up of turbidite sediments (Photo 2.7).

The Shizui and Taihuai Subgroups are made up of two independent volcano-sedimentary cycles and two subcycles occur in each of them. Compared with the Shizui Subgroup, the Taihuai Subgroup has higher Na content and more acidic rocks in the upper part of its sequence. The basic members contain both tholeiite and calc-alkaline series lithologies. The intermediate-acid rocks are mostly comprised of andesitic, dacitic to rhyolitic rocks, with a keratophyre sequence also present. The volcanic cycles tends to pass from tholeiite to the calc-alkaline series.

Some of the ultramafic rocks, with a $\Sigma REE=17.69-22.28$ ppm, $\Sigma LREE/\Sigma HREE=0.32-0.36$. The REE pattern is a smooth curve, with depleted LREE and a minor positive anomaly of Eu. This pattern corresponds to basaltic komatiite or olivine komatiite (Guo Jinjing 1989). In contrast, amphibolite in the lower Shizui Subgroup shows a $\Sigma REE=23.42-96.15$ppm, and $\Sigma LREE/\Sigma HREE=0.46-1.23$. The REE pattern is smooth except for a weak positive Eu anomaly. The mafic meta volcanics of greenschist facies in the Taihuai Subgroup have an average ΣREE of 39.70 ppm and a $\Sigma LREE/\Sigma HREE$ of 3.34, with slightly enriched LREE, a smooth curve and a slight positive anomaly (Guo Jinjing 1989); as a whole these parameters are consistent with Archean TH_1 type tholeiite (Condie, 1981) and correspond to modern ocean ridge to island arc tholeiite. On the <FeO>:MgO versus SiO_2 diagram and the <FeO>/MgO to <FeO> diagram, most points fall into the oceanic tholeiite area. The amphibolite in the upper-middle level of Shizui Subgroup however, yields an even ΣREE of 85.52 ppm, $\Sigma LREE/\Sigma HREE$ of 4.65 and an even $(La/Lu)_N$ of 4.15, with an insignificant Eu anomaly, corresponding to Archean TH_2 type tholeiite (Condie 1981) and equivalent to the present calc-alkaline series (and island arc) tholeiite.

The meta-sedimentary rocks in both the Shizui and Taihuai Subgroups have low REE abundance (85.92-114.08) and $\Sigma LREE/\Sigma HREE$ value (11.28-11.67), with no Eu anomaly (Eu/Eu* =0.99). In addition to reflecting an Archean feature (McLennan et al. 1979; Nance and Taylor 1976, 1977), these values indicate that the Archean supracrustal rocks were prominently composed of mafic and felsic volcanics (Taylor and McLennan, 1981). Some metasedimentary rocks display a REE pattern indistinguishable from those of present island arc calc-alkaline series, suggesting the rock - forming materials were derived from a similar volcanic series.

The REE pattern of the Gaofan Subgroup is marked by negative anomaly of Eu, (Eu / Eu* =0.6, $\Sigma LREE / \Sigma HREE=7.31$) and a higher REE abundance (202.18), implying a large granite area as the source area, which is very different from the underlying sedimentary strata. The similarity of the REE pattern to that of the Upper Archean shows that it was strongly influenced by the underlying potassic granite.

The average K_2O / Na_2O of the whole Wutai greenstone belt is 0.82, due to a large mass of low K, Mg and Fe volcanic rocks. The K_2O/Na_2O ratio of sedimentary rocks increases from the lower towards the upper part of the sequence, thus showing an increasing maturity from bottom to top.

The granitic rocks are dominated by sodium-rich types emplaced after the development of the Taihuai Subgroup. The sodium-rich granites, including tonalite, trondhjemite and granodiorite, were intruded in the early stages of plutonism whereas potassic granite occurred in the later stages. This reflects an evolving trend typified by the calc-alkaline series. Petrochemical data indicate that the former is the product of partial fusion of amphibolite and wacke sedimentary rock. However the REE chemistry resulted from the further differentiation of the partially melted magma. Hence, the evolution of the volcanic magma is closely related to tectonic development.

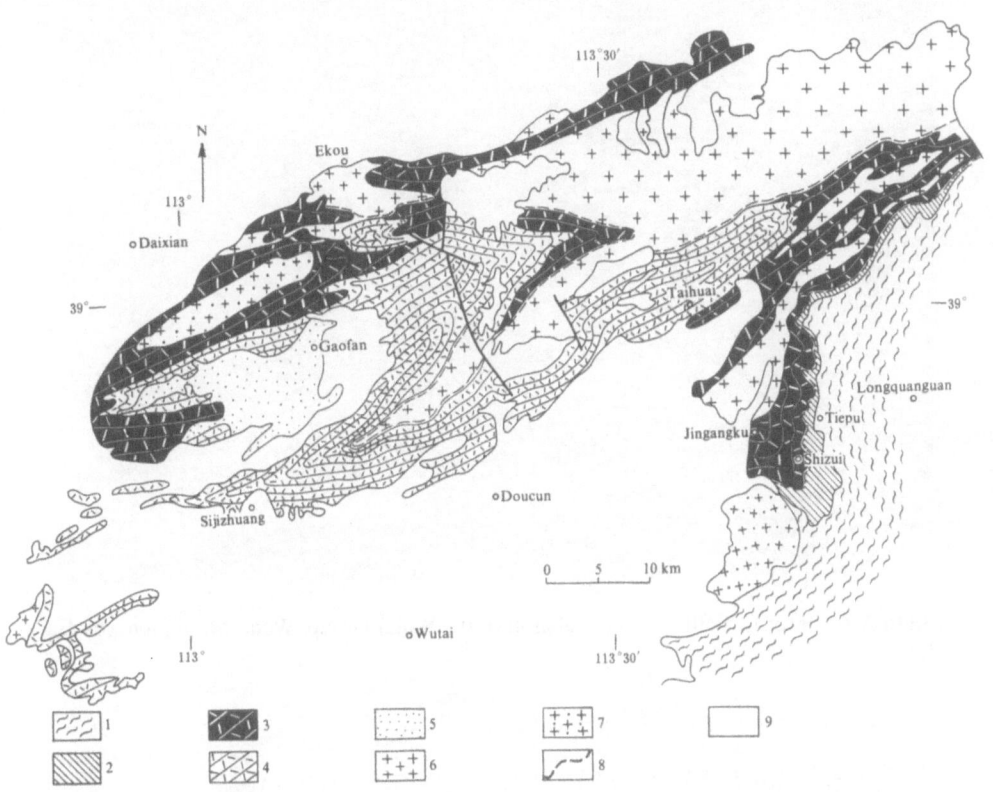

Fig. 2.14. Geological map of the Archean granite-greenstone terrane in Wutai Mt. (Modified after Bai Jin et al. 1986): 1. Granulite-gneiss terrane; 2. Banyukou Fm.; 3. Shizui Subgroup (no Babyukou Fm.); 4. Taihuai Subgroup; 5. Gaofan Subgroup; 6. tonalite-trondhjemite-diorite suite; 7. K-granite 8. ductile shear zone; 9. post-Archean

The most significant structural characteristics of the Wutai greenstone belt is the occurrence of recumbent folds and accompanying nappe-ductile shear-belt. The four unconformities, three within the greenstone belt and one on top of it, represent the three phases of the Wutai Orogeny and divide the Wutai Group into three structural stages (Bai Jin 1986). The early generation folds in each stage are recumbent ones, indicating strong deformation. Meanwhile the superimposed late generation deformation, share common patterns with the preceding deformation in spatial distribution and orientation. The early generated recumbent folds of the Shizui and Taihuai structural stages, together with the thrust nappe-ductile shear belt parallel to the penetrative-axis plane-schistosity, reveals lateral shearing and severe shortening. Apart from the strong deformations, all the three phases of the Wutai Orogeny caused distinct superimposed deformation such as transposition of foliation, compounding, refolding and other displacements of previous structural features in lithologies affected by preceding structural stages (Bai Jin 1987). Where early fold-axis plane-

Photo 2 .6. Deformed pillows in metabasalt of the Wutai Group. Wutai Mt. region. (by Bai Jin)

Photo 2 .7. Metaturbidite of the Wutai Group, from top to bottom: horizontal lamination (D, ~5cm thick), convolute lamination (C, ~10cm), horizontal lamination (B, ~5cm); down again, the lithologic units B and A of another turbidite bed. There are sandy fragments within the A unit, in Wutai Mt. region. (by Bai Jin)

schistosity has formed, the later folds would grow along the schistosity, demonstrating the flexural-slip folding mechanism. The direction of the structural lines of the 3 structural stages are presumed respectively to be NW, NS and NE, hinting at a clockwise migration. This is the most striking characteristics of the structures of the Wutai greenstone belt and shows the strong degree of mobility of the belt and the adjoining terrane (Bai Jin 1986). The ductile shear belt between the paleo-continental shelf sediments (represented by the Banyukou Formation in the lower Shizui Subgroup) and the paleo-ocean crust (represented by the ultramafic and mafic volcanics) is the tectonic boundary between the Archean greenstone belt and the adjacent high-grade region. The dramatic difference in the rock assemblages on each side of the boundary reveals a long displacement along the oceanic-continental boundary.

The distribution of the Wutai Group facies zones demonstrates the following pattern: the sub-greenschist facies area of the Gaofan Subgroup in the southwest part of the belt constitutes the core facies zone. To the northwest, northeast and southeast there developed in turn a low temperature greenschist facies chlorite-biotite zone and an amphibolite facies staurolite-kyanite zone. A marginal sillimanite zone is also present. The temperature rise from the core to the outskirts is also reflected by the paleo-temperature measurements. For instance, the biotite-garnet of the amphibolite facies in the northern part of the belt had a metamorphic temperature rising from 500°C to 580°C northeastwards. In the northeast, retrogressive metamorphic temperature determined by oxygen isotopes varies from 411°C to 434°C. Towards the northeast end of this region, mineral analysis yields metamorphic paleotemperatures as high as 670°C. Despite the sparsity of data, the fall of retrogressive metamorphic temperature from 429°C to 420°C towards the southeast is deemed significant. By combining the lateral temperature trend with the vertical one uncovered by regional profiles, the three-dimensional palaeothermal mode of the Wutai Group can be discerned as a "geothermal trough". i.e. it descends from the outer part to the centre (Fig. 2.15c). The present orientation of the "geothermal trough" is the final product of the latest superimposed deformation (Bai Jin 1986). Since regional metamorphism is closely related to tectonic setting and evolution (Miyashiro 1973), the "geothermal trough" was no doubt closely situated near to a subduction zone.

The Hengshan complex, to the northwest of the Wutai greenstone belt, is mainly made up of granulite facies grey gneiss. In contrast the Fuping Group, unconformably underlying the Banyukou Formation of the Wutai Group and exposed southeast of the basin, is made of metasedimentary rocks of amphibolite facies, with somewhat different tectonic orientations. Aeromagnetic anomaly maps of the Hengshan complex show a high linear positive anomaly whereas the Fuping Group displays a large area of negative anomaly with a few scattered positive spots. These geological and geophysical differences imply that the old land blocks on both sides of the greenstone basin are less likely to be a uniform massif. The tectonic rock assemblage, regional structural style and paleo-geothermal pattern jointly reflect an environment of paleo-island arc trench for the Wutai greenstone belt.

The amphibolite of the Shizui Subgroup has yielded a whole-rock Sm-Nd isochron age of 2599.16±41.49 Ma (2σ) with $\varepsilon_{Nd}(T)=2.4063\pm0.3721$. Thus the igneous protolith was derived from a depleted mantle source. However, in the amphibolite-facies metamorphic-process, water disrupted Sm-Nd isotope systematics, and therefore the isochron cannot represent the age of formation (Bai Jin 1990). Meanwhile a Rb-Sr isochron age of 2573±47 Ma (2 σ) with an intial $^{87}Sr/^{86}Sr$ ratio of 0.7021 has been found. In addition, Liu Dunyi et al. (1985) reported a U-Pb concordia age of 2508±2 Ma from metamorphic zircon within the leptynite. These values are consistent with the age of metamorphism and are supported by other U-Pb zircon ages of 2522+17/-16 Ma, 2530 ±30 Ma, 2560±6 Ma and 2507±16 Ma (Liu Dunyi et al. 1985; Bai Jin, 1986). In addition, Li Huimin (1989) reported a single zircon U-Pb concordia age for the granite in the belt of 2607±36 M (2σ). All these mentioned age data for both metamorphic rocks and granitoids are obviously related to the pronounced and extensive tectono-thermal event which occurred at the end of Archean time.

In recent years Tian Yongqing and coworkers published a monograph on the Geology and Gold Mineralisation of the Wutaishan-Hengshan greenstone belt (see Tian Yongqing et al. 1991). According to their studies they subdivided the Wutai Group into two parts instead of three. The lower part of the Group was termed the Shizui Subgroup and the upper part the Gaofan Subgroup. They pointed out that when compared with typical greenstone belts of the world, the Wutai belt has some special features:

1) The bottom of Wutai Group consists of terrigenous clastic rocks, with no associated komatiites.

2) In the volcanic rock series of the Shizui Subgroup, the proportion of sedimentary rocks in terms of thickness is 22%, whereas in the Gaofan Subgroup sedimentary rocks are limited.

3) There are two volcanic eruption cycles in the Wutai Group, which constitute two greenstone formations. The two cycles differ chemically and lithologically. The lower greenstone formation is rich in andesite with less basalt, lack of rhyolite and low Na_2O content (3.5%). In contrast, basalt is dominant in the upper greenstone formation. Rhyolite tends to increase upwards whereas the amount of andesite decreases. In the lower part of both cycles, the mafic rocks are dominant with the presence of BIF. Generally, the rocks are Na-rich; most of them are classified as high alkali tholeiite, a few belongs to alkali basalt and spilite. Pillow, amygdaloidal and vesicular sturctures can be observed. The REE signature is of TH_2 type (ie calc alkaline or continental basaltic affinity). The wholerock I_{Sr} value is very low (0.702 - 0.705), and $\varepsilon_{Nd}(T)$ is about +2.5, which show that the volcanics are derived from mantle melting.

4) The chemical composition of the volcanic rocks remarkably expresses a bimodal distribution with few intermediate compositions. The geochemistry also shows the influence of fractional crystallisation and crustal contamination.

5) The granitoid rocks developed in the greenstone belt are of various ages, sources and origins: ie - they are the products of different evolutionary stages of the greenstone belt rather than being formed as a single phase of late intrusions.

After detailed analysis of the formation and evolution of the Wutai greenstone belt, Tian et al. (1991) inferred that it has formed in a continental rift environment, which no ocean crust being generated. It developed ensialically (Kröner, 1981) in four stages.

Bai Jin (1986) considered that the Wutai Group has experienced the following processes during its evolution:

1. In the later period of the Late Archean, differentiation between continental crust and oceanic crust had already occurred on some scale. The Banyukou Formation, a continental shelf formation in the lower part of the Shizui Subgroup, was deposited on the continental margin.

2. In the early period of the first phase of the Wutaian orogenic cycle (2.6-2.35 Ga) the oceanic crust, represented by oceanic tholeiite and ultramafic rocks, was obducted and it overrode the shelf sediments.

3. Later, the oceanic crust was subducted, and in this geological setting there occurred an island-arc environment. Finally the Shizui Subgroup, a volcano-sedimentary formation in the lower part of the Wutai Group, was formed.

4. In the late period of the first phase of the Wutaian orogenic cycle, the oceanic crust was strongly subducted again and the sea basin contracted; as a result, the Shizui Subgroup underwent strong horizontal compression and shearing. This gave rise to deformation including recumbent folding and amphibolite-facies grade metamorphism.

5. The subduction of oceanic crust was inhibited and the volcanic arc uplifted, resulting in erosion and then renewed subsidence. This led to the uncomformable deposition of the volcano-sedimentary Taihuai Subgroup on the erosion surface.

6. In the second phase of the Wutaian orogenic cycle, the oceanic crust was intensely subducted for the third time and the sea basin contracted further; as a result, the Taihuai Subgroup also underwent horizontal compression and shear, thus generating recumbent folds and thrust nappes and bringing about greenschist-facies metamorphism. Meanwhile, the underlying Shizui Subgroup was subjected to superimposed deformation and retrogressive metamorphism. At a late stage, diapiric emplacement of sodic granite and potassic granite took place in succession.

7. Subduction waned and the volcanic arc was uplifted once more. Subsequent erosion followed by renewed subsidence resulted in the deposition of the Gaofan Subgroup (a turbidity current formation in the upper part of the Wutai Group) on the unconformable surface.

8. In the third phase of the Wutaian orogenic cycle, collision occurred between the Hengshan block and the Fuping block so that the Gaofan Subgroup underwent horizontal compression and shear. This resulted in recumbent folding and subgreenschist-facies metamorphism. Hence the underlying terrain underwent superimposed deformation and a certain degree of retrogressive metamorphism.

The periodic subduction of oceanic crust and partial melting of the sinking plate gave rise to a polycyclic tholeiitic and calc-alkaline volcano-sedimentary formation and calc-alkaline granitic plutons. Together these lithologies constitute a granite-greenstone terrane. Continous subduction of oceanic crust caused the sea basin to shrink continuously and ultimately this resulted in continental collision. Finally, about 2.5 Ga ago, the Wutai sea basin closed, forming a uniform continental crust.

Today the Late Archean-Proterozoic granite-greenstone terranes represented by the Wutai Group in the Mt. Wutai area may be regarded as having a tectonic environment which can be understood in terms of their spatial distribution. This is also the case with the Dongwufenzi Group in the western sector of the northern margin of the North China Craton. In the Mt. Wutai area (Fig. 2.14), post Middle Proterozoic strata is represented by gentle cover sediments which are either weakly magnetised or non-magnetic in character. Therefore, the magnetic anomalies in these areas covered by these strata are all the reflection of their basement Early Precambrian terranes. Gravity anomalies also essentially express a similar tectonic connotation. On the regional aeromagnetic anomaly map (Fig. 2.15D), the area where the Wutai Group is distributed is generally marked by a stable negative magnetic field distributed in a NE-ENE direction, surrounded by a regional positive magnetic field (on the northwest side) and a fluctuating magnetic field (on the southeast side). This feature is manifest on the gravity anomaly map too. There exist characteristic lines of NW-SE, nearly N-S and nearly E-W trends on the regional background of a NE-ENE trend. Hence, both regional gravity and aeromagnetic anomalies show that the Wutai Group was formed in a "basin" that had undergone polyphase deformation. This basin is located in a gradient zone where the crustal thickness changes dramatically in the Taihang Mountains. The northwest side of the "Wutai basin" is represented by the Hengshan complex, dominated by grey gneiss, whose metamorphic grade reaches granulite facies; the southeast side of the "basin" consists of the Fuping Group, dominated by metasedimentary rocks, whose metamorphic grade largely attains amphibolite facies (Fig 2.15B,C). The structural orientations of both sides are not consistent either (Fig. 2 .15A).

The Henshen Group shows similar relationships: on the regional aeromagnetic anomaly map (Fig. 2 .15D), the Hengshan complex is marked by a relatively high positive anomaly zone, while the Fuping Group is shown by a large negative anomaly zone, which suggests that the "Wutai basin" is situated just at a tectonic boundary. The blocks on both sides might not have resulted from rifting of a single massif (Bai Jin et al. 1986).

Dengfeng Greenstone Belt. The Archean terrane on the south margin of the North China Platform is predominantly made up of a granulite-gneiss region represented by the Taihua Group and the greenstone belt marked by the Dengfeng Group. The two groups are bordered by a south-dipping shear-zone, along which the Taihua terrane overthrust onto the Dengfeng terrane. Zhang Guowei et al. (1985) suggested that the two were generated at the same time but in different environments, whereas Kröner et al. (1988) envisaged that at least a part of the Taihua terrane is older than the Dengfeng terrane.

Within an area of 65 X 50 km², the greenstone belt occurs as N-S stretched stripes of relics of various sizes within tonalite, trondhjemite and granodiorite intrusions (Fig. 2.16). The proportion of greenstone belt material to the granitoid is 1:4.

According to Zhang Guowei's division, the Dengfeng Group is composed in ascending order of the Guojiayao Formation, Jinjiamen Formation and Laoyanggou Formation. The Guojiayao Formation contains amphibolite, hornblende plagioclase gneiss, biotite leptynite and minor magnetitic quartzite and leptite, reflecting a protolith equivalent to a set of basic to intermediate-acid volcanics and tuffs. The formation makes up the main body of the greenstone belt and is subdivisible into two parts: the lower part is ultramafic and mafic volcanics and the upper part is made up of bimodal volcanics. The ultramafic volcanics occur in the basement of the basalt and have components corresponding to basaltic komatiite. The volcanics are broken up by the intrusive granite and only exposed in scattered fragments. The bimodal volcanics largely consist of a mafic to felsic volcanic cycle interbedded with sedimentary rock. Intermediate volcanic rocks such as andesite occur only sparsely. There were at least ten such bimodal cycles, with single cycles ranging from 0.5-15 m thick. The mafic volcanics are tholeiitic and contain pillow structures, whereas the felsic ones are mainly dacitic and rhyolitic calc-alkline series.

The Jinjiamen Formation is composed of hornblende mica schist, mica quartz schist, staurolite mica quartz schist and a little magnetite quartzite. Together with the Laoyanggou Formation, it makes up the sedimentary series of the upper greenstone belt. The protoliths of the formation are pelitic sedimentary rocks and a small amount of banded iron formation, with sparse metabasic volcanics at the bottom of the BIF. Rhythmic beds are seen in the sedimentary sequence, revealing a turbidite-featured flysch formation. The Laoyanggou Formation is principally built up by meta-conglomerate which represent a molasse formation bearing granitic gravels of the underlying beds.

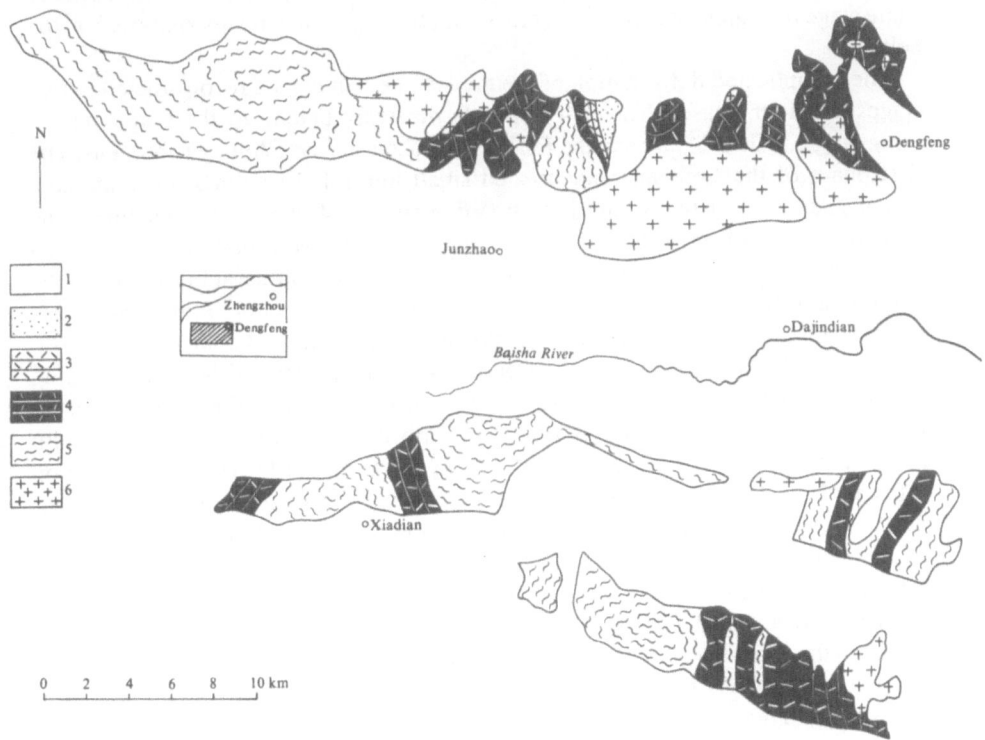

Fig. 2.15. A synthetic analysis of the tectonic setting of "Wutai basin" (after Bai Jin et al. 1990)

Zhang Guowei et al. (1985) suggested that the chemical composition of the mafic volcanics is of typical quartz-bearing tholeiite, characterized by low K_2O (0.14-0.4%), high MgO (8.07-10.6%) and medium Al_2O_3 (14-15%) as well as CaO (9-12.5%). The trace elements in felsic volcanics are roughly similar to Condie's F_I island arc type. In respect to REE, the mafic volcanics are weakly enriched in LREE, with smooth or slightly depleted HREE and no Eu anomaly. The $(Ce/Yb)_N$ shows some LREE/HREE fractionation, and the REE pattern corresponds to Condie's (1981) TH_2 type. According to Chen Haoshou et al. (1980), the amphibolite in the Dengfeng greenstone belt yields ^{87}Sr / ^{86}Sr ratios of 0.61999 and 0.7021, indicating a possible mantle source for the basic volcanics. In addition, Li Shuguang (1986) reported a $\varepsilon_{Nd}(T)$ of 2.2±0.4, which supports a depleted mantle source for the basic volcanics. However, a further $^{87}Sr/^{86}Sr$ ratio of 0.7124 implies that the magmas may have been crustally contaminated.

The whole greenstone belt reached a low amphibolite facies metamorphic temperature of 610°C (determined by oxygen isotopes), whereas the temperature demonstrated by mineral couples of hornblende-plagioclase and almandite-biotite are 486°C and 470-503°C respectively, with a geothermal gradient of spanning 17-20°C km. The b value of 9.024-9.036 from mica indicates a pressure of no less than a medium compressive type (Zhang Guowei 1985; Ma Xingyuan et al. 1981). Finally a late stage retrogressive greenschist facies metamorphism was superimposed on the belt.

The complicated deformation of the greenstone belt can be reconstructed in two stages and four phases. The two stages respectively belong to the Late Archean Songyang 2.5 Ga orogeny and the Early Proterozoic 1.7 Ga Zhongyue orogeny. The first phase of the Archean stage created small intrafolial isoclinal folds, an axial planar gneissic schistosity, and E-W trending strings of domes and synclines. The second phase built up roughly E-W trending, closed, overturned linear-folds and axial plane schistosity, mullion and sausage structures and mineral lineations. At the beginning of the Early Proterozoic NNW-oriented close isoclinal folds were superimposed on the Archean E-W trending structures and generated a ductile shear-belt. The late Early Proterozoic deformation is largely characterized by NNE-trending linear superposed folds, fractures and crenulation (Zhang Guowei et al. 1985). In the type region, the gneissosity was folded to form the second generation folds with horizontal or gently dipped hinges (Ma Xingyuan et al. 1981). The first generation folds are intrafolial and isoclinal, with axial planar gneissosity. If the effects of second generation folds are removed, then it is clear that the first generation folds are recumbent. This is important for the study of the dynamics of early deformation of the greenstone belt.

Chen Haoshou et al. (1980) obtained a whole-rock Rb-Sr isochron age of 2562±17 Ma from the meta-intermediate-basic rocks and Li Shuguang et al. (1986) reported a whole-rock Sm-Nd isochron age of 2509±16 Ma for the amphibolite and meta-acid volcanics and regarded it as the age of the protolith. In addition, Kröner et al. (1988) reported a zircon SHRIMP age of 2511±4 Ma for the metadacite and regarded it as the age of crystallization. Some zircon crystals have inner cores yielding an age of 2945±44 Ma (Kröner et al. 1988) and indicate that the xenolith crystals are derived

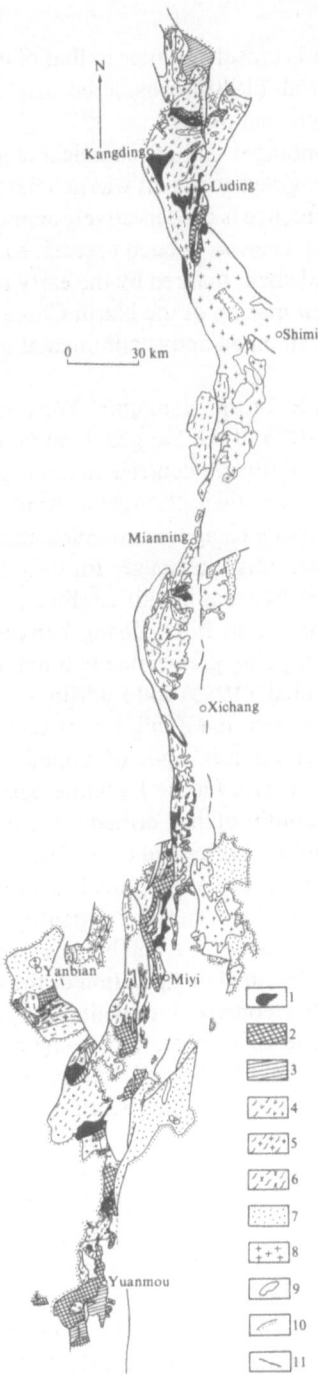

Legend:
1
2
3
4
5
6
7
8
9
10
11

Fig. 2.16. Geological map of the Dengfeng greenstone belt and gneiss terrane in Henan Province. (modified after Zhang Guowei et al. 1985): 1. Post-Archean; 2. Laoyanggou Fm; 3. Jingjiamen Fm; 4. Guojiayao Fm; 5. felsic gneiss; 6. early Proterozoic granite

from older rocks. This age is closely similar to that obtained from the Taihua complex. Therefore, Kröner et al. (1988) considered that older crust certainly existed underneath the Dengfeng greenstone belt.

Based on the rock assemblages and geochemical characteristics, Zhang Guowei et al. (1985) infered that the greenstone belt was generated in an environment resembling a modern rift. This inference is conservatively supported by Kröner et al. (1988), who thought that more evidence was needed to reach such a conclusion. Taking into consideration the horizontal shear marked by the early recumbent folds and the tectonic setting of the southern margin of the North China Craton, we deduce that the belt was developed in a basin on an active continental margin.

Kangdian Greenstone Belt. In the Kangdian (Western Sichuan-Eastern Yunnan) region of the southwest margin of the Yangtze Craton, extending from 28° 31'-30° 31' N and 101° 20'-102° 40'E, there occurred an old metamorphic complex. Huang Jiqing (1954) termed the region then "Kangdian Axis", and one decade later, Lee Chunyu (1963) named the rocks Kangding complex and suggested that it was Early Proterozoic in age. A number of Archean ages for the complex has been obtained in recent years and a greenstone belt was identified (Fig. 2.17) and named the Kangdian greenstone belt (Pan Xingnan et al. 1987; Zhang Yunxiang and Luo Yaonan 1988).

The rock mass of the Kangdian greenstone belt makes up the Kangding Group. According to Feng Benzhi et al. (1986) and Lu Minjie (1986), the Kangding Group is divided in ascending order into the Zanli, Lengzhuguan and Xiasuo Formations. The 5300 m thick Zanli Formation is a pile of amphibolite and hornblende-plagioclase leptinolite, interbedded with biotite leptynite, sparse leptite and minor basic granulite at its base. The protoliths of the Formation were basic volcanics interbedded with minor intermediate-acid volcanics and tuffs. Blastophitic and blastoporphyritic textures are often seen in the massive amphibolite and in some parts phenocryst-bearing and phenocryst-free rocks occur intermittently. This is taken as evidence of volcanic cycles. In addition, blastopillow and blastoamygdaloidal structures are also found in some places. The chemical composition of the amphibolite-dominated basic volcanics denotes that they consist of tholeiite and calc-alkaline basalt, with no boundary between them. The REE allocation is characterized by enriched LREE $((La / Yb)_N=3.57-8.23)$ and slight Eu negative anomaly (Eu / Eu*=0.80-0.98), close to Condie's (1976) TH_2 type calc-alkaline basalt or continental tholeiite. Furthermore, the ascending trend of transition from tholeiite to calc-alkaline series indicates a palaeo-environment corresponding to the present island arcs. The 2300 m thick Lengzhuguan Formation contains mainly biotite plagioclase leptynite, hornblende plagioclase leptynite and leptite, intercalated amphibolite, and in some places with graphite schist and staurolite sillimanite mica schist. The protolith is reconstructed as mainly intermediate-acid volcanics and tuff and partly pyro-greywacke with minor amount of sedimentary rocks. The thick interbedded volcanics occur principally in the lower part of the sequence and the beds decrease and thin upwards. The intermediate-acid volcanics have medium content of SiO_2 (57-67%), high K and Na ($Na_2O+K_2O=5.11-7.12\%$) and $Na_2O / K_2O=1.38-5.23$. This chemistry shows affiliation to andesitic-dacitic calc-alkaline volcanics. Andesite is a large proportion of the sequence and has an enriched LREE and a negative Eu anomaly, indicative of an island arc environment. The Xiasuo Formation is made up largely of thick-bedded biotite leptynite and leptite, with local occurrence of amphibolite. This changes upward into biotite schist, mica quartz schist tremolite diopside marble and olivine marble. No top has been found to the Formation and the exposed thickness measures up to 6300 m. The formation suggests a protolith of acid volcanics and tuff with SiO_2 amounting to 75-77%. Distinct blastoporphyritic or blasto-crystal fragments with tuffaceous texture is seen in the volcanics. Further up sequence, tuffaceous clay rock, clastic and carbonate rocks increase in abundance. The appearance of the graphite schist, mica quartz schist and marble respectively represents protoliths of pelitic-sandy and carbonate sediments of a shallow marine environment, whereas the graphite schist may originally have been deposited in a carbonaceous lagoon. In summary, this volcano-sedimentary megacycle is demonstrably a greenstone belt sequence, but unfortunately komatiitic rocks have so far remained undiscovered.

Within the N-S stretching zone metamorphic grade has mostly attained amphibolite facies. However, high greenschist facies occurs at both ends of the zone, and emergent granulite facies rocks occur locally in the broad central area otherwise occupied by amphibolite facies rocks. The PT conditions, calculated by various methods and in different parts of the stretching zone, varies considerably: the central area reveals a metamorphic temperature of 540-600°C, pressure ranging from 0.35-0.41GPa, and a geothermal gradient of 34-41°C/km. However in the southern part of the belt the temperature is 550°C, the pressure is 0.53GPa and geothermal gradient 30°C/km, close to medium-pressure facies series (Lu Minjie 1986). As far as these studies are concerned, no rational explanation for the variation in metamorphic facies has been made. Taking into account the parallel distribution of the metamorphic mineral associations to the early foliation and their syntectonic character, we envisaged that the primary orientation of the metamorphic facies zone was consistent with the early stage tectonic direction, i.e. striking N-S.

Metamorphosed granitoids are widespread in the Kangdian region, which was accordingly named the Kangding complex or Kangding gneiss. These rocks are often exposed in the lower-middle level of the Kangding Group and are mostly intrusives in the core of domes and in the axes of antiforms. They mainly fall into two catagories: hornblende plagioclase gneiss and granitic gneiss.

Usually exposed in large areas, the hornblende plagioclase gneiss rock-bodies are marked by a grey to light-grey colour and gneissic structure. Mineralogically they contain plagioclase, (40-60% An_{30-40}), hornblende (10-20%) quartz (7-25%) and minor biotite (2-10%). Where amphibolite is in contact with the gneiss, it is often separated and brecciated into relics and xenoliths shaped as lenticles and irregular masses. Some of the boundaries between the masses and the granitic intrusives are clear, whereas others show gradual change and still others appear as ghosts, with their attitude coincident to the gneissic schistosity. Inclusions of hornblende two-pyroxene plagioclase gneiss and granulite are seen in some hornblende plagioclase gneiss. The gneiss has a chemical composition of SiO_2 (54.9-63.5%), Al_2O_3 (14.5-18.5%, mostly >15.5%), <FeO°>MgO (4.37-11.66%) and $Na_2O + K_2O$ (2-4%). The REE signature of the rocks is characterized by ΣREE=110.4-157.9ppm, $(La/Yb)_N$=3.81-

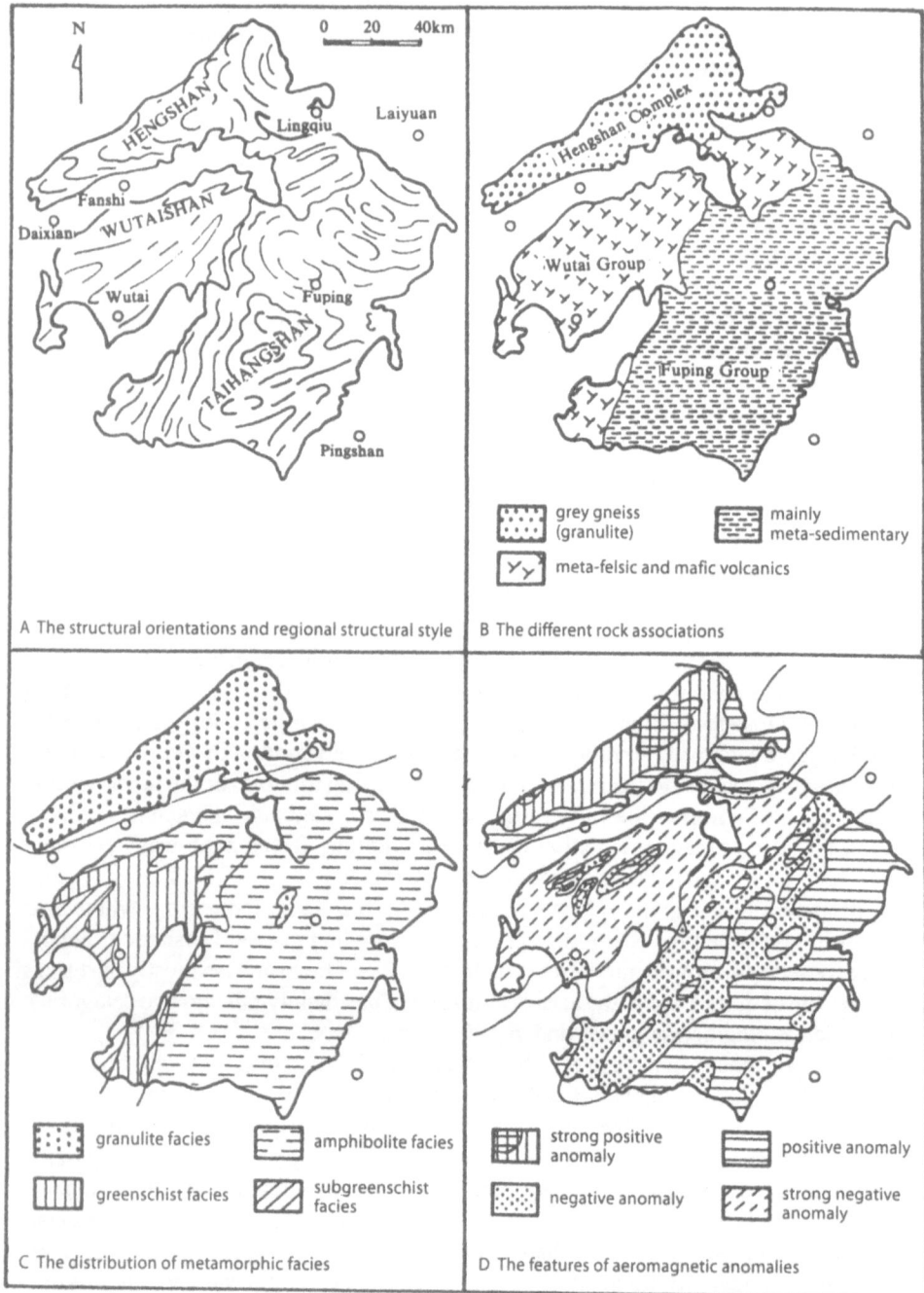

Fig. 2.17. Geological map of the Archean of the Kang-Dian region. (after Lu Minjie 1986).

12.29 and a weak to intermediate negative anomaly of Eu (Eu / Eu*=0.45-0.80), similar to those of the metabasic rocks in the region.

The granitic gneiss also occurs extensively in the region and has gneissose and schistose structure developed to varying degrees. The gneiss schistosity largely developed in the margin of the rock bodies and xenoliths of biotite leptynite and mica schist frequently appear on the fringes of them. Mostly light-grey colour, the rock bodies comprise oligoclase, quartz, biotite and minor microcline, and have a chemical composition of SiO_2 (66.5-76.5%), FeO*+MgO=2.29-4.12%, Al_2O_3=12.0-16.3%, N_2O+K_2O=5.61-7.93% and Na_2O / K_2O=1.25-3.66. On the whole it corresponds to oligoclase granite, partly to tonalite and granodiorite. The REE components of the rocks are characterized by low ΣREE (82.92-102.94 ppm), smooth patterns, a (La / Yb)$_N$ =6.49-12.36 and a normal Eu ratio (Eu / Eu*=0.86-1.12). These characteristics are similar to those of the xenolithic and migmatized biotite plagioclase leptynite. Since the REE signatures of the granitoids are identical to the relics of metamorphic rocks enclosed in it and this feature is shared by the Early Precambrian TTG suites throughout the world, it is considered that the granitoids are formed by the anatexis or partial melting products of the metamorphic rocks.

The Kangdian greenstone belt is tentatively dated as of Archean age by U-Pb concordia age of 2478 Ma for the regenerated zircon in the augen migmatitic gneiss (Wu Maode 1988), whole-rock Rb-Sr isochron age of 2.4 Ga from the gneiss in the central part of the belt (Yuan Haihua 1985), and a Pb-Pb age of 2957±304 (1σ) Ma for the migmatized hornblende pyroxene gneiss and hornblende two-pyroxene gneiss.

It is worth noting that although the Kangdian greenstone belt occurs in a N-S extension zone, although the structural direction of the belt is not necessarily N-S oriented. In fact, the structural styles demonstrated by the belt include prominent folds and domes with axes striking NE-SW to E-W and with the schistosity as a deformation plane (Lu Minjie 1986). Intrafolial folds are developed and have hinges plunging westward. Isoclinal and closed folds are frequently observed as well as recumbent folds (Feng Benzhi et al. 1986). Therefore, the E-W oriented structures are unlikely to be of the earliest generation and the enveloping surface covering the approximately E-W directed folds indicate a nearly N-S orientation for the first generation. The present N-S direction is influenced by the giant N-S trending Anning River fault developed since Middle Proterozoic times (Zhang Yunxiang. et al. 1988).

2.6.2 Rock Assemblages and Sequences

The sequence of basic volcanics-intermediate-acid volcano-sedimentary rocks is commonly found in the greenstone belts of China. Like their counterparts elsewhere in the world, Chinese greenstone belts have their large cycles subdivisible into several small cycles, each comprising a series from basic to intermediate-acid or from volcanic to sedimentary, which reflects the fact that the greenstone belts were to a large extent constructed by multiple volcanic eruptions. In the Luxi belt as many the 10 such small cycles are distinguishable, while two large and four small cycles were established in the Wutai greenstone belt.

At the base of the volcanic cycle of the greenstone belt, ultramafic rocks are widespread but poorly developed. They are usually thin-bedded (minimum thickness 0.5 m) or lenticular in form - a strongly distinguishing trait of the greenstone belt in China (e.g. in the Dengfeng greenstone belt). Most of the ultramafic rocks have chemical components equivalent to basaltic komatiite, in spite of the lack of spinifex texture. The only exception occurs in the 382 m thick Luxi greenstone belt which not only has both basaltic komatiite and komatiite compositions but shows clear spinifex texture. Although the universal shortage of ultramafic rocks in China can be superficially explained by the unfavourable geotectonic background, this is not an entirely satisfactory explanation: large numbers of ultramafic xenoliths occurring in granitoid (e.g. the Zunhua greenstone belt) and in ductile shear-belts situated at the boundary between the ultramafic rocks and the underlying rocks (e.g. the Wutai greenstone belt). This clearly suggest another possibility: namely, that just as it resulted in the smaller scale of the greenstone belts in China, the lack of ultramafic rocks results from the emplacement of granitoids and deformation during the closure of the basins into which the greenstone belts were originally emplaced.

The basic volcanics are dominated by TH_2 type tholeiites and a rapid upwards transition from the TH_1 type to the TH_2 type is frequently seen. This indicates contamination of the basic magma by crustal material during emplacement. The only exception to this trend is the overwhelming TH_1 type of the Wutai greenstone belt. The volcanics of the greenstone belts fall in two series, i.e. calc-alkaline and bimodal series and the latter is predominant.

The sedimentary rocks in the upper part of the greenstone belts vary considerably. Some contain terrigenous turbiditic or even carbonate rocks, whereas others have nothing but tuffaceous sediments. This is probably because of the short life of the paleo-basins within the belts;- few allowed abundant normal sediments to be deposited and later erosion may also have been a significant factor.

2.6.3 Metamorphism

High grade metamorphism is another characteristic of the greenstone belts in China. The ten greenstone belts mentioned range from greenschist facies to granulite facies, but are prominently amphibolite facies metamorphic grade. Two of them have basements of granulite facies, i.e. the Alax and the Jiaodong belts which were regarded as high-grade regions by Dong Shenbao et al. (1986). They usually consist of tremolite amphibolite (i.e. high amphibolite facies grade) in the lower part of the sequence. Low amphibolite facies often occurs in the middle to lower part, and is marked by the amphibolite without monoclinic pyroxene and with the occurrence of sillimanite, kyanite, and staurolite in the pelitic rock interbeds. High greenschist to greenschist facies is also common in the middle part, but low greenschist facies is exclusively found in the sedimentary rocks of the upper part of the sequence. In contrast, greenschist facies is also observed in retrogressive metamorphic zones and shear belts.

In the greenstone belts where greenschist facies metamorphism occurred (e.g. the Wutai, Dongwufenzi, Alax and Kangdian greenstone belts), distinct progressive metamorphism and clear facies zones are preserved, which reveal the paleo-geothermal state. For instance, the "geothermal trough" demonstated by the Wutai and Dongwufenzi greenstone belt is one of the significant criteria for determining the tectonic environment. In comparison, the seemingly "homofacies" greenstone belt may be a preserved "root" zone of the greenstone belt. The superimposed regional metamorphism is again a characteristic of the greenstone belts: Archean greenstone belts display the multiple-stages of thermo-tectonic events.

2.6.4 Deformation Styles

All the Archean greenstone belts have gone through several stages of structural deformation. The folding style is similar to that found in the granulite-gneiss terranes: i.e:- first generation folds are clearly intrafolial, closed, isoclinal or rootless folds. However, regional scale folds cannot be reconstructed. Parallel to the axial plane of the first generation folds developed penetrative schistosity and gneissosity defined by syntectonic metamorphic minerals. This deformation was accompanied by ductile shear zones and even imbricated structure. The second and third generation folds took the previous foliation as their deformation plane and usually formed closed isoclinal folds. In some cases crenulation in the hinge area as parallel to the axis plane; and others created strike-slipping ductile shear zones. The superimposition of folds of different generations may inherit the former axis but develop a new plane or create both new axes and planes (Photo. 2 .8) as well as dome-basin structures (Bai Jin 1987). The folds of the third generation assumed the form of open folds with vertical axial planes and sets of brittle fractures. It is apparent that this transition from ductile to brittle deformation marked the completion of cratonization in Late Archean times.

While the well-preserved greenstone belts (such as the Wutai and Dongwufenzi) demostrate complete synforms, most of the deeply metamorphosed greenstone belts emerge as relics or isolated islands in an ocean of granitoids and few intact outlines of regional folds are recognizible. After subtracting the interference effects of later folding, structural analysis shows that the style of first generation folds was recumbent. This suggests the formation of a nappe ductile shear-belt, a dynamic process of shear-compression and horizontal shortening for the whole of the greenstone belts. Such an idea is the key to the understanding of the tectonic setting of the greenstone belts.

2.6.5 Tectonic Setting

As explained the greenstone belts differ substantially from the granulite-gneiss region in their rock assemblage, and some of them (the Wutai and Dongwufenzi) have different regional structural direction from that of high grade region. Apart from those surrounded by intrusive granitoids, the contacts between the greenstone and the granulite-gneiss regions are marked by ductile shear-belts. These shear zones are parallel to the primitive plane schistosity and the metamorphic facies zones of the greenstone belts. Although showing a history of later reactivation, it is likely they formed the tectonic boundary when the greenstone basin closed.

The rock assemblages of the greenstone belts fall into two types. Type 1 is the bimodal volcano-sedimentary formation with oceanic tholeiite as is basic element (e.g. as seen in the Zunhua, Dongwufenzi, Dengfeng belts); type 2 is a volcano-sedimentary formation evolving from tholeiite to calc-alkaline series (e.g. as seen in the Qingyuan, Huadian, Alax, Western Shandong, Jiaodong, Wutai, and Kangdian belts). Those with a distinct progressive metamorphic facies zone (Wutai and Dongwufenzi) reveal a "trough"-shaped isothermal plane, and some (Dongwufenzi) even record an anticlockwise PTt track. Several of them (Wutai, Dongwufenzi, Huadian and Zunhua) were situated on regional geophysical anomaly boundaries or gradient belts. Taking into account their tectonic positions on the protoplatforms, the greenstone belts with type 1 rock-assemblage were generated in an environment equivalent to the present fore-arc and back-arc basin, or basins on active continental margins. Those with type 2 rock assemblages, however are thought to have occurred in modern island arc or trench type environments.

Photo 2. 8. Dis-axial displanar superimposed folds (similar to Ramsay's type 2) in sericite-chlorite schist of the Wutai Group. The F_1 folds are tightly isoclinal ones. Its hinges run nearly parallel to the picture. F_2 folds show themselves by a vertical axial plane. Its hinges are approximately perpendicular to the F_1 fold hinge. With measurement of the enveloping surface of the F_2 fold system, the recumbent pattern of F_1 folds can be recognized. Vertical view taken towards the east (by BaiJin)

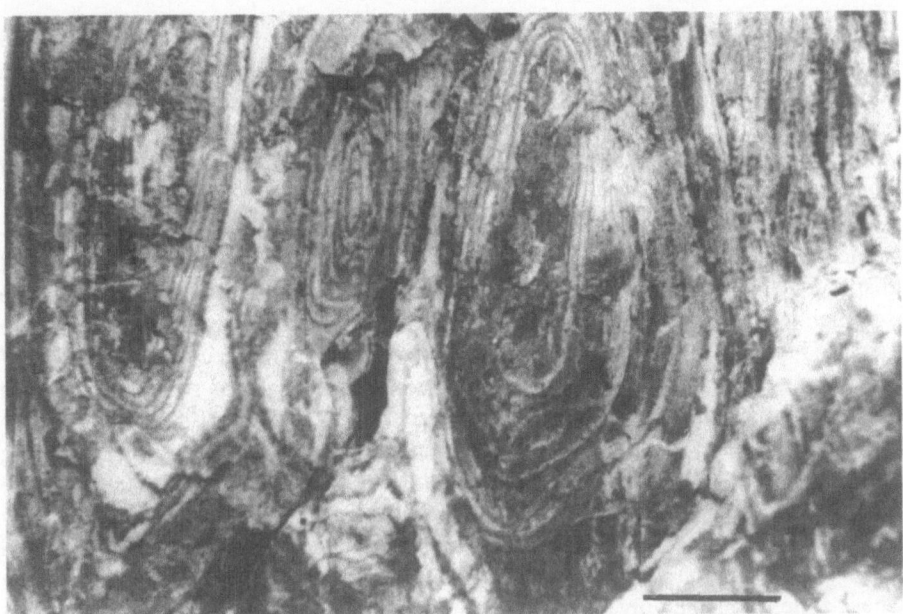

Photo 2 .9. Minor dome and basin structures in the BIF of the Wutai Group, elongated along the axial plane of an F_2 fold. Vertical view taken towards the west. Scale bar in 5 cm (by Gao Yaodong)

Apart from the Kangdian greenstone belt on the west margin of the Yangtze Craton, all the greenstone belts occur on the periphery of and within the North China Craton. This does not exclude the possibility that more greenstone belts occur underneath the post-Archean sedimentary cover. If those exposed in North China are a guide, then some belts are bound to be located around the bulges of deep magnetic boundaries (Fig 2 .7). If these magnetic anomalies represent terranes formed prior to Late Archean (see section 2.2.6), then potentially it are greenstone belts which weld the terranes together. Although no intact oceanic crustal relicts have been discovered, such an association could be explained if the greenstone belts are the relicts of oceanic crust sandwiched between terranes during orogenesis.

2.6.6 Geochronology

Isotopic ages yielded by the Archean greenstone belts are concentrated between 2.5-2.9 Ga. In all the ten greenstone belts described, only the Alax belt yields an uncertain Rb-Sr age of 3.2 Ga (Yang Zhende et al. 1988). Therefore, the Chinese greenstone belts are generally younger than 3.0 Ga and belong to Windley's (1977) juvenile type or Glikson's (1976) late greenstone belt type. The presently reported ages are mostly obtained by the U-Pb method for zircon from tonalite-trondhjemite. However, some Rb-Sr, Sm-Nd and U-Pb ages >2500 Ma have also been analysed from the lower basic volcanics, but they mostly record the age of metamorphism. Owing to the tremendous disruption exerted by the strong tectono-thermal events at the end of the Archean times, there is little chance of finding older age data from the Archean terranes of China. However, the possibility of greenstone belts older than the present ones being found cannot be discounted. The late Archean events also make it difficult to analyse the ages of earlier metamorphism or the diagenetic age of metasedimentary rocks.

The Gaofan Subgroup, largely composed of phyllite, in the upper Wutai greenstone belt yields a whole-rock Rb-Sr isochron age of 2175 Ma (Bai Jin 1986) and displays a REE pattern and K-Na content similar to that of the Early Proterozoic fine sedimentary rocks. This led Wang Hongzhen (1986) and Wang Qichao (1988) to regarded it as Early Proterozoic in age. However, if it is regarded as a component of a greenstone belt and considering that its structural deformation style is identical to the underlying volcanics, we classified it into the same unit.

In the next chapter, another greenstone belt in the North China Craton, the Qinglong greenstone belt, will be described. Despite the isotopic ages around 2.5 Ga that have been reported for it, it is regarded by most Chinese researchers as of Early Proterozoic age. This point of view is supported by structural analysis which reveals a single stage of deformation marked by isoclinal overturned folds and accompanying ductile shear-zones, considerably simpler than the Archean deformation. However, it is still possible that later study will prove it to be of Archean age.

The age data obtained so far seem to suggest that both the Archean greenstone belts and granulite-gneiss terranes (with a few exceptions) are confined to the interval 2.5-2.9 Ga. With regard to the dated samples, many granitoid samples of the granulite-gneiss region which yielded ages >2.5 Ga are petrologically similar to those intruded in the greenstone belts. For example, no reliable age has so far been obtained from the granulite-gneiss region in the western section of the north margin of the North China Protoplatform, but an age older than 2.5 Ga has been reported for the granitoid that is both enclosed and intruded in the Dongwufenzi greenstone belt. In contrast, the Dongwufenzi belt has a simpler deformation and low greenschist to low amphibolite metamorphic characteristics. Its contact with the granulite-gneiss region is marked by a ductile shear-zone. Thus the greenstone belt would appear to be younger than - and formed in a different environment to - the granulite-gneiss terrain. However, ages of 2511Ma and 2945 Ma for the volcanics of the Dengfeng greenstone belt have been reported (Kröner et al., 1988). The first age is taken to be the crystallisation age of the volcanics. However, the second age is from a zircon xenocryst, and this reflects the age of the granulite-gneiss terrane corresponding to the Taihua Group. The reported age data are insufficient to completely exclude the idea of simultaneous formation of both the greenstone belts and some of the granulite-gneiss regions. Furthermore the deduction that most of the granulite-gneiss terranes (including older supracrust rocks with greenstone characteristics), were formed before the greenstone belts should be acceptable. However, it is still uncertain whether the greenstone belt develop on a base of a pre-existed granulite-gneiss terrains or along the margins of them.

3 Early Proterozoic Crust

Bai Jin and Dai Fengyan
Tianjin Institute of Geology and Mineral Resources
Tianjin, China, 300170.

3.1 Distribution and General Features

Early Proterozoic rocks in China are mainly concentrated in north-western China and the North China Craton, closely associated - and constituting the basement with - Archean crustal units. They are distributed in the northern margins of the Yanshan and Yinshan ranges, in the meridional belt of eastern Shandong to Liaoning peninsulas (the Jiao-Liao belt), in the Jin-Yu belt from Shanxi to Henan Provinces and along the Longshoushan belt of the Alxa block. Outcrops are also situated along the northern and southern margins of the Tarim Craton further west (Fig 3.1).

Early Proterozoic basement rocks are located in the interior and on the western margin of the Yangtze Craton. Along the western margin of the craton, they are represented by the Dahongshan Group of central Yunnan Province. Within the vast area dominated by mobile belt tectonics, rocks of this age usually crop out along the upwarped sections of the belts. These occur, for example, within the Qinling-Dabie belt in between the North China and Yangtze Cratons and also in the Qilian, Kunlun, Tianshan and other mobile belts.

The stratigraphic, tectonic and petrochemical data for crustal thickening at the end of the Archean shows that a marked increase in lithosphere stability had taken place by the start of Early Proterozoic time. These newly stabilised plates were extensive, thick and rigid. Consequently, the style of deposition, deformation and intrusive processes are reflective of a more modern tectonic regime. Early Proterozoic crust typically adjoins and partly surrounds Archean cratonic nuclei. The highly variable patterns and compositions of Early Proterozoic mobile belts show the beginning of a remarkable history of crustal evolution (Goodwin, 1991). However, the transition from Archean to Proterozoic tectonic styles was undoubtedly diachronous: the high grade and low grade (greenstone) metamorphic belts typical of Archean tectonics continued to be formed during Early Proterozoic time.

The eastern and central sections of the North China Craton are host to the Jiao-Liao (Shandong-Liaoning) mobile belt (Bai Jin et al., 1990) and the Qinglong mobile belt respectively. The Jiao-Liao belt is stratigraphically represented by the metavolcanic-sedimentary formation of the Liaohe Group, and the Qinglong mobile belt by the Shuangshanzi and the Qinglonghe Groups. Following their formation, the fragmented craton was welded by a collision belt.

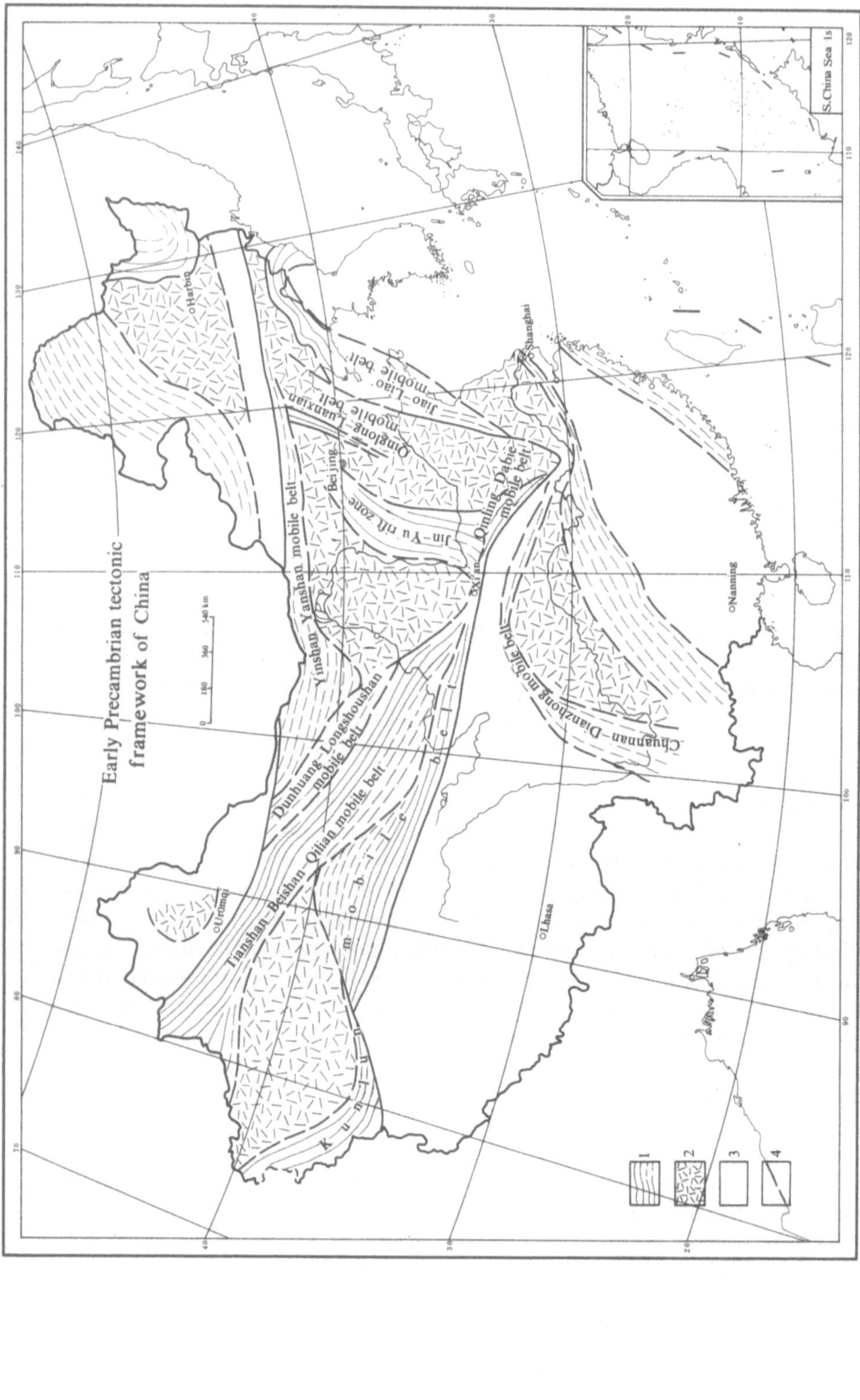

Fig 3.1. Early Precambrian tectonic framework of China: 1. Early Proterozoic mobile belts; 2. Archean craton; 3. Oceanic crust; 4. Boundary between tectonic provinces.

Photo 3.1. Horizontally exposed quartz sandstone of Changeheng System Changzhougou Formation, unconformably overlying the metabasalts of the Early Proterozoic Gantaohe Group, northern Taihang Mountains (Photo by Wu Tieshan).

Across the North China Craton, mountain ranges are host to a number of volcano-sedimentary sequences associated with belt formation: the previously described Hutuo Group is situated within the Wutai Mountains; the Lanhe Group, Yejishan Group and Heichashan Group in the Lüliang Mountains; the Zhongtiao Group and Danshanshi Group in the Zhongtiao Mountains. The Songshan Group occurs within the Jin-Yu rift Province and is also developed on the cratonic basement, where it is represented by a miogeoclinal formation. The Hongqiyingzi Group and the Qinling Group-Hong'an Group are situated on the northern and southern margins of the North China Protoplatform respectively. These form the marginal mobile belts formed during accretionary processes on the edges of the protoplatform.

The group of mobile belts discussed above were formed in succession to each other during Early Proterozoic time. Judging from their position and contact, the Jiaoliao collision belt and the Qinling mobile belt were probably formed before the marginal accretion belt. The Shanxi-Henan rifted Province, however, was probably developed during the accretion process. This was generated by the E-W extension and N-S compressional forces induced by the inward convection of the asthenosphere. Finally, the termination of the Lüliang movement resulted in the closure of the mobile belts and the formation of a protoplatform (photo 3.1).

3.2 Rifting, Reworking and Marginal Accretion of the North China Craton

3.2.1 Liao-Ji Mobile Belt

The Early Proterozoic rock sequence in the southern Liaoning-Jiln region can be divided into two suites of differing metamorphic grade. The two suites assume the style of a multiple fold zone stretching E-W and NE-SW. In the eastern Liaoning, they are jointly termed the Liaohe Group (Zhang Qiusheng et al., 1988). In southern Jilin, the former is named the Ji'an Group and the latter the Laoling Group. In regional terms, these Early Proterozoic metamorphic rocks, combined with the Early Proterozoic Motianling System in northeastern Korea, constitute an Early Proterozoic mobile belt lying between the Archean Longgang and Langlin blocks (Fig 3.2).

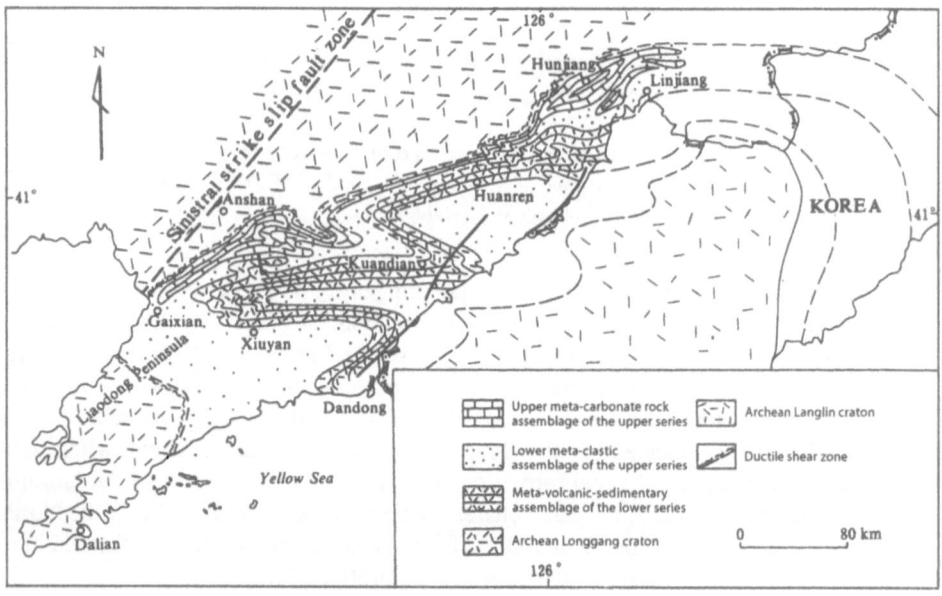

Fig 3.2. Structural map of the Liao-Ji mobile belt (after Bai Jin, 1991).

The first suite is a metavolcanic-sedimentary rock series. The lower horizons of the suite is composed of metavolcanics, pyroclastics and tuff, sandwiched between meta-argillaceous-arenaceous rocks and some magnesian carbonate rocks. The upper part of the suite is graphite-rich, and consists of meta-argillaceous-arenaceous and carbonate rocks, including schists, gneiss and marble. The volcanic rocks were erupted in a marine environment, and consist of a bimodal suite of intermediate-acid volcanics and intercalated basic volcanics. This bimodal suite occurs together with argillaceous-arenaceous deposits and carbonate rocks. The thickness of the lava increases from north to south (presumed original direction) and a maximum thickness of over 1000 m is reached in the Chengjin Series of the lower Motianling System. The lava is accompanied by stratified peridotites, which are possible relics of paleo-ocean crust in the mobile belt. The basic volcanic rocks are rich in Na_2O (2.20-4.40%) and Fe (Σ FeO=9.05-16.67%), with average TiO_2 =1% and a narrow range of $\Sigma FeO/MgO$ values. These features are similar to those of oceanic tholeiites, but on the AFM diagram they demonstrate both tholeiitic and calc-alkaline series affinities. The data points are concentrated around those of the Newfoundland Bay and the Luz Bay ophiolites, and are chemically similar to Mid-Atlantic Ridge tholeiites (Zhang Qiusheng et al. 1988). The gently sloping REE pattern and Sm/Nd values are also suggestive of mid-ocean ridge tholeiitic basalts. The intermediate-acid volcanics have a wide range of SiO_2 content (60.46-75.76%). These rock types are Si-oversaturated, showing a Na-rich to K-rich magmatic evolution trend. Intermediate volcanic rocks have an average Na_2O content of 6% (occasionally as high as 12%) and acid volcanic rocks a K_2O content between 6.6.-9.26%.

The second suite consists of metasedimentary rocks conformably overlying the metavolcanic rocks of the first suite. Lithological units include thick, immature metaterrigenous clastic rocks which occur in the lower parts of the sequence, and thick metacarbonate rocks at higher levels. In general, the rocks display turbiditic characteristics, clear stratification and constitute a sedimentary megacycle.

Within the eastern section of the Liaoning region, the Liao-Ji belt is host to some complex tectonic features, with three recognised episodes of deformation. The first deformation phase resulted in bedding plane deformation and produced closed, isoclinal, inclined and recumbent folds. These structures commonly preserved a penetrative axial-plane schistosity, widespread schistosity-bedding intersections and a N-S structural direction. "Tongue-shaped" or sheath folds of outcrop to regional scale are also developed in connection with progressive shear deformation. Judging from the present shape and direction of lineation, these structures were formed by an E-W low-angle napping mechanism.

The second deformation phase had the schistosity S_1 as deformation plane and created nearly horizontal, close to open folds that developed on the folds of the first phase in coaxial superposition, accompanied by axial plane bending cleavage. In places where the "tongue-shaped" folds developed, incoaxial, or even cross superposition of the two phases are found. The general direction of the structural lines is still nearly N-S, which, combining with fold styles, indicates that the Liaohe Group went through another E-W lateral compression.

The third deformation phase severely reformed the previous folds and formed E-W stretching, vertically plunging fold structures at various scales, reflecting a N-S horizontal compression. This phase virtually changed the N-S extending tectonic lines and rotated them to a nearly E-W direction. In comparison to the post-Early Proterozoic structural layers, the third phase probably occurred significantly later. This third phase of deformation may have resulted from sinistral shearing on the adjacent NNE-trending Tan-Lu fault during N-S compression.

The major metamorphism of the Early Proterozoic supracrustal rocks in the Liao-Ji region is synchronous with the first phase of deformation. Within the metavolcanic suite, metamorphic grade is generally of amphibolite facies and the metamorphic lithological units include leptynite, leptite, amphibolite, minor magnetite quartzite, schist, gneiss and magnesian marbles. Abundant columnar crystalloblasts of silli- manite occur in argillaceous rocks metamorphosed at high amphibolite facies. The temperature-pressure conditions of metamorphism are rather wide, with T = 500- 700°C and P = 0.2-0.6 GPa. Geochronological studies on this suite have yielded a variety of age dates: Jiang Chunchao et al. (1986) report K-Ar ages ranging 1958- 2270 Ma from biotite and phlogopite in lower series metavolcanic rocks. In addi- tion, Liu Hongyong et al. (1981) reported a Rb-Sr age of 2206±229 Ma, and a U-Pb zircon age of 1956 Ma for leptynites within the metavolcanic sequence. However, these dates are older than the K-Ar ages of 1565-1750 Ma previously determined for these rocks (Wang Jiyuan and Diao Naichang 1982 unpublished). Furthermore, rocks (Wang Jiyuan and Diao Naichang 1982 unpublished) obtained two different Sm-Nd isochron ages from stratigraphically separated amphibolites within the metavolcanic suite. The stratigraphically higher amphibolite yielded an age of 2063.24 ± 37.92 Ma (2σ) εNd (T) = 4.92 ± 0.32, whereas an amphibolite from a lower horizon yielded an age of 2214.20 ± 55.86 (2σ) Ma, εNd (T)= 3.77 ± 0.43. These values are inter- preted to approximately reflect the age of the Early Proterozoic magmatism and metamorphism of the lower rock series, while the younger values resulted from later tectono-thermal events.

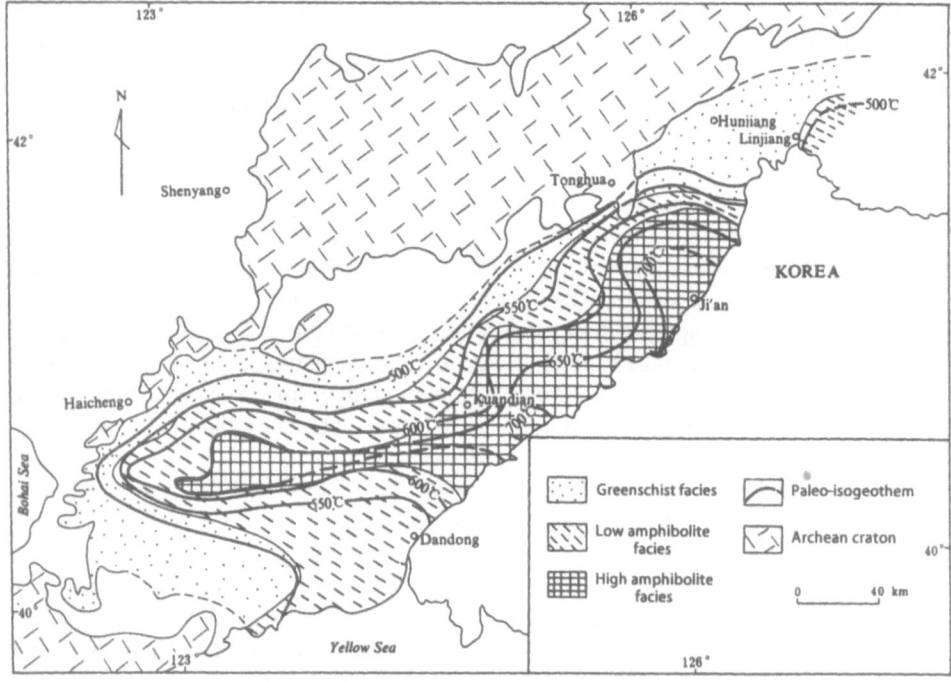

Fig. 3.3. Map showing the metamorphic facies and the paleo-isotherms of the Liao-Ji mobile belt (after Bai Jin 1991).

In comparison, the overlying metasedimentary suite reached greenschist to low amphibolite facies. Meta-argillaceous rocks contain staurolite, garnet and biotite, and reached phase and the temperature-pressure conditions of T = 430-600°C, P = 0.2-0.6 GPa. Metacarbonate rock types within this sequence yield the well-known Liaodong magnesite and talc deposits. As with the lower metavolcanic suite, a variety of ages have been obtained in geochronological studies using different dating methods: Pb-Pb and Sm-Nd studies yielded ages of 2210 Ma, 1892 Ma and 1869 Ma respectively (Zhang Qiusheng, 1988; Wang Kuiyuan et al., 1990). Within the Anshan-Benxi area region, Chen Yuwei et al. (1981) obtained a whole-rock isochron Pb-Pb age of 1977±49 Ma (σ) from the metasedimentary suite and suggested the maximum age of rock deposition to be 2.0 Ga. In comparison, Rb-Sr (wholerock isochron) dating of the metasedimentary sequence in southern Jilin (Jiang Chunchao, 1987) produced slightly younger ages of 1860±124 Ma (σ) and 1728±70 Ma (σ). Hu Guowei (1989) obtained Rb-Sr isochron ages of 1875±216 Ma (2σ) and 1768±351 Ma (2σ) from schists of the upper series and second generation metabiotite respectively. In this case, the older age is thought to represent the timing of regional metamorphism and deformation, and the latter date is indicating a later thermal event. A summary of the age dates obtained suggests that the metavolcanic and metasedimentary rocks of the Liao-Ji belt were originally deposited between 2.3 and 1.9 Ga.

The range in temperature-pressure conditions in both metamorphic suites are interpreted to show an inhomogeneous paleothermal state (Fig 3.3). Vertical zonation in metamorphic conditions is demonstrated by the generally higher grade attained by the metavolcanic rock suite in comparison to the overlying metasedimentary suite. Clear metamorphic facies changes also occur horizontally within the same rock series. In addition, although as a whole both suites belong to a medium pressure facies series, the lithology of some localities show evidence of low temperature-high pressure metamorphism.

The strike of the metamorphic belt is parallel to the earlier folds, i.e. generally N-S, but interference by later folding has caused it to assume an S-shaped structure. Metamorphic grade increases eastwards to low/high amphibolite facies. As the original strike of the mobile belt was N-S, the present NE-SW facing curvature may have resulted from sinistral shear on the western margin of the belt. Originally, the metamorphic grade probably increased westwards, with the distribution of the metamorphic belt forming a paleothermal trough. However, the western side of this trough was cut off by a paleosubductive zone.

The metamorphic history of the Liao-Ji belt has been further elucidated by the electron microprobe analysis of garnet crystals (Spear et al., 1986). The clockwise pressure-temperature-time (PT-t) track obtained demonstrated two thermal-dynamic cycles, the first falls in the range of high P-low T (0.55-0.75 GPa, 450-525°C) and the second in medium P, with a slight rise of T (0.4-0.62 Pa, T450-540°C). The PT-t path indicates the rock has been in a "thermal valley" environment, showing the typical PT-t track of a continental collision belt (Thompson and England 1984).

Following regional deformation and metamorphism, post-tectonic magmatic intrusion resulted in the variable degrees of contact metamorphism. Intrusive rocks include both S-type and I-type granitoids. Early granitoids are S-type and are spatially associated with the lower metavolcanic rock sequence, constituting an Early Proterozoic igneous rock belt bordering the Archean Langlin block. Age dating of these intrusive rocks by Jiang Chunchao et al. (1986) yielded Pb-Pb isochron and U-Pb concordia ages of 2093±22 (2σ) Ma and 2053±69 (2σ) Ma. Late- stage granitoids have the geochemical character of collision granites and are classified as I-type (Yan Yaoyang 1990). At some localities, igneous intrusion resulted in only low-grade thermal effects, and hence caused retrogression of the regionally metamorphosed lithologies. Alternatively, at other localities high grade contact metamorphism was superimposed on low-grade regional metamorphic rocks.

In summary, the environment of mobile belt formation may be inferred from the various petrological, structural, metamorphic and geochronological studies outlined above. On the basis of the chemical bimodality of the metavolcanic rock suite, Zhang Qiusheng et al. (1988) and Jiang Chunchao et al. (1987) suggested that the early stages of belt evolution took place in a rift-valley environment. However, from the analysis of regional structures, rock assemblages, geothermal state and thermal-dynamics, it is reasonable to conclude that Liao-Ji mobile belt was originally formed in an oceanic basin adjacent to a subduction zone (Fig.3.4a). As the tectonic stress regime changed from one of extension to shear-compression, the basin began to close. Formation of the tectonic belt finally occurred during the continental collision that welded the Archean Longgang block and the Langlin block. This development process is summarised in Fig. 3.4 and below.

A. Convergent stage (Fig. 3.4b). A primary N-S orientation is presumed for the mobile belt. Oblique subduction then caused low-angle nappe-shearing and formed recumbent bedding folds with N-S hinges, created local "tongue-shaped" folds (sheath folds) and a ductile shear zone on the boundary surface between the stratified rocks and the Archean basement. Metamorphism is synchronous with the deformation which produced the penetrative axial schistosity S_1.

B. Collision stage (Fig. 3.4c). With the subduction of the continental blocks replaced by oblique collision, lateral compression became predominant to generate the vertical schistosity fold F_2, which is coaxial to F_1.

C. Superimposition stage. Probably later than the collision stage, this stage again saw a N-S lateral compression derived from the sinistral sliding of the old Tan-Lu fault zone. This resulted in the formation of large plunging vertical folds in the western part of the belt close to the fault zone. To the east the folding gradually diminished and only wide gentle folds were developed. Even further to the east, no folding at all is developed in the Motianling System in Korea. The resultant fold mountains became slightly denuded during this terminal stage of belt development. Thereafter molasse formations developed which became weakly deformed and metamorphosed. Finally, Mid-Late Proterozoic stable cover sediments were deposited on top of the molasse deposits.

3.2.2 Qinglong Mobile Belt

This belt, represented by the Early Proterozoic Qinglong greenstone belt is located within Hebei Province between 40°05'-40°30' N and 119°-119°15' E. The belt is strongly folded and metamorphosed to high greenschist-low amphibolite facies. In addition, it is fragmented by a major Proterozoic NNE-trending fault zone. The east and west sides of the belt are also fault-bounded, and lie adjacent to terranes of Archean age. In contrast, the north and south ends are covered unconformably by the Mid-Proterozoic Changcheng System (Qian Xianglin et al. 1985; Bai Jin and Yang Chunliang 1984).

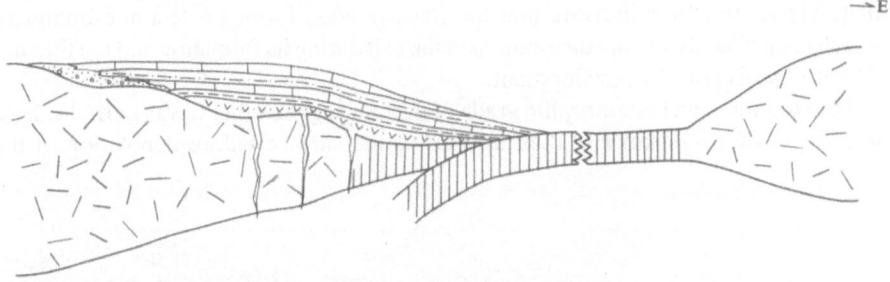

A. Formation of volcanic–sedimentary series

B. Convergent stage

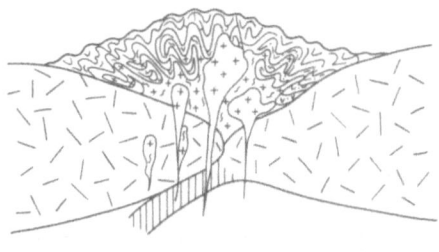

C. Collision stage

Fig. 3.4. Schematic diagram showing the tectonic evolution of the Early Proterozoic Liao-Ji mobile belt (after Bai Jin 1991)

The fault system which cuts the belt was named the Liaocheng-Lankao crustal fracture by Huang Jiqing et al. (1980). It extends to the northern border of the North China Protoplatform, southwestwards through the North China Plain, and probably to the southern border of the protoplatform. It is still an active seismic zone to the present day, and was responsible for the catastrophic Tangshan Earthquake in 1976. The stratigraphic boundaries of the greenstone belt are parallel to this fracture zone. Notable features include a deep crustal ductile shear zone which developed along its edges, and syntectonic metamorphic belts parallel to the Liaocheng-Lankao fault belt. These structures indicate that the fracture zone formed a tectonic boundary which controlled the spatial distribution of the belt during its formation and the thermo-dynamic process of its development.

The structure and metamorphic grade of the belt suggests that it was buried at least to a depth of 15-20 km and then uplifted to the surface before deposition of the

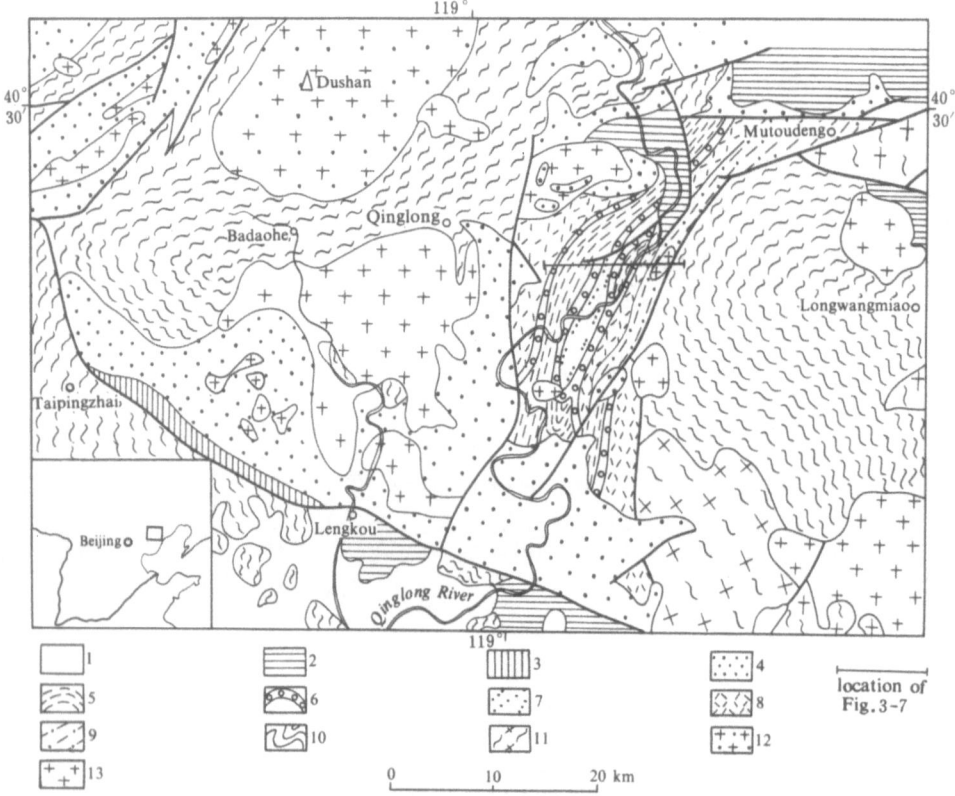

Fig. 3.5. Geological sketch map of the Early Proterozoic of the Qinglong region, Hebei Province (after Bai Jin et al. 1984b): 1. Cenozoic; 2. Mesozoic; 3. Paleozoic; 4 Middle Proterozoic; Early Proterozoic Qinglonghe Group: 5 Boluotai Formation; 6. Zhangjiagou Formation; Shuangshanzi Group: 7. Xiabaicheng Formation: 8. Luzhangzi Formation; 9. Ciyushan Formation; 10. Archean; 11. Archean granitoids; 12. Proterozoic granite; 13. Mesozoic granite

sedimentary rocks of the Changcheng System. It is therefore probable that most of this greenstone belt has either been eroded or alternatively destroyed during fault movement. Hence the original scale of the Qinglong belt was probably considerably larger than the area which was preserved and subsequently exhumed.

Rock Assemblages. The Qinglong greenstone belt can be subdivided into two Groups: the lower Shuangshanzi Group and the upper Qinglonghe Group (Bai Jin and Yang Chunliang 1984b). The Shuangshanzi Group is 3000 m in total thickness and consists of metavolcano-sedimentary lithologies. The metavolcanic rocks constitute 70% of the Group and show a variety of rock facies and clear rhythmic layers which show little along strike variation. The Group can be further subdivided into three large volcano-sedimentary cycles. The petrology of these cycles is summarised below in stratigraphical order. Where known, the original lithologies are given in parentheses:

Cycle (1) - biotite amphibolite (intermediate volcanics), blasto-porphyritic mica-quartz schist (acid volcanics), calc-sericite schist.

Cycle (2) - amphibolite (pillow lavas), blasto- porphyritic biotite- plagioclase leptynite (intermediate-acid volcanics), carbonic -sericite phyllite.

Cycle (3) - amphibolite (basic volcanics), blasto-porphyritic biotite-plagioclase leptynite (intermediate- acid volcanics), two-mica-plagioclase leptynite.

In the SiO_2,-$Na_2O + K_2O$ diagram of Middlemost (1980), the biotite amphibolite, blasto-porphyritic biotite-plagioclase leptynite and blasto-porphyritic sericite schist plot within the basalt, andesite and rhyolite fields respectively. Basalts and andesitic rocks dominate, followed by dacites and rare rhyolitic rocks. They constitute an obvious calc-alkaline series in the Fe_T-Na_2O+K_2O - MgO diagram. The average TiO_2, K_2O, Na_2O and Fe_T compositions of the metabasaltic rocks is 0.95%, 0.5%, 2.88%, and 10.33% respectively, which are characteristic of island arc tholeiites (Jakes and White 1972). These basaltic rocks also plotted into the field of island arc tholeiite in the diagrams of Ti x 100 - Zr -Sr/2 proposed by Pearce (1975). However, in the Cr-FeO/MgO and Ni-FeO/MgO diagrams, they fall into or adjacent to the island arc, active continental marginal and abyssal tholeiite fields (Sun Dazhong 1984). Dating of the metabasic volcanics using the Rb-Sr whole-rock method obtained an isochron ages 2217±43 Ma and 2228 ±136 Ma (Lu Gongyi et al., 1987) These ages are considered to represent the timing of metamorphism.

Metasedimentary rocks include two-mica - plagioclase leptynite, schist and phyllite. The precursors to these lithologies are greywacke, pelite and pelitic siltstone. These rock units are sometimes calcium-rich and show extensive graded-bedding. Locally, small-scale turbidite-type cross-bedding is present. In addition, some phyllites contain notably higher contents of organic carbon (reaching 0.56%) and are termed carbonic phyllites. Chemically, the average $Al_2O_3/(Na_2O + K_2O)$ ratio in the metasediments is 2.5, indicating a low degree of maturity. A U-Pb geochronological study of two (probably detrital) zircon grains taken from the metasiltstone rocks by Shen Qihan et al. (1990) obtained an average age of 2497 ± 2 Ma using the evaporation method of analysis.

The Qinglonghe Group is a turbidite formation and represents the upper assemblage of the greenstone belt. It is over 1000 m thick in exposure and unconformably overlies the Shuangshanzi Group. A notable layer within the Group is the 50-170 m thick layer of metabasal conglomerate. This composite paraconglomerate also contains some features of lump conglomerate (Pettijohn 1975) in which the gravels are well rounded, but poorly sorted, showing that gravels of different size are randomly mingled. The metaconglomerate is basally cemented, with the pebbles making up about 50-70% of the rock. The cements are mainly two mica (biotite)-plagioclase leptynite, two mica schist and a little amphibolite. The protolith sequence of these cementing lithologies was dominated by pelitic-psamitic and pelitic-calcareous sediments intercalated with small amounts of volcanic materials.

The metaconglomerate layers are intercalated with thin and graded layers of plagioclase leptynite and less mature two mica (±garnet) schist, which show features of turbidite sedimentation. The pebbles are mostly derived from the underlying high-grade metamorphic rocks, and are dominated by meta-intermediate-acidic volcanics, with only a few meta-basic volcanic pebbles are encountered. The bias towards meta-acid compositions is evidently likely to be due to the poor preservation potential of metabasic clasts. Considerable quantities of granitic pebbles, coexisting with less-mature leptynites, may indicate a process when the sedimental basin was rapidly depressed, and the crystalline rocks forming the floor and the steep walls of the basin were speedly eroded (Pettijohn 1975). The pebble composition, pebble size and cements of the metaconglomerate vary regularly (Fig. 3.6), i.e. the iso-grain lines of the pebble are roughly parallel to the original layering, and the grain size becomes smaller inwards. This further evidence not only suggests that a fracture zone (primarily a ductile shear zone) controlled the greenstone belt, but also that the sea floor turbidity current responsible for depositing the sequence flowed from east to west. Above this metaconglomerate layer, there are a series of biotite (±muscovite)-plagioclase leptynite, (garnet) two mica-quartz schists intercalated with cummingtonite-magnetite-quartzite and magnetite-bearing cummingtonite. Graded layers of varying thickness are developed, and commonly this alternation of leptynite and schist occurs in layers of over tens of metres to a hundred metres thick: this is also a characteristic of turbidite sedimentation. A zircon U-Pb concordia age of 2512+57/-47 Ma has been obtained from the leptynites, which probably indicates the age of the detrital zircon crystallization (Shen Qihan et al. 1990). In addition, a Rb-Sr isochron age of 2082±314 Ma has been gained from the same rocks, but its error seems too large (Sun Dazhong 1984).

Deformation. The structural pattern in the Qinglong greenstone belt is generally a NNE-trending, isoclinal, overturned fold system (Fig. 3.7). The penetrative axial plane schistosity is well-developed and dips WNW. The schistosity is mostly in concordance with primary bedding, and the projection of its poles constructs a trend of 120° with a plunge of 40°-50° in attitude (Gao Yadong 1984). However, at the hinge zones of folds, the bedding commonly reverts in dip and cuts the uniform attitude of the schistosity. Regional folding is accompanied by the development of ductile shear zones; these are especially developed along the boundaries of the greenstone belt

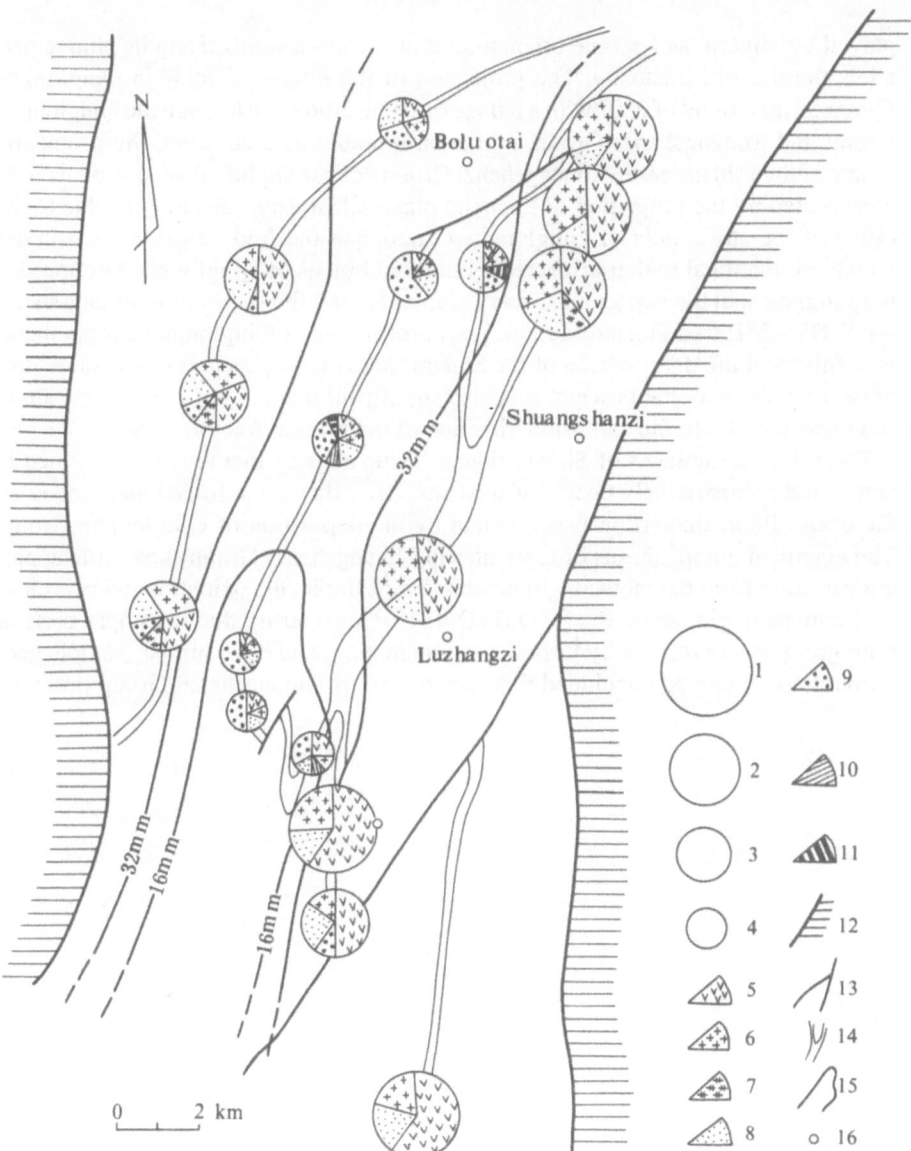

Fig. 3.6. Composition and grain size of Qinglonghe Group basal conglomerate pebbles. Mean diameter of gravel: 1. 64-236 mm; 2. 32-64 mm; 3. 16-32 mm; 4. 4-16 mm; 5. metavolcanics 6. metagranite; 7. leptite; 9. vein quartz; 10. schist; 11. BIF; 12. Archean rocks; 13. fault; 14. metacongloerate; 15 isogram of gravel diameter; 16. village

and highly-deformed tectonic rocks such as mylonites are quite common. The ratio of the maximum, intermediate and minimum widths of the pebbles in metaconglomerate rocks (expressed as axes a:b:c) is generally, 6:3:1, while in the nearby ductile shear-zone, the ratio changes to 10:3:1. The stretching lineation displayed by mineral and pebble orientation is also well-developed and its plunge parallels the dip of schistosity. The projection of the hinges of folds in Qinglonghe Group forms a trend of 230° with a plunge of 20° in attitude. Although the Qinglonghe Group and Shuangshanzi Group are unconformable to each other, the projection points of the fold hinges in Shuangshanzi Group deviate slightly from the great circle constructed by the projection of the axial plane schistosity. This may be due to the effect of the stress field of Qinglonghe Group, and the fold hinges have attitudes which are identical to that projected by the fold hinges of Qinglonghe Group. This may suggest that the two groups were deformed under the same compressing-shearing WNW-ESE trending stress field. The deformation of Qinglonghe Group therefore followed the deformation of the Shuangshanzi Group as part of the same progressive deformational process. It is also significant that the long axis of the stress field was parallel to the extension direction of the present fracture zone.

The rock assemblages of Shuangshanzi Group suggest that they were formed in active and comparatively deep ocean basins. After they were folded and eroded on the ocean floor, this group was overlain by the deposition of Qinglonghe Group. The chemical compositions of leptynites in Shuangshanzi Group show little apparent variation from the metaconglomerates: for example, the oxidation index is close to the metaconglomerate $R_1 = Fe_2O_3/FeO = 0.45$. Considering that the upper parts of both groups were formed by turbidite sedimentation, and have similar isotopic geochronology, it can be concluded that the erosion of Shuangshanzi Group probably

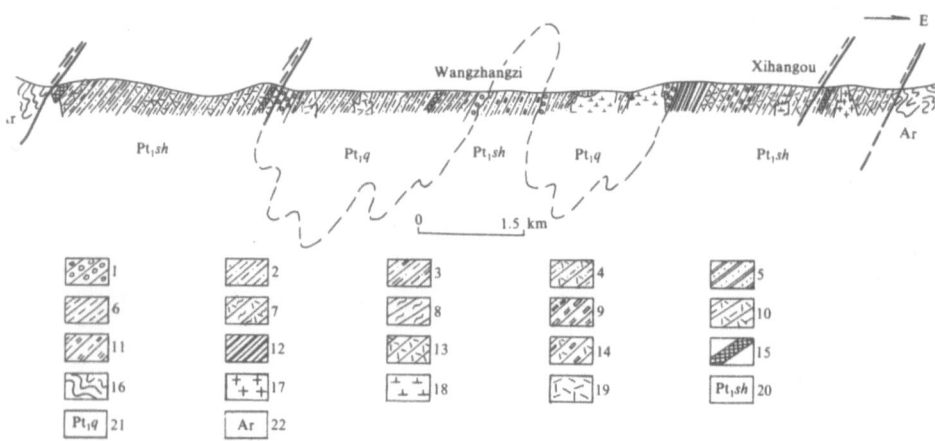

Fig. 3.7. Geological cross-section through the Shuangshanzi Group and the Qinglonghe Group (after Bai Jin and Yang Zhunliang 1984): 1. Metaconglomerate; 2. biotite leptite; 3. two-mica leptite; 4. biotite-hornblende leptite; 5. light leptite; 6. biotite schist; 7. Amphibole - chlorite schist; 8. chlorite schist; 9. blastoporphyritic sericite-quartz schist; 10. biotite-hornblende schist; 11. two-mica quartz schist; 12. phyllite; 13. metabasic volcanics; 14. porphyritic amphibolite; 15. BIF; 16. gneisses; 17. granite; 18. meta-acid-intermediate intrusion; 19. metabasic intrusion; 20. Early Proterozoic Shuangshanzi Group; 21. Early Proterozoic Qinglonghe Group; 22. Archean; 23. ductile shear zone

took place on an ocean floor, and the time interval between the completion of the erosion and the deposition of Qinglonghe Group was very small.

Metamorphism. The Qinglong greenstone belt is metamorphosed to high greenschist facies - low amphibolite facies grades (Jin Wenshan and Wang Wuyun 1984). Evidence of metamorphic grade is given by metapelites in both the Shuangshanzi and Qinglonghe Groups: in the Shuangshanzi Group metapelites are plagioclase + quartz+ biotite ± muscovite ± garnet ± staurolite. Additionally, metabasites in this Group are composed of plagioclase + hornblende ± biotite. The appearance of staurolite in metapelites and of extensive green hornblende plus plagioclase (Anorthite content >30%) in the metabasites indicates that the metamorphism of this Group reached low amphibolite facies. In comparison, the Qinglonghe Group contains no staurolite in metapelites and has plagioclase + hornblende + epidote in metabasites, indicating it belongs to high greenschist facies.

According to garnet-biotite geothermometry, the metamorphic temperature has been calculated to be 510 °C for Qinglonghe Group, and 548 °C for Shuangshanzi Group. It seems that the metamorphic temperature of this greenstone belt decreases from both sides inwards, displaying a concave isothermal plane whose long extension is in accordance with the regional tectonic lines. Geobarometry studies have concentrated on systematically collected samples of muscovite of muscovite schist and mica leptynite in the two groups. The b_0^* values of muscovite have been measured by an X-ray diffractometer. In a percentage distribution diagram, the muscovite b_0 values of Qinglonghe Group are mainly concentrated between 8.993-8.998 Å with a maximum of 37.5%. These measurements indicate that the two groups have been subjected to different metamorphic pressures. There are two concentrations of muscovite b_0 values in Shuangshanzi Group: one is between 9.002-9.006 Å with a maximum percentage of 44%, and the other between 9.015-9.017 Å. The cumulative frequency curves of muscovite in the two groups are sharp, the averages are 8.996 Å and 9.009 Å for the Qinglonghe and Shuangshanzi Groups respectively. The two muscovite b_0 concentrations in Shuangshanzi Group may be the reflection of two pressure types. Hence the metamorphic conditions reflected from mineral studies are therefore consistent with those from mineral assemblages: 500-600°C, 0.55-0.7 GPa for Shuangshanzi Group and 500 °C, 0.4-0.6 GPa for the Qinglonghe Group.

Tectonic Setting and Evolution. The Qinglong greenstone belt has been assumed to represent a paleo-rift (Bai Jin and Yang Chunliang 1984a; Bai Jin 1985). With the recent advances of studies on the geology and geophysics of this greenstone belt, it is necessary to reconsider its tectonic setting and evolution.

As previously mentioned, the Qinglong greenstone belt is distributed within a long-lived NNE-trending fracture zone. This fracture zone is displayed by a linear negative aeromagnetic anomaly interrupted in some places by E-W trending positive anomalies. The tectonic lines of the greenstone belt cut those of adjacent Archean terranes, indicating that the belt was juxtaposed onto the Archean cratons. Correspondingly there is a regional negative magnetic anomaly extending for 20 km, dipping to the west. To the east of the greenstone belt, there is the ancient Shanhaiguan granitic terrane which shows NE-trending low positive areomagnetive anomalies. However, to the west of the greenstone belt, there is the Malanyu metamorphic complex of medium-high metamorphic grade which shows high E-W trending positive anomalies. It is evident that these two were not initially united. This greenstone belt, although incompletely preserved, can be reconstructed to be a linear overturned synclinorium dipping west, associated with ductile shear zones. The metamorphic grade tends to decrease from the edge to centre, forming a "thermal trough".

To summarize:- from the rock assemblages, regional structural patterns, paleo-geothermal regimes and geophysic anomalies, it can be concluded that this greenstone belt formed in island arc setting. The belt was subjected to deformation and metamorphism with the convergence of the blocks on two sides and eventually survived in the collision zone between continent and island arc. The evolutional process is postulated as follows (Fig. 3.8):

1. After 2.5 Ga, the Archean craton was extended and separated in a NNE direction, resulting in the formation of an ancient ocean of an unknown size. Subsequently, subduction occurred along the west coast of the ocean, leading to the formation of a subaqueous island arc.

2. Volcanic eruptions, together with turbidite deposition in periods of volcanic quiescence, resulted in the formation of the Shuangshanzi Group. The Group experienced a tectonothermal event related to subduction at 2.2 Ga. This event resulted in the amphibolite facies metamorphism of this Group.

3. Erosion of the sea floor lead to the formation of a layer of composite paraconglomerate. This was followed by the deposition of the upper turbidite series, i.e. - the Qinglonghe Group. As subduction continued, the collision between the island arc and continent occurred at 2.0 Ga. As a result of this collision, the two groups were deformed during progressive metamorphism. Eventually the belt was deformed into an isoclinal, overturned fold system which was subsequently subjected to ductile shearing. The folds are mostly moderately inclined, rather than recumbent (thrusting) type displayed in converging plate boundaries. This may imply that this greenstone belt was the root of the collision zone.

During deformation, a thermal event causing the high greenschist facies metamorphism of the Qinglonghe Group occurred simultaneously and formed a "thermal trough". This was then followed by the intrusion of basic dykes and granodioritic porphyrites. The separated cratons were ultimately welded along this collision zone during the Lüliang movement. The Mid-Proterozoic Changcheng System was deposited on the gentle eroded surface formed after cratonic collision.

3.2.3 Shanxi-Henan Rifted Province

The Shanxi-Henan Rifted Province occurs in a region covering 34°15' - 39° longitude and 110° 45' - 114° 50' latitude. This area makes up most of Shanxi, the western margin of Hebei and western Henan areas, together with the Wutai, Taihang, Lüliang, Zhongtiao and Songshan Mountain ranges. In geological terms, these regions are host respectively to the Hutuo Group, the Gantaohe Group, the Heichashan Group-

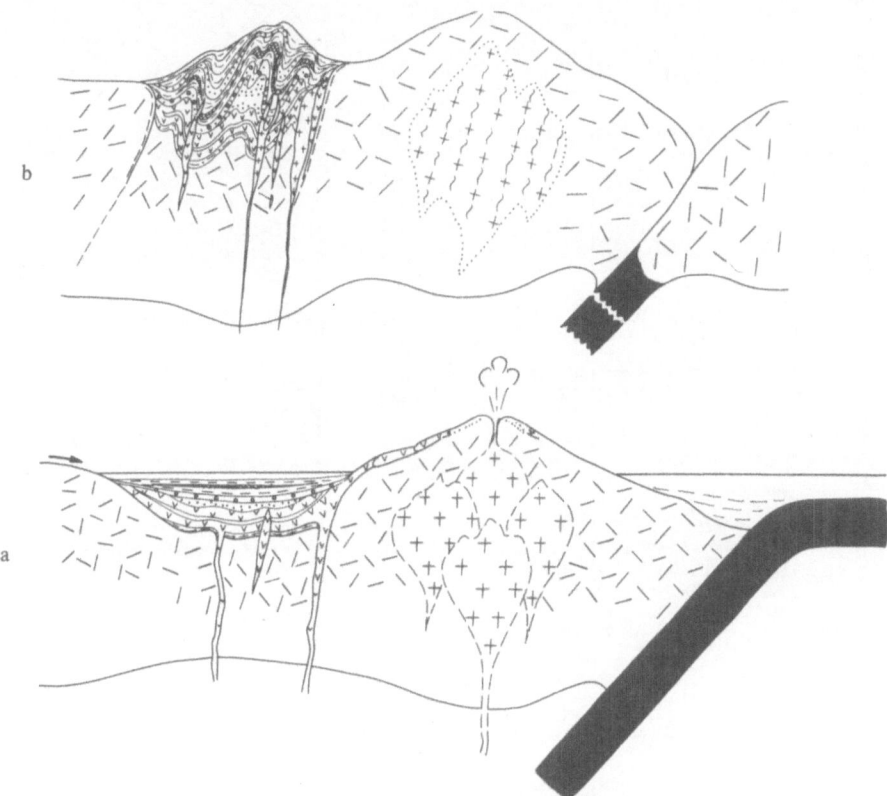

Fig. 3.8. Schematic model showing the evolution of the Qing-Luan collision zone

Yejishan Group-Lanhe Group, the Zhongtiao Group and the Songshan Group. These Groups all consist of volcano-sedimentary, miogeosynclinal deposits formed in the Early Proterozoic and largely controlled or confined by thrust faults (Fig. 3. 9). The similarity of environment, thermal dynamic setting and regional structural style strongly suggests that they are the remnants of aulacogen formations developed on a single Archean cratonic basement.

Rock assemblages. Most of above-mentioned groups are composed of clastic, argillaceous and carbonate rocks with lesser amounts of continental tholeiitic basalts.

In the Lüliang Mts. (Fig. 3.10), the succession is comprised of 52.4% medium-coarse clastic rocks (conglomerate and sandstone), 19.6% argillaceous sediments, 10.7% carbonate and 22% volcanics (mainly basic lava). The maximum thickness of the sequence is greater than 9500 m. The Early Proterozoic succession can be divided ascendingly into the Lanhe, the Yejishan and the Heichashan Groups. The Lanhe Group consists of cyclic terrigenous clastic sedimentary and carbonate formations containing metaconglomerate, feldspar quartzite, phyllite and crystalline dolomite interbedded with metabasic volcanics. The base of the Group is repre-

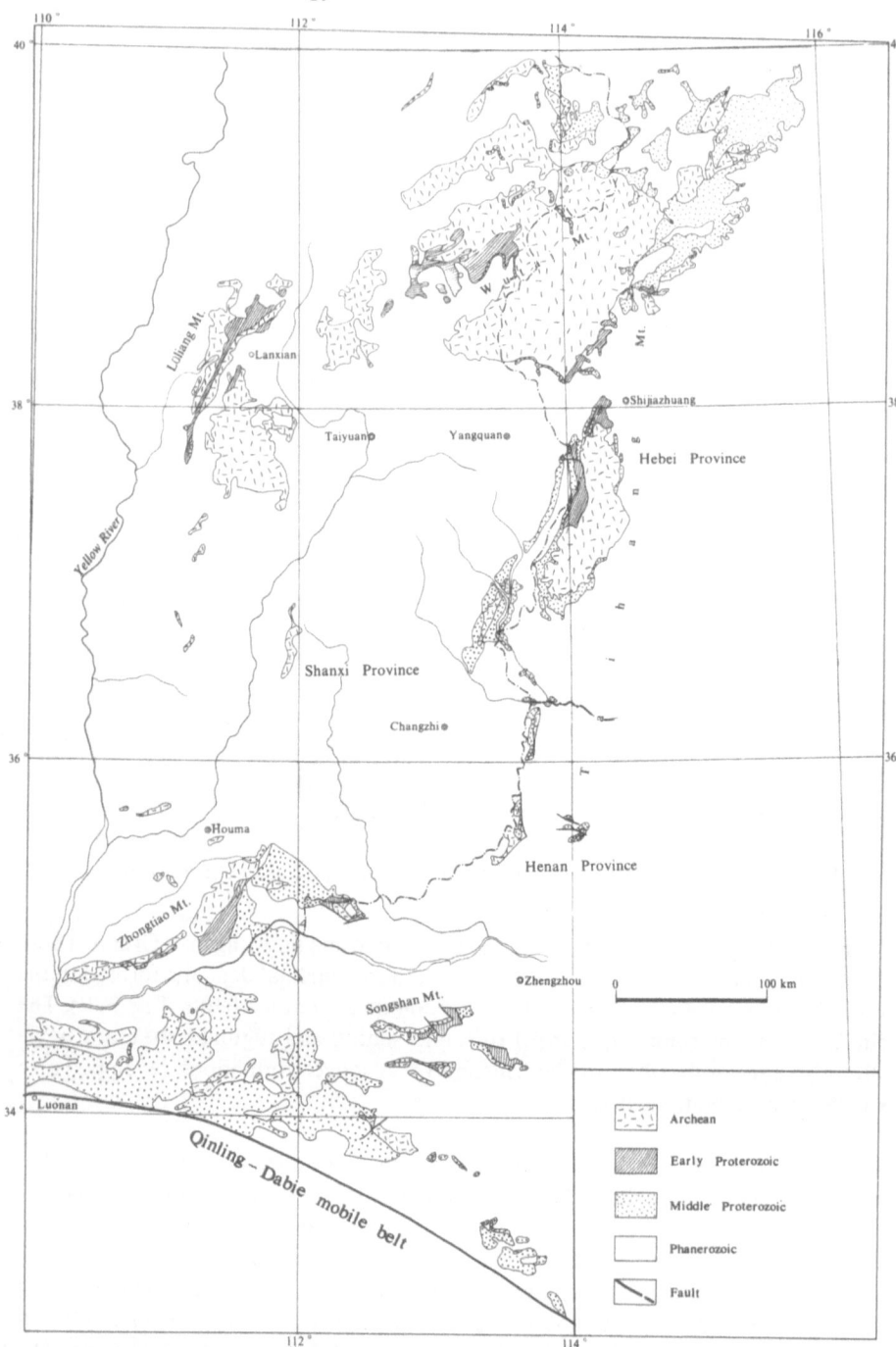

Fig. 3.9. Simplified geological map showing the distribution of the Early Proterozoic rocks in Jin-Yurift Province (compiled from the 1:500 000 geological maps of Shanxi, Hebei and Henan Provinces): 1. Archean crystalline basement; 2. Early Proterozoic metamorphic rocks and their structural trend; 3. Middle Proterozoic strata; 4. Phanerozoic cover; 5. northern boundary of the Qinling-Dabie mobile belt.

sented by unsaturated sandy conglomerate. Such a sequence reflects rapid deposition and sedimentary differentiation in the early stages of formation of a rift valley. The lower, middle and upper parts of the sequence are dominated by coarse-grained clastics, carbonates and finer clastics respectively. The lithologies reflect the instability of crust from the onset of rifting to the later flexural downwarp of the crust. Tectonic activity is accompanied by frequent fluctuation of sea level and variable orders of cyclic deposits: for example, the Yejishan Group is a graded bedded sequence dominated by dark grey and black fine clastics. It represents a flysch sequence, deposited in an abyssal environment during the downwarp stage of evolution of the basin. Basic volcanics with minor rhyolites are developed in the middle part of the Group. These volcanics are mostly metamorphosed to amphibolitic rocks (e.g. amphibolite, amphibolic leptynite, amphibole schist) with preserved vesicular structure. According to their composition the metabasic volcanics can be divided into metabasalt and metabasaltic andesite, which are characterized by high TiO_2 content (1.33%), Al_2O_3 (14.76%) and $K_2O + Na_2O$ (4.89%), as well as meta-alkaline basalt that is characterized by an even higher TiO_2 (1.55%), Al_2O_3 (15.06%), CaO(7.31%) and K_2O+Na_2O (5.39%). On the AFM and FeO^*-MgO-Al_2O_3 discrimination diagrams these basalt types both fall into the tholeiite and continental basalt areas respectively. The overlying Heichashan Group is a molasse formation composed of coarse clastic rocks. The deposition of the Group reflects the end of the sedimentary basin.

In summary, it is evident from the Early Proterozoic stratigraphy in the Lüliang Mountains that an intracontinental neritic environment existed on the western margin of the Early Proterozoic rift zone. The frequent variation of lithology in the sequence, bearing several gravel-sand-mud cycles, reveals considerable change in water depth.

The Early Proterozoic stratigraphic column in the Wutai Mountains is termed the Hutuo Group (Fig. 3.11). This Group is mainly composed of polycyclic coarse clastics, argillaceous and carbonate rocks and has a thickness as great as 10,000 m. The sequence comprises 9% conglomerate, 20.8% arenaceous rocks, 26.9% argillaceous rocks, 40.7% carbonate rocks and >2.5% volcanics. The Group is divided in ascending order into the Doucun, Dongye and Guojiazhai Subgroups.

The base of the Doucun Subgroup is of polygeneous metaconglomerate typified by poor sorting, basal cement, moderate roundness and varied thickness. These features indicate that since its orogenic crumpling the Wutai Group went through a protracted denudation, followed by the rapid deposition of cover rocks. This cycle was accompanied by uplifting and downwarping caused by an earlier rifting episode. The lower Doucun Subgroup is complex, being largely composed of terrigenous, argillaceous and arenaceous clastics with mechanical bedding structures (disturbed, wavy and graded bedding) and emergent sedimentary structures such as mudcracks and pseudocrystals of salt stone. This sequence reflects the crustal mobility which occurs during the initiation of the sedimentary basins. In contrast the upper Doucun Subgroup is developed in a more kinetically stable condition. It is largely composed of small reefs of biotic carbonate mixed with considerable arenaceous and argillaceous materials. These sediments were deposited in the mature (stable

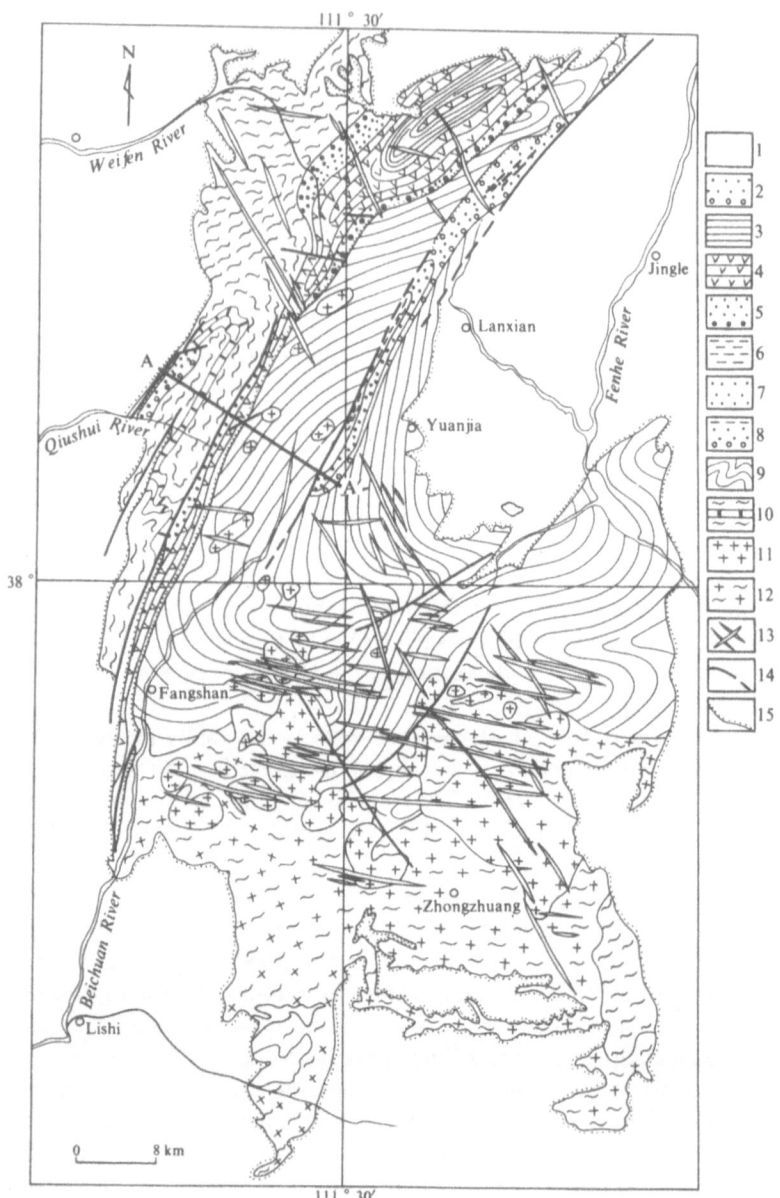

Fig. 3.10. Geological sketch map of Precambrian area in the Lüliang Mountains (modified after the 1:200 000 geologic map of Shanxi Province): 1. Phanerozoic cover; 2. Heichashan Group, metaconglomerate and quartzite; Yejishan Group: 3. Changdaogou Formation, phyllite; 4. Bailongshan Formation, metabasic volcanics; 5. Yangliushuwan Formation, metaconglomerate feldspathic quartzite; Lanhe Group: 6. Luanshicun Formation, phyllite with crystalline dolomite and metabasalt; 7. Shiyaowa Formation, feldspathic quartzite; 8. Qianmazong Formation, metaconglomerate, feldspathic quartzite phyllite; 9. Late Archean Lüliangshan Group, metamorphosed volcanic-sedimentary series; 10. Early to Middle Archean Jiehekou complex, metamorphosed supracrustals; 11. Archean granites; 12. Archean gneisses, migmatitic granites and gneisses; 13. basic dyke swarm; 14. fault; 15. unconformity. - Location of A-A' in Fig. 3.15

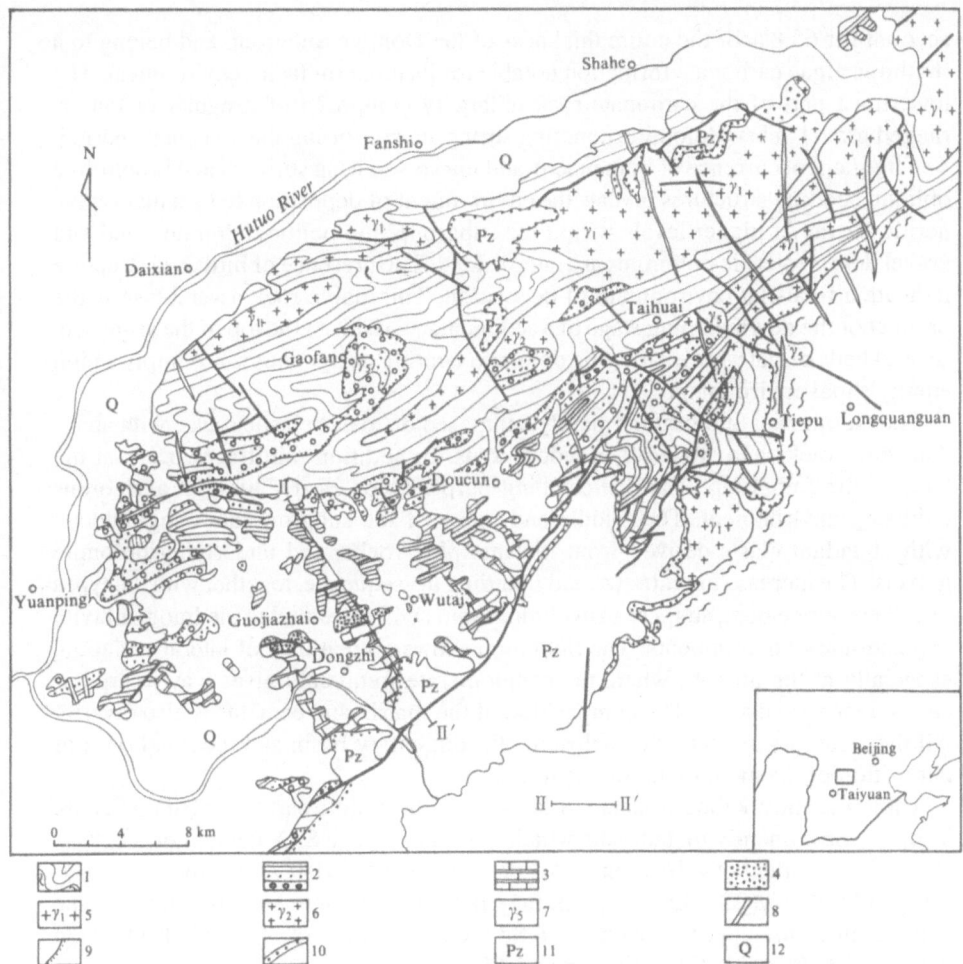

Fig. 3.11. Simplified geological map of the Wutai Mts. region (modified after Bai Jin et al. 1986).
1. Late Archean-Early Proterozoic Wutai Group; Early Proterozoic Hutuo Group: 2. Doucun Subgroup; 3. Dongye Subgroup; 4. Guojiazhai Subgroup; 5. Archean granite; 6. Proterozoic granite; 7. Mesozoic granite; 8. basic dyke swarm; 9. unconformity; 10. Middle Proterozoic; 11. Paleozoic; 12. Quaternary. Location of II-II' in Fig. 3. 15

downwarping) stage of basin development. The top of the subgroup consists of tholeiitic continental flood basalts, revealing the increased depth of the basin as downwarping progressed.

The Dongye Subgroup covers the old weathered crust of the Doucun Subgroup with a deposit of feldspar-quartzite. The Subgroup consists of seven sedimentary cycles made up of argillaceous and carbonate rocks. Lithologies are slaty dolomites, oolitic dolomites, fine-grained micritic dolomites, stromatolitic micritic dolomites and siliceous laminated micritic dolomites, with micritic support. The carbonate rocks account for 65.8% of the entire thickness of the Dongye Subgroup and belong to a rhythmic algae carbonate formation notable for its bioherm facies environment. The lowermost part of the carbonate rock is largely composed of irregular or fan-arranged gravel debris dolomite, denoting strong storms during the diagenetic period. The frequent occurrence of both macro-and micro-scouring surfaces and broom-like oblique bedding structures reveals the environment of deposition to be a metastable neritic basin. At higher levels within the subgroup, the oolitic dolomite, sand and gravel debris micritic dolomites are interbedded. The presence of birds eye structure indicate a tidal-flat environment of deposition. This stable rock assemblage is the product of the downwarping stage of a sedimentary basin. At the top of the sequence, several beds of slate-gravel micritic dolomite-bearing terrigenous sands imply a high energy shoal environment.

The Guojiazhai Subgroup is mainly composed of coarse clastic rocks, with abundant mudcracks and stonesalt pseudocrystals. In addition rain traces occur at the base of the Subgroup. The eye-catching purple color clearly hints at a vigorous oxidising environment. The middle and upper of the subgroup are conglomerate with abundant clasts derived from metamorphic rocks, and underlying dolomite gravels. The increase in grain size and rounding up-sequence, together with the types of calcareous cement, suggest an evolution from swift torrential to piedmont alluvial depositionary environments. The Subgroup shows a great deal of lateral variation, especially at the bottom, where the conglomerate frequently gives way to argillaceous quartz sandstone. The composition of the matrix and cements is also diverse. All these factors indicate the enclosure of sedimentary basin and the rapid rates of deposition of the piedmont molasse formation.

Tholeiitic and alkaline volcanic rocks occur in the middle part of the Hutuo Group. They are prominently metabasalts which have an average SiO_2 content of 50.2%, a K_2O+Na_2O content of 4.14 % and a K_2O/Na_2O ratio of 0.01-0.71. On the TiO_2-K_2O-P_2O_5 and FeO^*-MgO-Al_2O_3 diagram they fall into the continental basalt area. The metabasalt in the upper horizon has a SiO_2 concentration of 46.48%, K_2O+Na_2O value of 5.60% and K_2O/Na_2O ratio of 1.46. These values are very close to those of alkaline basalt (Bai Jin et al. 1986; Wu Jiashan et al. 1986). The SiO_2 content decreases upwards, whereas K_2O+Na_2O and K_2O/Na_2O increase. This suggests volcanism developed from shallow to deep sources and an evolution from sea basin to cratonic environments. The REE contents of the metabasalt are characterized by the following: ÝREE=94-123 ppm, with evident fractionation of light and heavy REE, {ÝLREE/ÝHREE=5.06-5.93, $(La/Lu)_{cn}$=4.83-5.71}. In contrast, the REE concentrations of the meta-alkaline basalt in the upper horizons are greater and their signature more fractionated {ÝREE=152.7 ppm, ÝLREE/ÝHREE = 6.80, $(La/Lu)_{cn}$ =6.59}. The increase in REE concentration and LREE/HREE ratio relative to the underlying tholeiites correlate well with the increase in K_2O/Na_2O ratio.

Dating of zircon within the metabasalt yields a U-Pb concordia age of 2366+103/-94 Ma (Wu Jiashan et al. 1986), which is regarded as the age of basalt eruption. Alternatively, a single zircon grain Pb-Pb concordia age of 2483±1 Ma (2σ) has been obtained from the metagranite intruded into the lower Hutuo Group (Bai Jin et al. 1990). This indicates an age for the lower limit of the Hutuo Group considerably older than 2.4 Ga.

In the Taihang Mountains region to the east, the Early Proterozoic stratigraphy consists of the 4000m thick Gantaohe Group and the 130 m thick Dongjiao Group. The protolith of the Gantaohe Group is a volcano-sedimentary sequence: the sedimentary rocks are composed of coarse clastics, argillaceous and carbonate rocks which are interbedded with volcanic rocks. Proportionately the Group consists of 29% clastics, 16% argillaceous rocks, 14.3% carbonates and 40.7% volcanic rocks. The coarse metaclastics include red unsaturated sandy conglomerate and feldspar quartzite, developed with steep cross-bedding and accumulated heavy minerals in the middle-lower part of the Group. The meta-argillaceous rocks are dark in color and contain rare ripple and mud cracks. This suggests a deep sea reducing environment. No stromatolite is found in the carbonate rocks.

The volcanic rocks in the Gantaohe Group are largely basic volcanic agglomerate, lava and tuffite, which are metamorphosed into chlorite schist and amphibole schist with preserved relic doleritic and amygdaloidal structures in the lower part of the sequence. The volcanic eruptions are multi-cyclic, with the early stage being explosive and the late stage resulting in the deposition of stratified lavas. The mean composition of the basic volcanics are: TiO_2=1.08%, Al_2O_3=15.34%, FeO* =11-15%, MgO=6.67%, $K_2O + Na_2O$ =3.81%. They also possess high K, Rb, Sr and Ba contents, high Rb/Sr and Ba/La values, and low Ir, Nb and Y contents. On the AFM diagram they mostly fall in the tholeiite area and on FeO*-MgO-Al_2O_3 diagram largely in the continental basalt area, with a few falling into calc-alkaline basalt category. The ÝREE is 87.4 ppm, with high LREE concentrations and evident REE fractionation (La/Yb=4-8 and Eu/Eu*=0.6-1.3). These parameters denote a continental tholeiite composition (Wu Jiashan et al. 1988; Cui Zhengkun 1984). Two types of zircons are found in the volcanics: large, dark grains interpreted as xenocrysts and light, small crystals crystallized from the magma. The single zircon Pb/Pb age of the former type is 2.5 Ga and that of the latter type is 2.3 Ga (Wu Jiashan et al. 1988).

The Gantaohe Group contains several conglomerate-sandstone-mudstone and conglomerate-sandstone-mudstone-carbonate rock cycles. The general features of the group indicate an unstable coastal-neritic environment.

Unconformably overlying the Gantaohe Group, the Dongjiao Group was derived from cycles of clastic protoliths. In ascending order, it consists of phosphate siliceous rocks and slate, feldspar quartzite (rich in hematite and magnetite bands), and cross-bedded metaconglomerate interbedded with meta-quartz keratophyre, metadacite and metatrachyte-rhyolite. This sequence is suggestive of the inversion stage in a sedimentary marine trough.

In the Zhongtiao Mountains region in the south, the Early Proterozoic is divided into the 6000 m thick Zhongtiao Group and the 250 m thick Danshanshi Group, (Fig. 3.12). The Zhongtiao Group is a typical miogeosynclinal rock series, consisting proportionately of 3.3% conglomerate, 22.2% sandstone, 44.7% argillaceous rocks, 27.3% carbonate rocks and minor basic volcanics. The volcanics are characterized by 45-53% SiO_2, are Na-rich and belong to the calc-alkaline series. They fall in the tholeiite area of the AFM diagram and the intraplate area of the Ti-Zr, La-Ce,

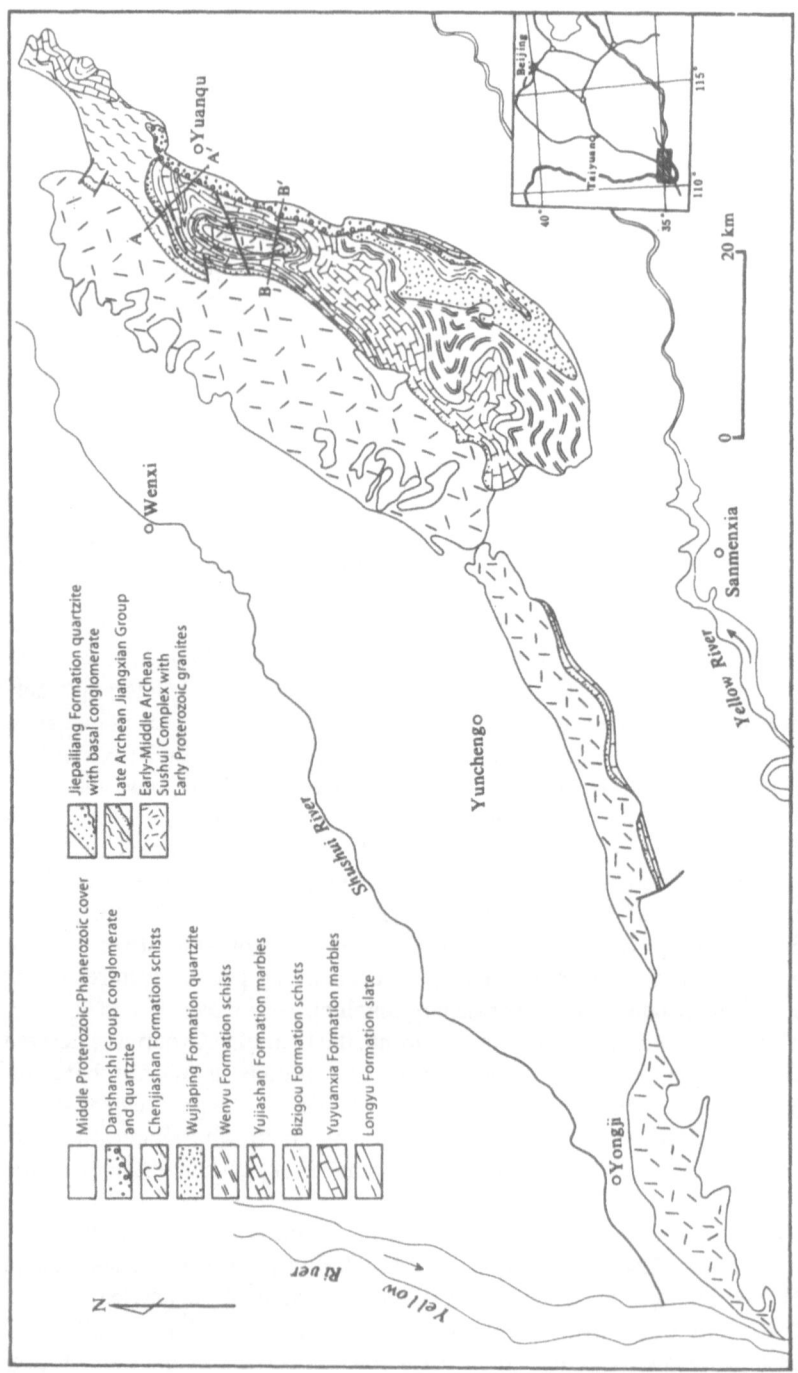

Fig. 3.12. Geological sketch map of the Early Precambrian of Zhongtiao Mt. region, Shanxi Province (modified after the 1:500 000 geologic map of Shanxi Province): 1. Middle Proterozoic-Phanerozoic cover; 2. Danshanshi Group conglomerate and quartzite; 3. Chenjiashan Formation schists; 4. Wujiaping Formation quartzite; 5. Wenyu Formation schists; 6. Yüjiashan Formation marbles; 7. Bizigou Formation shists; 8. Yüyuanxia Formation marbles; 9. Longyü Formation slate; 10. Jiepailiang Formation quartzite with basal conglomerate; 11. Late Archean Jiangxian Group; 12. Early-Middle Archean Sushui Complex with Early Proterozoic granites. Locations of A-A' and B-B' in Fig. 3.16

Ti / 100-Zr-Y and Zr / Y-Zr diagrams (Sun Dazhong et al. 1988). Their REE compositions are marked by a negative Eu anomaly, low fractionation and smooth REE patterns. The Zhongtiao Group consists of three arenaceous-argillaceous-carbonate cycles, which reflect a stable intracontinent neritic environment. The Danshanshi Group unconformably overlies the Zhongtiao Group. The basal conglomerate consists of variable components, derived from the underlying Zhongtiao Group, and reveals an alluvial genesis. Upward the conglomerate becomes quartzite and basal type conglomerate with arenaceous cement, and the gravel contents can be as less as 10-20%, representing a molasse formation stage.

Fig. 3.13. Geological sketch map of the Early Precambrian in Songshan region, Henan province: 1. Syncline and overturned syncline; 2. anticline and overturned anticline; 3. early Proterozoic basic rock; 4. Proterozoic granite; 5. Xiaohuayu Formation phyllite, schist and marble; 6. Miaopo Formation quartzite; 7. Wuzhiling Formation schist and phyllite; 8. Luohandong Formation quartzite; 9. Archean crystalline basement; 10. Sinian to Phanerozoic cover

Age dating of uraninite and brannerite from the marble within the Zhongtiao Group yield a U-Pb concordia age of 1829.6+34/-32 Ma. A range of K-Ar ages of 1.5-1.8 Ga from biotite, amphibole and wholerock samples have also been analysed, and are regarded as the age of metamorphism of the Zhongtiao Group. Additionally, a tuffite yielded a single zircon Pb-Pb evaporation age of 2104±5 Ma. Hence it is considered that the upper limit of the Zhongtiao Group is no younger than 1.8 Ga and the lower boundary is approximately 2.1 Ga (Sun Dazhong et al. 1988).

In the Songshan region in the southeast fringe of the Shanxi-Henan region (Fig. 3.13), the Early Proterozoic stratigraphic pile is called the Songshan Group (Ma Xingyuan 1981b). This Group is mostly composed of feldspar quartzite, phyllite, quartz schist, and dolomitic marble, with the base made of metaconglomerate and sitting unconformably on the Archean basement. The preserved thickness is no more than 2000 m. The protolith is a typical craton rock assemblage largely comprising terrigenous, arenaceous, argillaceous and dolomitic carbonate sediments. Proportionately the lithologies consist of 1.8 % conglomerate, 65% arenaceous rocks, 29.7% argillaceous rocks, 3.5% carbonates, and minor volcanic rocks. The sedimentary sequence exhibits clear cyclicity, with oblique bedding and ripples. These features reflect an epi-continental environment. The presence of facies changes is illustrated by perfectly sorted, uniformly very thick platform quartz sandstone formations and a flysch formation with well developed graded bedding. The facies changes reveal that during the deposition of the Songshan Group the basement was constantly changing between stable and mobile conditions. Such an alteration coincides with the marginal aulacogen position of the Group.

The above descriptions show that these five Early Proterozoic regions have rock assemblages rather similar in composition and environment of formation. In terms of sedimentary formation, all of them have cyclic changes from coarse clastics at the base to sandstone, argillaceous rocks in the middle and carbonate rocks at the top of the sequence. The environment-indicating sedimentary structures occurred in all the rock columns and similar stromatolites are found in each of the carbonate series. Dominantly continental tholeiite volcanic rocks are developed in three areas: the Lüliang, the Wutai and the northern Taihang Mountains regions. All these volcanic rocks are characterized by the same chemistry and tectonic settings.

Structural Deformation Styles. During the Lüliang movement at the end of Early Proterozoic the structural evolution of the Shanxi-Henan region was dominated by E-W lateral compression between the Archean land blocks, forming gigantic multiple folds and strike thrust faults (Fig. 3.14). This deformationary episode finally resulted in the closure of the sedimentary basin. It is clear from the structural patterns in the remaining Early Proterozoic strata that the general structural trend is N-S. However, the Hutuo Group in the Wutai Mountains and the Zhongtiao Group in the south are oriented NE-SW so making the whole structural line display an "S" pattern. This pattern is produced by later clockwise shearing along the north and the south margins of the North China Protoplatform.

In the Lüliang Mountains the Early Proterozoic rocks exhibit a huge fan-shaped synclinorium striking NNE-SSW (Fig. 3.14). The axial plane of the folds on the west limb of the synclinorium dip to west and the folds are reversed eastward; meanwhile those on the east limb are tilted to the east. The Heichashan Group in the westernmost part of the area is a syncline with its axial plane trending west and two near-vertical limbs, occurring on the east of the thrust fault. The Yejishan Group lying in the central area is a narrow syncline with a vertical axial plane tilting to west and accompanied by subordinate folds. The Lanhe Group in the east has a west-dipped reverse syncline in the north combined by two to three very close isoclinal

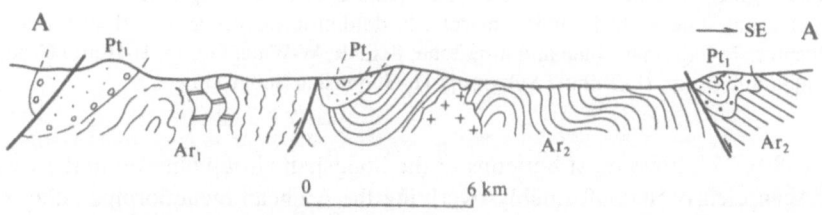

Fig. 3. 14. Cross-section through the three rifts in the Lüliang Mts. fold belt (after Ma Xingyuan et al. 1984). Ar_1. Archean Jiehekou Complex; Ar_2 - Late Archean Lüliang Group; Pt_1 - Early Proterozoic Hutuo Group. Location of A-A' in Fig. 3. 10)

synclines and an east-dipped reverse syncline. In the Wutai Mountains the Hutuo Group displays a gaint fan-shaped synclinorium with the axis trending NE 60° (Fig. 3.15). Well-exposed rocks reveal a perfect structural form characterized by non-parallel axial planes on the two limbs: the NW limb dipping to the north and the SE limb to the south. The folds constituting the synclinorium are all closed and have parallel plane lines, forming a typical linear fold zone. A set of large thrust faults are developed along the major axial plane of the synclinorium.

The Gantaohe Group in the Taihang Mountains contains a wide open synclinorium striking NNE and a near vertical axial plane, decorated with strike thrust faults of varying scales.

The structural line of the Zhongtiao Group exhibits an "S"-like pattern, forming a NNE multiple fold belt. The NE sector is dominated by rather close reversed folds having an axial plane dipping to the NE, whereas the southwestern sector has open folds with axial planes trending NW-SE (Fig. 3.16). As the Zhongtiao Group unconformably overlies the Archean rocks in the northwest, the higher horizons of the Group are largely absent in the northeast. However, the combination with rock outcrops and fold styles suggest that the whole fold belt may be an anti-fan-shaped synclinorium, crosscut by a series of large strike-slip faults.

The Songshan Group in the south forms a westward-dipping, N-S striking multiple linear fold belt. The eastern limb of the fold is reversed. The subordinate folds have variable shapes depending on their position to the main fold: open folds occur on the two limbs, while they grow closer and tilted, even reversed and down-facing in the core of the belt. They are accompanied by compression-shearing thrust faults (Fig. 3.17) which parallel the axes of the folds (Ma Xingyuan et al. 1981b; Liu RQ et

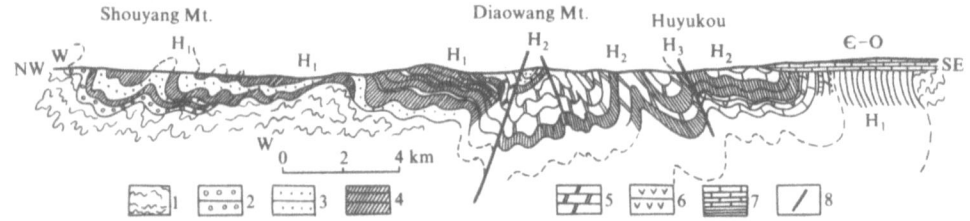

Fig. 3.15. Cross-section through the Hutuo fold belt (after Ma Xingyuan et al. 1984):1. Late Archean greenstone belt – Wutai Group; 2. Hutuo Group metaconglomerate; 3. quartzite; 4. phyllite and slate with dolomitic marble; 5. dolomitic marble and marble; 6. metabasic volcanics; 7. Cambrian shale and limestone; 8. fault; W-Wutai Group; H-Hutuo Group; H_1-Doucun Subgroup; H_2-Taihuai Subgroup; H_3 Guojiazhai Subgroup

al. 1980). The lowermost horizons of the Songshan Group outcrop in the west and are seen clearly unconformably overlying the Archean metamorphic suite. Meanwhile the uppermost horizon is observed in the east of the belt, covered by the Middle Proterozoic Wufoshan Group. Clearly, the incomplete outcrop of the Early Protero-

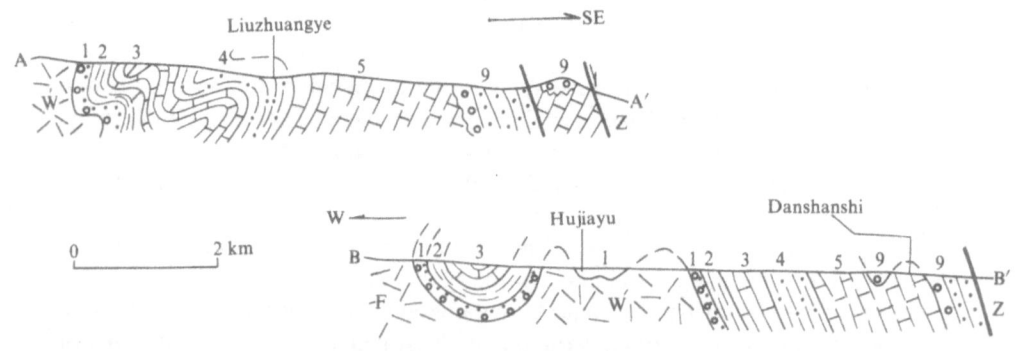

Fig. 3.16. Cross-sections through the Zhongtiao fold belt. (Location and legends in Fig. 3.12)

zoic occurs only in the western half of the original fold belt. It is thus possible that the whole Early Proterozoic fold belt is also a fan-shaped synclinorium.

The tectonic frameworks displayed by the meta-volcanic and meta-sedimentary formations in these mountain regions have remarkable similarity both in structural style and orientation. The Lanhe Group-Yejishan Group-Heichashan Group constitutes a NNE fan-shaped fold belt delimited on both sides by thrusts which parallel the axial planes; the Hutuo Group is also a fan-shaped fold belt striking NE-SW, and controlled by thrusts both in the centre of the basin and elsewhere; the Gantaohe Group-Dongjiao Group occur as a NNE-trending fold belt with vertical axial planes and a series of strike-slip faults; - hence the Zhongtiao Mountains may also be a fan-shaped fold belt. Finally the Songshan Group, if it were intact, would display a similar fan-shaped fold belt morphology. In general, the metasedimentary forma-

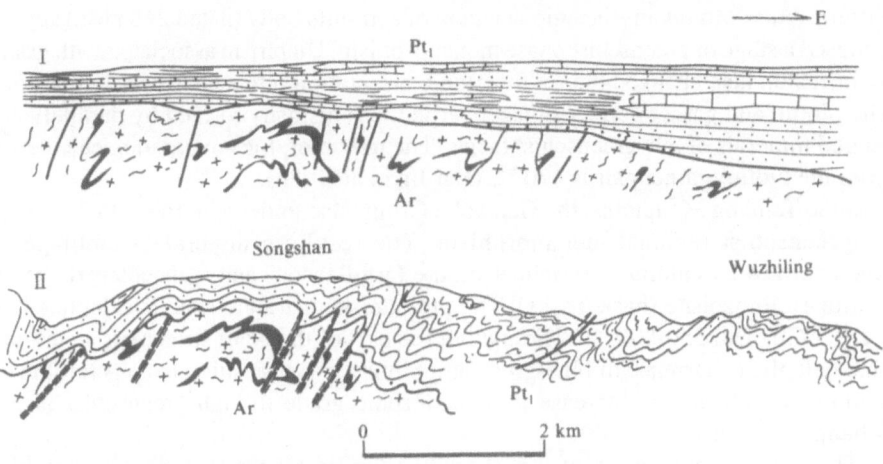

Fig. 3. 17. Schematic cross-section through the Songshan basin, Henan Province, showing I-Sedimentation of the Songshan Group; II-Structural section of the Songshan Group (after Ma Xingyuan et al. 1984): Ar-Archean Dengfeng Complex; Pt_1-Early Proterozoic Songshan Group

tions remaining in each of mountains of the Shanxi-Henan Aulacogen have yielded similar structural styles and orientation together with a fan-shaped fold morphology. This proves that the folds were developed on a rigid Archean cratonic basement. The fan-shaped fold system and accompanying thrust faults are caused by lateral compression from both sides of the sedimentary sequence which has a strong base but contains a suite of more incompetent deposits within them.

Metamorphism. The Early Proterozoic volcano-sedimentary formations mentioned above have gone through extensive syntectonic regional metamorphism to low greenschist facies, and in some places they even reach high greenschist to low amphibolite facies. Apart from the upper molasse formation, which is separated from the major miogeosynclinal formation by an unconformity plane, the major metamorphic facies zone can serve as criteria for studying the paleo-geothermal state in the sequences exposed within the various mountain ranges.

In the Lüliang Mountains, the Lanhe Group is metamorphosed to low greenschist facies, having a representative mineral association of sericite + chlorite + albite + quartz. The Yejishan Group in the central part of the region reaches high greenschist facies, marked by amphibole + plagioclase + epidote + biotite + almandine + quartz (basic rocks) and biotite + mica + plagioclase + almandine + epidote + quartz (argillaceous rocks). The biotite-garnet mineral pair yields metamorphic P-T conditions of 0.2-0.41 GPa and 470°-530 °C. The geothermal gradient is calculated at 30.9-64.9 °C and the depth of metamorphism at 7-17 km, belonging to low pressure metamorphism (Yang Wenkui et al. 1988).

The Hutuo Group in the Wutai Mountains experienced a regional metamorphism of lower subgreenschist facies, with a mineral association composed largely of chlorite + sericite + albite + quartz (Dong Shenbao et al. 1986; Bai Jin et al. 1986). In the eastern Wutai Mountains the emplacement of a granite body (dated 2483 Ma) super-imposed a stage of greenschist phase metamorphism. Diapirism associated with gran-ite intrusion also changed the ENE tectonic orientation into a nearly N-S direction. The biotite and garnet porphyroblast overlie the schistosity formed by parallely ar-ranged minerals of low greenschist facies. The metamorphic temperature calculated from the biotite-garnet pair is 470 °C (Bai Jin et al. 1986).

In the Taihang Mountains, the Gantaohe Group also underwent Early Proterozoic subgreenschist regional metamorphism. The resulting mineral assemblage is characterized by chlorite + sericite + biotite (argillaceous and arenaceous rocks) + biotite + clinozoisite (basic rocks). The pressure-temperature state is calculated as 0.2-0.7 GPa and 350-500 °C. Throughout the east of the NNE stretching metamor-phic belt, the occurrence of blue-green amphibole and a transition from actinolite to amphibole indicate the increase in metamorphic grade to high greenschist facies (Zhang Chunhua et al. 1988).

The metamorphic grade of the Zhongtiao Group ranges from low greenschist facies to low amphibolite facies. Thus the Group is divisible into a biotite zone, garnet zone and staurolite zone (Yang Wenkui et al. 1988). The metamorphic grade of this Group is evidently higher than Early Proterozoic sequences situated in other mountain ranges. If the interpretation of the structural style of the Group as a fan-shaped synclinorium is correct, then the northeast limb has a higher grade of meta-morphism. Liang Yingfang et al. (1988) suggested that the Group reveals a meta-morphic temperature of 450-500 °C and a pressure of 0.2-1.0 GPa.

The Songshan Group is metamorphosed to low greenschist facies and has a parage-netic mineral association of muscovite + chlorite + quartz + epidote + albite + mi-crocline. In terms of the mineral association it is divisible into a muscovite-chlorite subfacies and a biotite subfacies. The facies zone distribution is concordant to the regional N-S tectonic orientation. Metamorphic grade increases towards the east (Ma Xingyuan et al. 1981). Taking into account the regional structural spreading, the presently found facies zone shows a paleogeothermal plane that is likely a rem-nant on the eastern margin of a larger geothermal trough.

In general, the preserved Early Proterozoic volcano-sedimentary formations in the various mountain ranges commonly suffered from a low-medium pressure re-gional dynamic metamorphism which ranged between 350-500°C in temperature and 0.25-1 GPa in pressure, reflecting that these Early Proterozoic rocks were 15-20 km underground when the metamorphism associated with the Lüliang movement took place. Thus the Archean basement on which the Early Proterozoic sedimentary trough developed had risen 15-20 km even before Middle Proterozoic deposition occurred. Therefore, it is reasonable to infer that most of the regionally folded sedi-mentary rocks have been denuded away. The volcano-sedimentary formation that occurred in the Lüliang, the Wutai, the Taihang, the Zhongtiao and the Songshan Mountains regions were formed in the same tectonic environment. Taking the meta-morphic facies zones as a whole, higher geothermal values are found on the two sides of the aulacogen, which might constitute a paleogeothermal trough with indis-tinct relief. What is interesting is that such a geothermal distribution does not coordinate with that of an aulacogen: there must have been an utterly different thermo-dynamic process operating in the belt.

Dynamic Processes. In the strictest sense, the Shanxi-Henan region is not an oro-genic environment. However, the sedimentary formation was folded and metamor-phosed by lateral compression immediately after its deposition which suggests a reasonably dynamic mechanism of formation.

The rock association and regional structural styles of the volcano-sedimentary formation remnants of the Shanxi-Henan region indicate that they were developed in a cratonic environment subject to tensional stresses. With the nearly N-S com-pression-shearing stress state on both southern and especially the northern margin of the North China Protoplatform, the rifting was probably formed by a E-W tensional stress coupled with N-S compression. Rifting resulted in the formation of a trough and the beginning of deposition of miogeosynclinal sediments together with the erup-tion of basic lavas.

As shown in Fig. 3.15 and described in section 3.2.3, the regional structural orien-tation of the Hutuo Group is NE-SW. In the south, the southern part of the Zhongtiao Group is also reoriented NE-SW. This reorientation is possibly caused by the later dextral shearing that occurred on both the south and the north margins of the North China Protoplatform. This fold belt is covered by the horizontally-bedded Changcheng System, which indicates that the dextral shearing at both south and north end of the fold belt started in the early Early Proterozoic and had ceased by the terminal Lüliang movement.

3.2.4 Yinshan–Yanshan Mobile Belt

Owing to the collision between the Siberia-Mongolian Plate and the Sino-Korean Plate in Caledonian times and poor geochronological constraints, the classification of the Precambrian rock units along the over 2000 km long northern boundary of the North China Protoplatform has been considerably hindered. Presently classified as Early Proterozoic units are the Halaqin Group in Huhhot and the Hongqiyingzi Group located in northern Hebei, situated respectively in the western and eastern sections of the northern boundary of the craton.

The Halaqin Group is composed of intermediate grade metamorphic rocks. Com-monly occurring lithologies within the Group are biotite quartz schist, staurolite two-mica schist, staurolite garnet mica schist, biotite plagioclase gneiss, staurolite sillimanite garnet biotite schist, amphibolite and marble. Among them the amphibo-lite suggests a protolith of basic volcanics which respectively fall into the tholeiite and continental tholeiite areas in the $FeO^*-Na_2O-K_2O-MgO$ and the $FeO^*-MgO-Al_2O_3$ diagrams. The schists of the Group have protoliths of clastic and pelitic sedi-ments, and the marble originates from Mg-rich carbonate rocks. In short, the Group is a suite of sandy-pelitic-carbonate sedimentary rocks interbedded with continental tholeiite. This assemblage and the lack of intermediate-acidic volcanic rocks implies a stable faulted basin environment (Jin Wenshan et al. 1989). An intermediate-pressure low amphibolite facies metamorphism has affected the Halaqin Group: the metamorphic PT conditions are calculated at 0.5-0.7 GPa and 585-630 °C, with a geothermal gradient of 20-25 °C/km calculated from the garnet-biotite and garnet-staurolite mineral pairs.

The Hongqiyingzi Group is composed of largely metamorphic rocks of high greenschist to low amphibolite facies. Lithologies include biotite plagioclase gneiss, biotite plagioclase leptynite and hornblende plagioclase gneiss interbedded with amphibolite and marble. A series of zircon U-Pb concordia ages of ~2300 Ma have been reported for the Group. Lithofacies and geochemical studies (Regional Reconnaissance Party of Hebei Bureau of Geology and Mineral Resources 1989, unpublished) show that the protoliths of the amphibolite are basic volcanics which occur in the tholeiite area of the $FeO*-Na_2O + K_2O$ -MgO diagram and scatter in an area of oceanic island basalt. The leptynite in the lower part of the sequence has a protolith consisting of basic, intermediate and acid volcanics. Chemically, the basic volcanics have an average composition of SiO_2=50.84%, TiO_2=1.36%, Al_2O_3=18.53%, $FeO*$=9.11%, CaO=5.99%, MgO=4.12%, $K_2O + Na_2O$ = 7.13% and Na_2O=4.39%, which are analogous to olivine latite; the intermediate rocks have average components of SiO_2=57.10%, TiO_2=0.96%, Al_2O_3=17.29%, $FeO*$=5.58%, CaO=4.93%, MgO=1.96%, K_2O+Na_2O=8.23% and Na_2O=5.38% and are classified as latites. Lastly, the acid volcanics are of rhyolitic to dacitic composition, with average SiO_2=66.19%, TiO_2 =0.44%, Al_2O_3 =15.46%, $FeO*$=3.43%, CaO=3.04%, MgO=0.87%, K_2O+Na_2O=7.67% and Na_2O=3.68%. The chemistry of the volcanics reflects a calc-alkaline evolution trend. The protolith of the leptynite graphite-bearing garnet gneiss were sandy-pelite sediments, whereas other gneissic rocks have intermediate-basic pyroclastics and tuff as major protoliths. In summary the Hongqiyingzi Group was originally a volcano-sedimentary rock series composed of sandy-pelite carbonate rocks associated with basic to acid calc-alkalic volcanics, which reflect an active continental margin environment.

The P-T conditions of the low amphibolite facies regional metamorphism imposed on the Hongqiyingzi Group was 0.6-0.8 GPa, 600 °C, with a geothermal gradient of 21-28 °C/km, corresponding to intermediate pressure facies series (Dong Shenbao et al. 1986). The volcano-sedimentary rock series also suffered from a N-S compression-shearing in Early Proterozoic times which resulted in E-W striking regional overturned multiple folds. The gneissic schistosity acted as a deformation plane and associated ductile shear belt (Regional Reconnaissance Party of Hebei Bureau of Geology and Mineral Resources 1989, unpublished).

In the Yanshan region to the east, an old metamorphic series of postulated Archean age was termed the Dantazi Group (Cheng Yuqi et al. 1982; Wang Qichao, 1989). However, recently Hu Xuewen and Zhu Yingxi et al. (1989) detached a Proterozoic unit from the stratigraphy and correlated it with the Hongqiyingzi Group to the west. The regional structural styles are demonstrated by ENE directed ductile shear rock-slices separated by highly strained mylonite belts, accompanied by northward overturned brachy-antiforms and interfolial folds. The gneissic schistosity acted as a deformation plane during folding.

The newly identified Hongqiyingzi Group outcrops north of the Archean Dantazi Group. In sharp contrast to the Dantazi Group, it is mainly made up of a suite of amphibolite, marble, leptynite and migmatized biotite (hornblende) plagioclase gneiss in high greenschist to low amphibolite facies. A biotite leptynite within the Group yielded a zircon U-Pb concordia age of 2333+7.7/-5.4 Ma. Basic volcanics predominate among the protoliths of the sequence, with minor intermediate-acid volcanics and sandy-pelitic rocks. The geochemical parameters indicate that the major part of the basic volcanics are analogous to the modern island arc tholeiite. Structurally, the region exhibits prominent E-W striking and N-plunging (50-60°) linear multiple folds and unusually well-developed ductile shear-belts represented by mylonite zones (Hu Xuewen and Zhu Yingxi 1989).

3.2.5 Qinling-Dabie Mobile Belt

The Qinling and Dabie Mountains on the southern border of North China Protoplatform stretch southeastwards from 34°N and 105°E, southeastwards to 30°N and 116°E. The Early Proterozoic volcano-sedimentary series in this long, narrow belt has been greatly altered or even consumed during the post Mid-Proterozoic tectonic events, especially during Caledonian, Indosinian and Yanshanian movements. Therefore, it is very difficult to study the Early Proterozoic crustal evolution of this belt. Fortunately, by means of rock assemblages, regional structural patterns and paleogeothermal regimes provided by the various volcano-sedimentary Groups (e.g. Qinling and Hong'an Groups), we can also investigate the Early Proterozoic history of this belt.

Central Western Section of the Qinling-Dabie Mobile Belt. The Qinling Group is located in the central western section of the Qinling-Dabie mobile belt. It has experienced multiple-phase deformation and metamorphism usually followed by magmatism. In tectonic terms, it acted as the metamorphic core complex in the Qinling orogenic belt. The Group is composed of a series of gneisses, marbles and amphibolites of medium metamorphic grade. According to the recent studies of You Zhendong et al. (1987), the Group is divisible into three main units: the lower horizons of the Group consists of biotite-plagioclase gneiss, amphibolite, garnet-bearing-biotite-monzonic gneiss, garnet-sillimanite biotite gneiss and minor amounts of marble. The middle part of the Group consists of calc silicates, marble, garnet-sillimanite gneiss and biotite gneiss. Finally the upper part of the Group consists of graphite-bearing, thick-layered marble locally intercalated with diopside amphibolite and sillimanite gneiss. The biotite (two mica)-plagioclase gneiss and sericite (or biotite)-quartz schist are metamorphosed mostly from argillaceous and psammitic terrigenous sediments, and partly from intermediate-acidic volcanic rocks. The average chemical composition of the meta-sediments is: SiO_2=65.98%, TiO_2=0.74%, Al_2O_3=15.56%, Fe_2O_3=0.59%, FeO=5.56%, MgO=1.61%, CaO=2.75%, Na_2O=2.59%, K_2O=3.04%, and Al_2O_3/Na_2O =6. They are less mature, poorly sorted

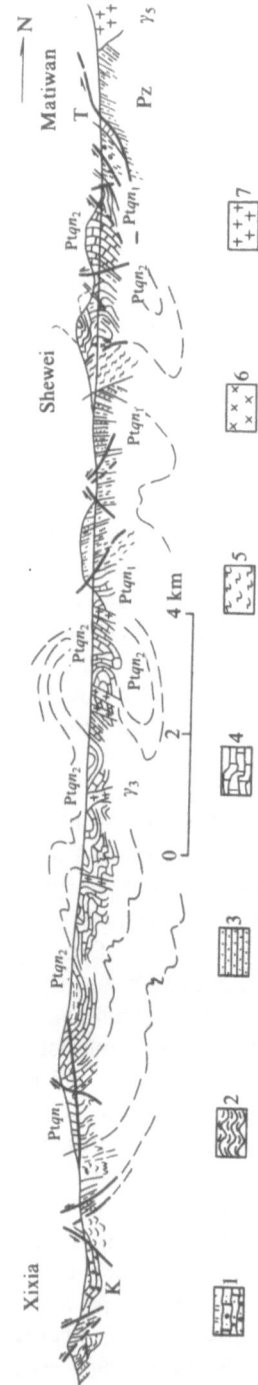

Fig. 3.18. Cross-section of Qinling Complex from Xixia to Motianling Henan Province (after You Zhendong et al. 1990): 1. Cretaceous red beds; 2. gneiss; 3. biotite-leptynite; 4. marble; 5. ductile shear-zone; 6. Caledonian gabbro; 7. granite; T-Triassic fault-depressed metamorphic zone; $Ptqn_1$-lower Qinling Complex; $Ptqn_2$-upper Qinling Complex; P_2-Paleozoic; K-Cretaceous

and found adjacent to source areas, showing features of being deposited in active tectonic settings (Pettijohn 1975).

The minor elements and REE geochemistry of some of representative metapelite and metapsammites from the Qinling complex indicate that they are post Archaean metasediments. The REE distribution pattern (Fig. 3. 19) of the metapelites is similar to that of PAAS (Post-Archean Australian average shale). The La/Th ratios are 2.817 and 2.7 respectively which are near to those of the Proterozoic metasediments (2.7±0.2). The plots of La-Th-Sc ternary diagram are in the field of post Archaean shale and near to the present-day upper crust (UC) but independent of the linear trend of Archaean two-component mixing (i.e. the lines on the ternary diagram, Fig. 3. 20).

There are two types of amphibolite within the Qinling complex: one type is derived from basic intrusions, has the simple composition of hornblende plus plagioclase and possesses clear intrusive contacts with country rocks; the other is derived from basic lavas and shows features of continental marginal tholeiites, the average chemical compositions of which is: SiO_2=49.81%, TiO_2=1.51%, Al_2O_3=13.84%, Fe_2O_3=1.66%, FeO=10.72%, MnO=0.21%, MgO=6.84%, CaO=9.80%, Na_2O=1.58%, K_2O=1.34% (You Zhendong et al. 1987). These data plot into the field of oceanic islands, but very close to that of continent in the FeO*-MgO-Al_2O_3 diagram (after Pearce et al.1977). Such a distribution is characteristic for continental margins. They show lower total REE content (ΎREE=67 ppm, the average of two samples), LREE enrichment {LREE=53.8 ppm, (La/Yb)$_N$=3.87-8.14} and a positive Eu anomaly (Eu*=1.42-1.75). Such REE patterns are between those of Hawaii tholeiites and alkaline basalts. Thus the lava may have been formed by a low degree of partial fusion in the upper mantle (An Sanyuan et al. 1988).

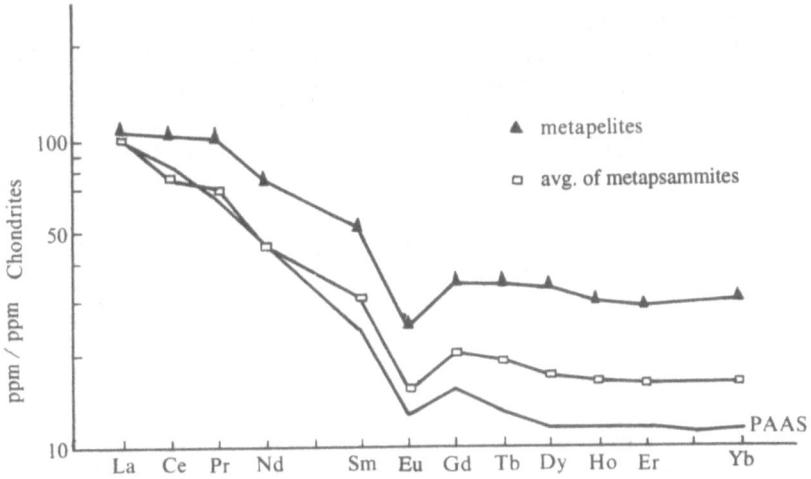

Fig. 3.19. Chondrite normalized REE diagram showing typical post Archean sedimentary REE pattern.

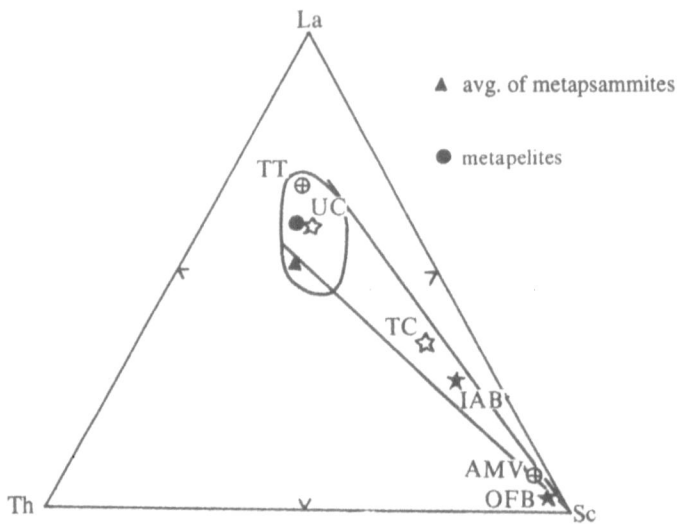

Fig. 3.20. Ternary La-Th-Sc diagram: UC-present-day upper continental crust; TC-present day total continental crust predicted by andesite model; IAB-island arc basalt; OFB-ocean floor basalt; TT-Archean felsic igneous rock; AMV-Archaean mafic volcanic, the delineated area-field of post-Archaean shales (McLennan et al. (1979).

Secondly, some amphibolites are interlayered with marbles and gneisses, and mostly have composite mineral assemblages such as garnet, diopside, biotite and scapolite as well as hornblende and plagioclase. Obviously, these amphibolites are formed from the metamorphism of marlites. Aluminium-rich pelites were the precursors to other meta-sediments such as sillimanite- and garnet-rich biotite gneisses, and dolomite-bearing marbles derived from magnesium-bearing carbonate rocks.

The rock associations of the Qinling Group are in several ways similar to the Late Archaean Taihua Group on the North China Platform (e.g. the ubiquitous graphitic marble). However, overall the lithologies and their geochemistry described above suggest that the Qinling Group does not belong to the Archean as many authors had previously proposed. Nor is the Group equivalent to the metamorphic Paleozoic strata in the adjacent orogenic belts because they contain entirely different rock assemblages and metamorphic style. Therefore the Qinling Group is likely to be an Early Proterozoic continental margin volcano-sedimentary deposit bordering the North China Platform (Wang 1982).

Geochronological studies have confirmed this interpretation: the garnet-bearing biotite gneisses have yielded a zircon U-Pb concordia age of 2226+173/-153 Ma and a whole-rock Th-Pb isochron of 2298 Ma; while the amphibolite has produced a whole rock Sm-Nd isochron of 1892 Ma. It has been therefore suggested that the Qinling Group was formed between 2250-1900 Ma (Geng Shufang 1989). In addition, three sets of isotopic ages, i.e. 1800 Ma, 990.68 Ma and 779.30 Ma have been obtained from the Group. They have been considered to reflect the Lüliang movement in the North China Protoplatform, and the subsequent Sibao and Jining movements in the Yangtze Protoplatform (You Zhendong et al. 1989).

In summary, the Qinling Group is a series of medium-high grade metamorphic rocks which have experienced extensive migmatization. The lower part of the Group was originally composed of basic explosive rocks and terrigenous argillites which gave way upwards to magnesium-rich carbonate rocks formed in closed basin environment. The entire sequence represents a volcano-sedimentary formation of an Early Proterozoic active continental margin.

During tectonic movements, the Qinling Group was subjected to low-amphibolite facies metamorphism (You Zhendong et al. 1987). The most typical metamorphic zones occur in the Xixia area. These are described below from south to north:

1. Sillimanite-muscovite zone: this zone occurs in metapelitic and felsic rocks and is characterized by the assemblage of sillimanite + muscovite. Its metamorphic conditions have been calculated to be 618-665 °C and 0.54-0.55 GPa.

2. Sillimanite-K-feldspar zone: this zone occurs in the core of the Qinling Group and is characterized by the occurrence of sillimanite + K-feldspar in metapelites. The P-T conditions have been calculated as 680-739 °C and P=0.62-0.70 GPa.

3. Andalusite-muscovite zone, characterized by the presence of andalusite + muscovite in metafelsic rocks. Calculated P-T conditions are T=640 °C, P<0.5 GPa.

The distribution of the three metamorphic zones probably indicates that the paleogeotherm tends to decrease from the periphery of the North China Protoplatform southwards whilst the pressure increases. Some surviving kyanites have been found in many rocks. It can inferred that the Group was deposited approximately syntectonically because most metamorphic minerals are oriented parallel to the axial plane foliations of early recumbent folds. The major mineral assemblages of the Qinling Group can be shown by ACF and A'KF diagrams which indicate that they belong to amphibolite facies (Fig. 3.21).

The Qinling Group experienced multiple-phase deformation. The first deformationary event occurred at the time of the Lüliang movement, and is characterized by the development of regional recumbent folds, followed by ductile shear zones (Fig. 3. 18; You Zhendong et al. 1989; Suo Shutian et al. 1987). According to Zhou Dingwu et al. (1989), the early phase deformation can be divided into two stages: the first stage usually led to the formation of regional recumbent folds where the original bedding acted as a deformation surface. Deformation led to the production of penetrative gneissosity, schistosity and lineation. The extensive, grey-white and banded gneiss and the discontinuously exposed, uniformly grey-white biotite-plagioclase gneiss are probably the equivalents of the migmatites and tonalitic-trondhjemitic intrusions of this stage. The large number of small recumbent folds and of the ptygmatic folds of felsic dykes may signify that the first stage deformation was ductile in character. The second stage of deformation resulted in the coaxially refolding of the early recumbent folds and produced closed folds overturned to the north and more recumbent folds. The penetrative axial plane foliation S_2 and lineation as well as flow-like ductile shear-zones were developed. A regional metamorphism of amphibolite facies took place simultaneously with these two stages of deformation.

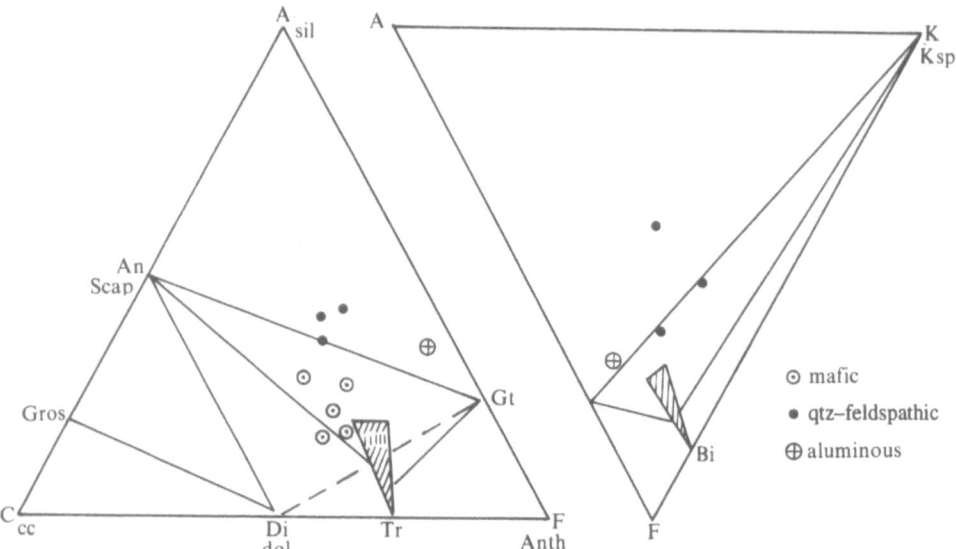

Fig. 3.21. The ACF and A'KF diagram of major mineral assemblages in Qinling Group.

The regional metamorphism is subdivided into two tectonometamorphic cycles, according to the distribution of metamorphic zones and the interrelation between the sequence of metamorphic mineral growth and deformation. These are the Proterozoic and Caledonian cycles, which is in consensus with the Chaidam-Qilian-north Qinling Proterozoic- early Paleozoic metamorphic terrain.

The Proterozoic tectonometamorphic cycle includes notable mineral growth during the D_1, and D_2 deformational episodes. The intergrowth of sillimanite and muscovite representing the regional foliation S_2 replace the kyanites formed during an earlier generation of deformation. The kyanites are sometimes replaced by intergrowths of andalusite, plagioclase, muscovite and quartz. Consequently, in all three regional metamorphic zones Proterozoic mineral growth is characterized by kyanite transformed to sillimanite and/or andalusite. Furthermore, the presence of plagioclase replacing garnet suggests a decompressive metamorphic reaction. The trend of the regional metamorphic zones cuts through the axial plane F_2 of the recumbent fold, which indicates that the final formation of the metamorphic zones are later than the end of the D_2 event. The quartz-feldspathic gneiss of Mashankou, Neixian, yielded a Rb/Sr whole rock isochron date of 990 Ma. It is suggested that the formation of the metamorphic sequence was prior to 1000 Ma.

The PT trajectory of the Qinling complex can be obtained using metamorphic reactions and the mineral geobarothermometry. In particular, garnet zoning in metapelites is especially useful in calculating relative geobarothermometry. These studies suggest the PTt path has a clockwise pattern (Fig 3.22 path M_1): there was a pressure and temperature increase up to the Ky-Ms-kfs field, which then passed

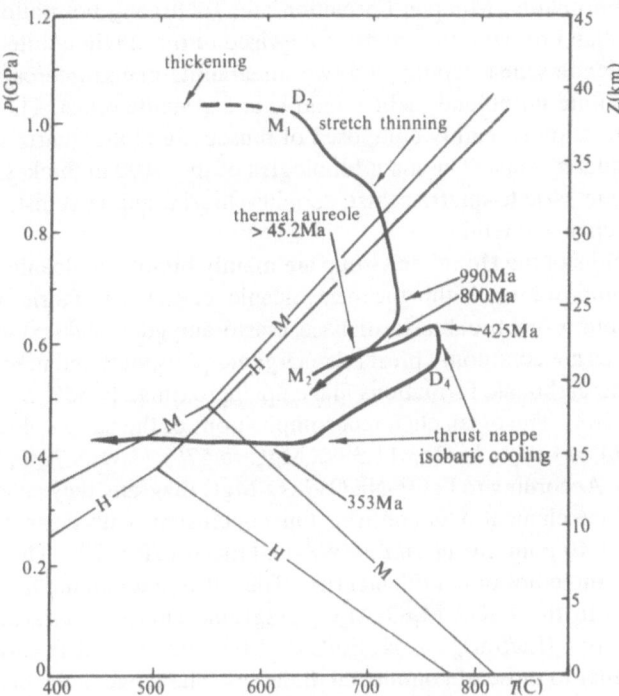

Fig. 3.22. The PTt path of the Qinling Group in western Henan Province, D_2, D_4 Proterozoic and Caledonian deformation episodes, M', M_2-Proterozoic and Caledonian tectonometamorphic cycle respectively, H, M. for Al_2SiO_3 polymorphic transformation by Holdaway (1971), Muller and Saxena (1977).

through the sillimanite + K-feldspar field. Following the peak of metamorphism, the complex passed through a decompressive process.

It is concluded that during the Proterozoic tectonometamorphic cycle the large scale folding of D_1 and D_2 deformation had caused the horizontal shortening and vertical thickening of the continental crust. This implies that the Qinling complex was located in a collisional environment. As the sialic crust began to become stretched, the pressure decreased due to crustal thinning, whereas the temperature rose up to the metamorphic peak. This was followed by a cooling process until the end of the tectonothermal event.

Eastern Section of the Mobile Belt. Located in the Dabie Mountains on the eastern section of the southern periphery of the North China Protoplatform (Fig 3.23), the Early Proterozoic Hong' an Group unconformably overlies the Archean Dabie Group. It is mainly comprised of epidote-albite-hornblende schist, chlorite schist, two mica-albite gneiss, leptite, muscovite-quartz schist, phosphate-bearing leptynite and marble. Based on "Regional Geology of Hubei Province" (Bureau of Geology and Mineral

Resources of Hubei Province, 1986), the Hong' an Group can be divided into the Qijiaoshan Formation, Muopan Formation and Ta' ergang Formation. The 500 m thick Qijiaoshan Formation is chiefly comprised of muscovite-albite gneiss, schist, albite-hornblende schist, leptite, and two mica-albite gneiss intercalated with garnet-bearing albite-hornblende schist, marble and graphite schist. The 2200 m thick Muopan Formation is mainly composed of muscovite-albite-quartz schist and epidote-albite-quartz schist. The main lithologies of the 2400 m thick Ta' ergang Formation include sericite-quartz schist, sericite-chlorite-quartz schist, muscovite-albite-quartz schist and leptite.

The protoliths of the Hong' an Group are mainly bimodal volcanic rocks of basic and acidic compositions. Within the metavolcanic rocks, blasto-fabrics such as blasto-porphyritic, blasto-gabbro, blasto-diabase, blasto-amygdaloidal textures and blasto-fluid structures are commonly preserved. Organic-phosphate and maganese-bearing sediments and carbonate formations make up approximately 30% of the lithologies (Zeng Jiaji 1983). The mean chemical compositions of the basic volcanics are: TiO_2 =1.66%, Al_2O_3=14.03%, FeO* =11.39%, MgO=5.57%, CaO=8.71%, Na_2O=2.71%, K_2O= 0.53%. According to FeO*-Na_2O+K_2O-MgO diagram, they can be considered tholeiites. Trace element data confirms this conclusion, with mean Sr = 355 ppm, mean Rb = 21.40 ppm, mean Ti/Y is 398, and mean K/Rb=282. These features are similar to the tholeiites of continental rifts. They also plot into the field of continental tholeiites in the FeO*-MgO-Al_2O_3 diagram. The REE characteristics are, ΎREE=132 ppm, $(La/Sm)_N$ =2.64, Eu/Eu*=1.03, and the LREE enriched patterns are quite similar to those of continental tholeiites. The range in the concentration of oxide components in the meta-acidic volcanics are, SiO_2 = 65.15-77.28%, (mean 73.16%); CaO = 0.27-3.33%, (mean 1.03%). Na_2O concentrations are greater than those of K_2O, and mean Na_2O + K_2O = 6.43%. The composition of this rock therefore resembles quartz keratophyre. The K/Rb ratio is 235, similar to that of rhyolite. In terms of REE contents, ΎREE content = 192.5 ppm, with $(La/Sm)_N$ =3.49, and mean Eu/Eu* = 0.90. REE patterns show LREE enrichment.

In contrast, the chemical features of the metapelites bear some resemblance to basic rocks, but the high K_2O contents (greater than those of Na_2O or FeO) may indicate a sedimentary protolith: SiO_2 contents range from 36.03 to 53.46%, (mean 45.46%); Al_2O_3 ranges from 9.41 to 16.34%, and TiO_2 content ranges from 1.03 to 5.90%. The metapelites have been interpreted as intermittent sediments deposited during basic volcanic eruptions. The metatuffaceous sandstones show silicic content from 68.30 to 80.6%, and Na_2O from 0.13-7.26%, similar to the compositions of acidic volcanic rocks. This indicates that the main constituents of the sandstones may be derived from acidic eruptions. In addition the Rb/K ratio of the metasediments ranges between 0.0012 and 0.005, and Sr/Ba ratio mostly between 0.02 and 0.476, which concide with those of terrigenous sediments.

Based on these characteristics and combined with the information that the dolomitic marbles in the upper part of Hong' an Group are the products of a shallow sea, it can be concluded that the sedimentary environment was intermediate between continental and oceanic. This may indicate tectonic evolution from a continental to

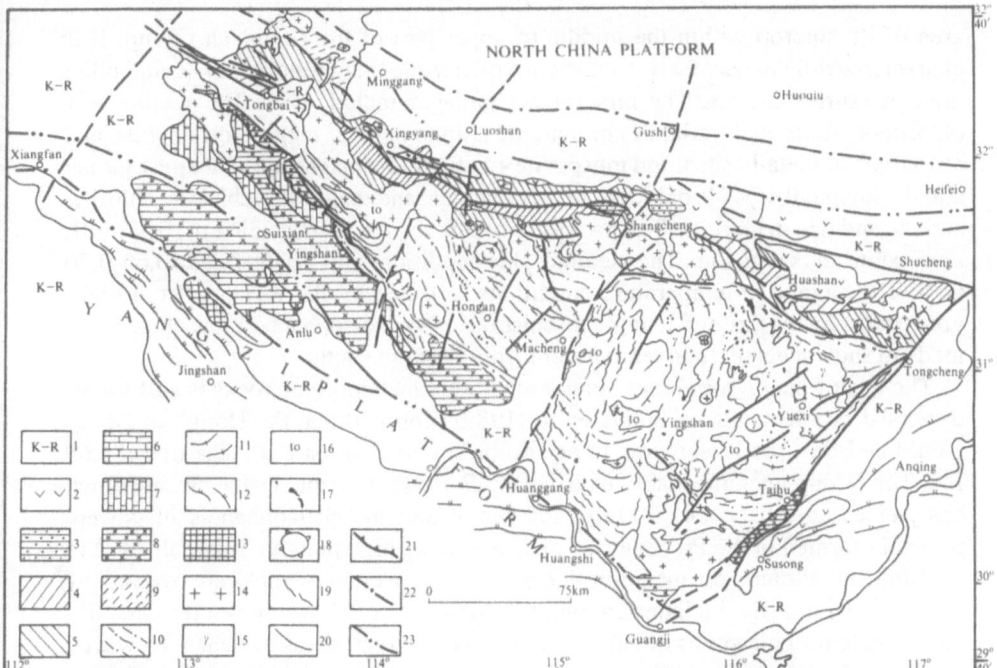

Fig. 3.23. Simplified tectonic map of Tongbai-Dabie Mountains (Compiled by Suo Shutain et al., 1992): 1. Cretaceous-Tertiary; 2. Jurassic, 3.Carboniferous of northern Huaiyang region; 4. Xinyang Group; 5. Lower Paleozoic of northern Huaiyang region; 6. Lower Paleozoic of southern Huaiyang region; 7. Sinian; 8. Mid-Late Proterozoic Sui Xian Group; 9. Kuanping Group; 10. Early Proterozoic Hong'an Group , Susong Group and Qinling Group; 11. Dabie complex; 12. Foreland fold thrust belt; 13. Exposed basement of the Yangtze Platform; 14. Mesozoic granite; 15. Old granite; 16. Tonalite, 17. Eclogite, ultramafic rocks; 18. Gneiss dome; 19. Ductile shear zone; 20. Fault; 21. Brittle-Ductile nappe; 22. Deformed margin of the orogenic belt; 23. Margin of the buried basement

oceanic environment (Liu Yaqin et al. 1989). The volcano-sedimentary nature of this Group may reflect formation in a rift environment.

Three metamorphic facies have been identified from the Hong' an Group. In order from north to south, they are high-greenschist facies, low-greenschist facies and glaucophane-greenschist facies. High greenschist facies occurs mainly in the Qijiaoshan Formation, and is characterized by the presence of hornblende + epidote + albite ± garnet ± quartz in basic rocks, and of epidote + muscovite + albite + quartz and garnet + epidote + muscovite + biotite + albite + quartz in acidic and pelitic rocks. Greenschist facies occurs chiefly in the upper part of the Qijiaoshan Formation and in some lithologies of the Muopan Formation. The mineral assemblages are chlorite + epidote + albite + quartz and actinolite + epidote + chlorite + muscovite + albite in basic rocks; and epidote+ muscovite + albite + quartz, and garnet + actinolite + epidote + muscovite + albite+ quartz in acidic and pelitic rocks. In addition a few rare biotites are found in the upper meta- acidic and pelitic rocks of this facies belt. Glaucophane-greenschist facies occurs mainly in the southern area of its outcrop within the middle to upper part of the Hong' an Group. It is characterized by the extensive occurrence of crossite, magnesioriebickite and other high pressure minerals. The mineral assemblages include actinolite + epidote + chlorite + albite and variable amounts of stilpnomelane, magnesioriebickite and muscovite in meta-basites; and muscovite + actinolite + magnesioriebickite + garnet and the incipient crystals of biotite in meta-pelites; and magnesioriebickite, chlorite, sericite and talc in marbles. The protoliths of this sequence are acidic tuff, basic tuff, lava, pelite and carbonate. The metamorphic conditions are: T=350 °C, P=0.65-0.70 GPa. The change in temperature conditions reflected by the variance from high-greenschist facies through low greenschist facies to glaucophane-greenschist facies tends to show a decrease in temperature from north to south.

The age of the high-pressure, low-temperature metamorphic rocks is still a controversial problem. Zhang Shuye et al. (1989) proposed that the Hong' an Group would be Mid-Late Proterozoic in age according to a zircon U-Pb age of 1485.69 Ma. The Group unconformably overlies the Early-Proterozoic Dabie Group which has yielded ages between 1900 to 2400 Ma. Hence the glaucophane-schists were probably formed in the Precambrian because all the high pressure minerals exist in the Mid-Late Proterozoic metamorphic rocks. Moreover, these rocks are overlain by Paleozoic strata, and the latter are unmetamorphosed or show uneven regional dynamic metamorphic zones of subgreenschist facies to greenschist facies. However, studies of the relationship between deformation and metamorphism of glaucophane schists (Liu Xiaochun et al. 1989) suggest that glaucophanes are well-orientated, parallel to the regional schistosity, and usually enclosed by secondary actinolite. Occasionally, the glaucophanes are replaced by biotite, while independent actinolites are commonly not orientated. These indicate that the glaucophanes are the primary minerals formed simultaneously with the rest of the metamorphic facies. If the early phase metamorphism of the Hong' an Group was related to the Lüliang movement, we should not exclude the possibility that this high pressure belt was initiated in Early Proterozoic time. However, because the Qinling-Dabie mobile belt is situated between the North China and Yangtze Protoplatforms, and has experienced multiple-phases of orogenesis since Mid-Proterozoic time, it is also probably overprinted by poly-metamorphism. Therefore, it may be the case that it is difficult to obtain older ages.

The regional structural patterns of the Hong'an Group constitute a large NW-SE trending composite fold system extending in accordance with the structure of the Qinling Group. There is extensive development of penetrative schistosity and a stretching lineation. Folds are mostly isoclinal and recumbent. Rootless folds are also common, trending parallel with regional tectonic lines (Zhou Gaozhi et al. 1989).

Nature of the Mobile Belts. In the Qinling-Dabie mobile belt on the southern periphery of the Archean North China Protoplatform, the Qinling and Hong'an Groups have, to some extent, contrasting rock assemblages and environments of deposition: the Qinling Group represents an active continental margin setting whereas the Hong' an Group reflects a continental rift environment. However, the two Groups are generally analogous in regional structural patterns and paleo-geothermal regimes. The early phase deformation in both groups is characterized by recumbent folds and later ductile shear-zones (Fig. 3.24), and in general this structural style might be formed in an extensional tectonic setting. However, the tectonic style and syntectonic meta-morphism of up to amphibolite facies would not be accomplished without shear-compression at considerable crustal depths. Moreover, the spatial distribution of metamorphic facies in both groups runs parallel to that of the whole mobile belt (especially in the eastern section of the belt). The paleogeotherm tends to decrease from north to south, which is different from the geothermal regimes in continental rift environments. Consequently, if rifting did indeed occur, it would have been in an active continental margin environment. Considering the preserved lithological record within the whole mobile belts, the Qingling and Hong'an Groups could represent fragments of continental margins close to an ancient subduction zone, that subsequently changed to the accretional zone on the southern margin of the North China Protoplatform (Fig. 3.23). Their tectonic styles and paleogeothermal regimes are probably the result of formation in a subduction zone environment. If some fragments of ancient oceanic crust had been preserved, they would have been destroyed by the successive alternating processes of compression and collision, extension and separation, and strike-slip shearing since the Mid-Proterozoic Era. Finally, if the model outlined above is correct, it means that the Archean craton, North China and Yangtze Protoplatforms were separated by an ancient ocean in Early Proterozoic time.

3.2.6 Dunhuang-Longshoushan Mobile Belt

The Dunhuang-Longshoushan mobile belt lies on the south-western margin of the North China Protoplatform, between 38°-40° latitude and 94°-104° longitude. The belt can be divided into the Dunhuang, Longshoushan and Beidashan Groups which are exposed consecutively from the northwest to the southeast.

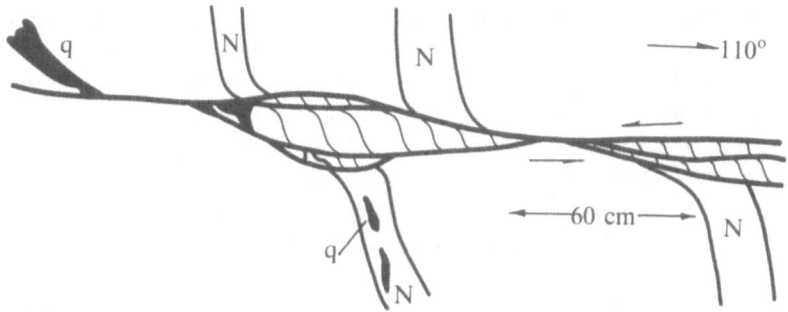

Fig 3.24. Parallel shear zones in the Early Proterozoic Susong Group: N-basic rock, q-quartz vein (By Suo Shutian).

The Dunhuang Group ranges in thickness between 4275-13532 m and is usually deformed into E-W trending fold belts, petrologically composed of biotite-plagioclase gneiss with biotite quartz schist, quartzite, (garnet-bearing) amphibolite and lenticular thin-bedded marble in the lower part of the Group. The concentrations of magnesia and graphite in marble increase upwards and gneiss and schist interbeds become more numerous. The middle part of the Group contains two-mica or garnet-bearing biotite quartz schist, (garnet) biotite amphibolite, amphibole-plagioclase gneiss and chlorite quartz schist, with interbedded garnetiferous plagioclase gneiss quartzite, leptynite and magnesian marble. The upper part of the Group consists of metarhyolite, dacitic breccia, tuffaceous lava, rhyolitic dacite and mica-quartz schist, amphibolite, epidote-chlorite schist, quartzite, and phyllite. Migmatism is widely seen in the middle to lower horizons, often represented by augen, sparse porphyritic migmatic gneiss, striped and banded migmatite and homogeneous migmatite.

The protoliths at the base of the Dunhuang Group is dominated by a set of marine terrigeneous clastic formations, sandwiched by (argillaceous) carbonatite and basic volcanic rocks. This changes upwards to magnesium-rich carbonatite interbedded with intermediate-acid volcanics. The upper part of the Group originally consisted of arenaceous-argillaceous clastics interbedded with intermediate-basic volcanics and carbonatite; and the top is largely acid volcanics with terrigenous clastics, forming two megacycles. The Group has undergone a low amphibolite to low greenschist facies metamorphism, with grade increasing from the limbs of the anticline to the core (Huang Dezheng et al. 1988).

The Longshoushan Group ranges in thickness between 7600-11000 m. It is distributed in NW-SE trending long bands which astride the Longshoushan region in northeastern Gansu. The lower part of the sequence is composed of banded and striped augen migmatite with migmatic amphibolite, migmatic gneiss and marble, changing upwards to magnesium-rich marble with biotite-plagioclase (or garnet-bearing two-mica) gneiss, garnet-bearing two-mica quartz schist as well as amphibolite. The carbonate contents increase northeasterly. In contrast, the middle part of the sequence is mainly made up of two-mica (biotite) quartz schist, amphibolite and plagioclase leptite. These units are interbedded with marble, varying thicknesses of dolomite, gneiss, quartzite and leptynite. Finally the upper part of the sequence is composed of leptite, quartzite, quartz schist, gneiss, schist meta-rhyolitic dacite, leptynite. The very top of the Group is made up of crystalline limestone.

The protolith of the Longshoushan Group was composed of intermediate-basic volcanics and marine terrigenous clastics-magnesium-rich carbonate formation overlain in turn by a marine carbonate-paraflysch formation and marine volcanic-sedimentary rocks- comprising terrigenous clastic formations. The metamorphism at the base of the Group is low-amphibolite facies, rising to middle high-greenschist facies in the middle of the Group and decreasing to low-greenschist facies in the upper part of the Group.

Amphibole-biotite gneiss within the Longshoushan Group yields a whole-rock Rb-Sr isochron age of 1949 Ma. Furthermore a number of muscovite / biotite K-Ar ages of 1700, 1715, 1719, 1741, 1786 Ma were obtained from the granite, granitic pegmatite, pegmatite and gneissic granite that intruded into the Group. Another whole-rock Rb-Sr isochron age of 2147 Ma is reported for the intrusive plagioclase magnetite (Bureau of Geology and Mineral Resources of Gansu Province, 1990). The Longshoushan Group is overlain unconformably by the stromatolite-bearing Dunzigou Group of the Jixian System (see next chapter).

The 3006-7022 m thick Beidashan Group has a base composed of biotite (± amphibole) plagioclase gneiss with biotite quartz schist, mica schist, amphibolite and lenticular marble. Strongly migmatized, it often occurs as banded, striped, ptygmatic and augen migmatite as well as some porphyrite migmatic biotite plagioclase gneiss. The middle of the Beidashan Group consists of marble, dolomite and amphibole (biotite plagioclase) gneiss containing two mica quartz schist, amphibolite interbedded with quartzite leptymite. The upper part is mainly two-mica quartz schist, amphibolite and two mica leptymite, interbedded with biotite plagioclase gneiss. Locally, leached, banded and striped migmatite are found associated with amphibole plagioclase gneiss. The protoliths of the Beidashan Group are similar to those of the Dunhuang Group: the lower horizons consist of an abyssal terrigenous-volcanic formation typified by intermediate-basic eruptions; the middle part is neritic-abyssal carbonate-terrigenous clastic formation (although thinner and less extensive than that of the Dunhuang region), with metasedimentary iron deposits; the upper part of the Group was composed primarily of neritic terrigenous clastics interbedded with strongly acidic volcanic rocks. The whole Group underwent high-greenschist to low-amphibolite facies metamorphism.

The three groups described above have very similar protoliths, metamorphic grade and facies zones, tectonic direction and other characteristics. The vigorous magmatism and flysch formation reflect an active continental margin environment. This mobile belt links with the Qinling-Dabie mobile belt to the southeast. Just as the existence of the Qinling-Dabie belt denotes the previous existence of oceanic crust lying between the North China and the Yangtze Protoplatforms, another large ocean may also have occurred between the Dunhuang-Longshoushan mobile belt and the Tianshan-Beishan-Qilian mobile belt on the north margin of the Tarim Protoplatform. It is possible that subduction might have take place during the Lüliang movement, which caused the collision of the plates on both sides.

3.3 Tarim Protoplatform

In recent years, great progress has been made in studying Early Precambrian geology. This includes the compilation of the metamorphic map of China (Dong Shenbao et al. 1986) and comprehensive studies on the regional geology and isotopic geochronology for the region. However, the original information was so slight that the present divisions of the strata and their ages could only be conjectured upon using the ages of overlying strata, rock sequences and the stratigraphic correlation with adjacent areas (Table 3.1). The tectonic evolution is still ambiguous. Nevertheless, according to the temporal and spatial distributions of regional structures, metamorphic zones and facies, and rock formations, a primary description of the tectonic settings can be made.

The Tarim Protoplatform occurs as a triangular plate, and is extensively covered by Mid to Late Proterozoic sediments and Phanerozoic strata. According to geophysical data and rarely exposed Archean rocks, it can be postulated that the basement is chiefly composed of Archean complexes. Early Proterozoic metamorphic rocks outcrop mainly in the northern and southern peripheries of the protoplatform and were probably related to successive orogenic episodes. The regional tectonic lines and metamorphic zones in the Early Proterozoic rocks are parallel to each other and, in turn, to the peripheries of the protoplatform (Fig. 3.25). It seems that the northern and southern peripheries of the protoplatform were mobile belts during Early Proterozoic times. These are the Tianshan-Beishan-Qilian mobile belt and Kunlun mobile belt respectively. The two belts were cut by the Altun lithospheric left-lateral shear fault (Ren Jishun et al. 1980), possibly during the Indosinian orogeny (Zhang Yifo 1982). This resulted in their subdivision into four sections: the Tianshan-Beishan section, Dakendaban-Qilian section, western Kunlun section and eastern Kunlun section.

3.3.1 Tianshan-Beishan-Qilian Mobile Belt

Tianshan-Beisha Section. The Tianshan and Beishan Mountains are located at 41°-45° N and 75°-95° E of the Xinjiang Uygur Autonomous Region and in 40-42° N and 92°-98° E of the boundary between Xinjiang and Gansu respectively.

The Early Precambrian metamorphic rocks in Tianshan are discontinuously exposed along the southern slope of the highest range of mountains, as a narrow east-west belt, stretching from the southern flank of the upper reaches of Bortala River, near Wenquan, on the boundary areas between China and Soviet Union, and the Yining and Zhaosu regions. The belt goes eastwards through the south part of Xinyuan, Narat, Balyguntay, Hejing, Turpan, the southeastern part of Shanshan, and the eastern and southern parts of the Hami, Weiya and Xingxingxia areas.

The Early Proterozoic metamorphic rocks in this region have been grouped into two sequences. The lower part of the sequence is mainly comprised of augen, injected biotite-plagioclase gneiss, granitic gneiss, migmatic gneiss, quartz schist and quartzite intercalated locally with banded magnetite-poor deposits. An example of this succession is found in the central-west of Tianshan Mountains, and also corresponds to the Early Proterozoic Koksu Group in Kuruktag. The suggested protolith is an intermediate-basic volcano-sedimentary formation, and metamorphism is dominated by medium-pressure low-amphibolite facies. Locally, progressive metamorphic zones are identifiable. Migmatization is well-developed.

The upper part of the sequence covers a wide area with protoliths including terrigenous clastics, carbonates, volcanics and volcano-sediments. This formation is subjected to low-greenschist to amphibolite facies metamorphism and migmatization as well as granitization at 1.8-1.9 Ga (Bureau of Geology and Mineral Resources of Xinjiang Uygur Autonomous Region, 1993). The rocks exposed in the Muzart River

Protoplatform	Region	Area	System	Group		
South Western Periphery of North China Protoplatform	Dunhuang-Longshoushan	Beidashan	Mesozoic Group	Beidashan Group		
		Longshoushan	Jixian System	Longshoushan Group		
		Dunhuang Anxi	Mesozoic Group	Dunhuang Group		
Tarim Protoplatform	Eastern Kunlun	Southern Periphery of Qaidam basin	Jixian System	Huhai Group Jinshuikou Group		Milan Group
	Western Kunlun-Altun	Altun	Changcheng System	Altun Group		
		Western Kunlun	Changcheng System	Elinkat Group Kongar Group	Karakox Group Bulungkol Group	
	Dacandaban-Qilian	Longshan	Changcheng System	Niutohe Group		
		Maxianshan	Jixian System	Maxianshan Group		
		Zoulangnanshan	Changcheng System	Beidahe Group		
		Yemananshan	Changcheng System	Yemanashan Group		
				Huangyuan Group		
		Northern Perphery of Qaidam Basin	Jixian System	Shalinhe Group Dacandaban Group		
	Tianshan-Beishan	Beishan	Jixian System	Beishan Group		
		Kuruktag	Changcheng System	Xingditag Group	Koksu Group	Daglagblag Group
		Central western Section of Tianshan	Changcheng System	Narat Group Muzart Group	Koksu Group	
		Area Stratigraphy	Overlying Strata	Upper	Lower	
				Early Proterozoic Group		

Table 3.1: Schematic correlation for the Early Precambrian Strata of Northwest China

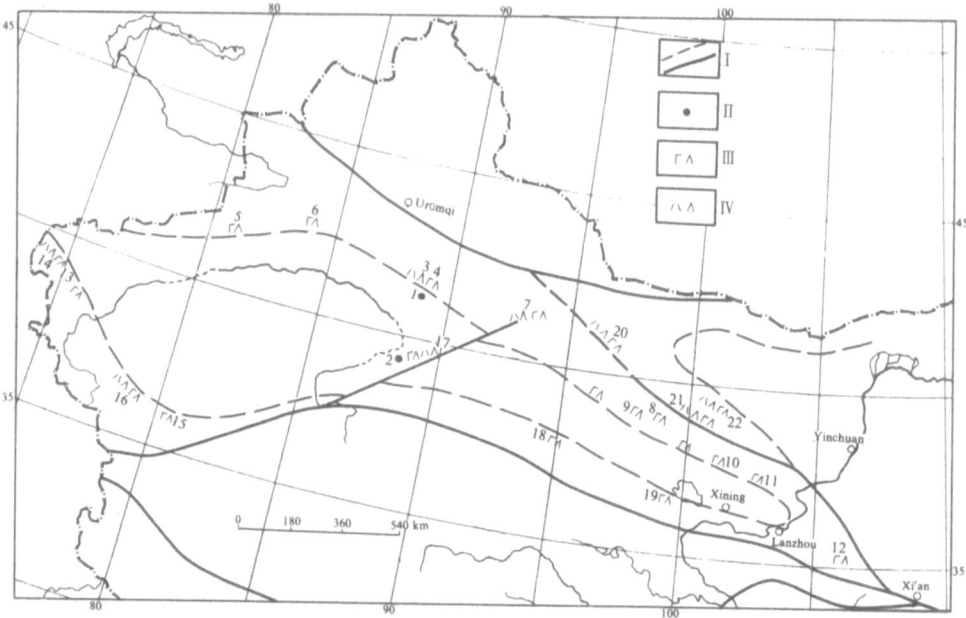

Fig. 3.25. Distribution of Early Proterozoic volcanics in the Tarim Protoplatform: I. Boundary between tectonic provinces and mobile belt; II. Archean exposure; 3. Early Precambrian intermediate-basic volcanics; IV. Early Precambrian inermediate-acid volcanics. 1. Daglak Bulak Group; 2. Miran Group; 3. Xingditag Group; 4. Koksu Group; 5. Muserthe Group; 6. Narat Group; 7. Beishan Group; 8. Beidashan Group; 9. Yemananshan Group; 10. Huanyuan Group; 11. Maxianshan Group; 12. Niutouhe Group; 13. Kongurshan Group; 14. Bulunkol Group; 15. Aliankat Group; 16. Karakax Group; 17. Altun Group; 18. Jinshuikou Group; 19. Kuhai Group; 20. Dunhuang Group; 21. Longshoushan Group; 22. Beidashan Group

area of the southern slope of Halik Mountain in Southern Tianshan are named the Muzart Group, whereas those in the northern slope of Halik Mountain and in the major range of the Narat Mountain are termed the Narat Group. The 5300 m thick Muzart Group is composed of a series of migmatites and gneisses with some schists occurring towards the top of the Group. The 5000 m thick Narat Group is also composed of various gneisses, migmatites, with minor amounts of dolomitic marble and schist.

Also notable is the Early Proterozoic sequence from the Aksu-Wushi area and from the southern slope of Halik Mountain. This sequence is a series of low-grade metamorphosed albite-muscovite-quartz schists, and glaucophane schists intercalated with various greenschists. Besides glaucophane, other high-pressure minerals such as crossite and stilpnomelane are present. Xiong Jibin, Gao Zhenjia and Peng Changwen (1985) have proposed the preliminary conclusion that this unit belongs to the high pressure glaucophane-greenschist facies.

The Kuruktag area in the eastern section of southern Tianshan is widely covered by a complete Early Proterozoic sucession. This succession is used as one of the Early Precambrian type sections in Xinjiang (from Xinjiang Volume, Regional Stratigraphy of Northwest China 1981). The sequence includes the Archean Daglagbulak Group, Early Proterozoic Koksu Group and late Early Proterozoic Xingditag Group. These Groups are covered unconformably by the Mid-to Late Proterozoic sequence corresponding to Changcheng, Jixian and Qingbaikou Systems. The Daglagbulak Group has unconformable contacts with the other two Groups. In contrast, the two Early Proterozoic Groups are juxtaposed by faulting (from Xinjiang Volume, Regional stratigraphy of Northwest China 1981; Gao Zhenjia and Peng Changwen, 1985; He Gaopin and Li Yingri 1988). Geochronological studies have shown that a blueish-quartz granite intruded into the Xingditag Group near Xingdi area has as U-Pb age of 2487±10.2 Ma by analysis of euhedral zircons using the evaporation method (Wang Yunshan and Chen Jiniang 1989). Moreover, three zircon U-Pb ages of 2150, 2000 and 2300 Ma have been obtained from the migmatitic gneiss (Yin Baoxiang et al., 1988).

The 4675 m thick Late Archean Daglagbulak Group occurs as faulted blocks in the central and western parts of Kuruktag as well as to the south of Xinggir Vilsage, where it is unconformably overlain by the Xingditag Group. It is comprised of various gneisses and migmatites intercalated with leptinite and hornblende schist as well as hyperthene-bearing migmatitic gneisses, garnet-biotite-plagioclase gneiss and hornblende two-mica-plagioclase gneiss. Considerable numbers of N-S trending closely spaced basic dyke swarms are commonly found in this Group. There was at least two phases of folding, of which the second phase was associated with many granitic intrusions along the tectonic contact of the Daglagbulak Group with the Mid-Proterozoic Bowam Group (Gao Zhenjia et al., 1989). The rocks are host to very strong deformation, schistosity, mylonitization, migmatization and retrogressive metamorphism. Pulse-like intrusions of various dykes are also common. The protoliths have been studied and classified as andesite, dacite and basalt (Wang Yunshan and Chen Jiniang 1989; Zhang Zhide 1990), which belong to the calc-alkaline series. The REE is typified by lower total REE content ($\acute{Y}$REE=37.19-59.39 ppm), highly fractionated LREE and HREE ($\acute{Y}$LREE/$\acute{Y}$HREE=14.2-16.9), and a positive Eu anomaly (Eu/Eu*=2.05-2.688) (Wang Yunshan and Chen Jiniang 1989). The original rock formation is dominated by intermediate-basic volcanics (with intermediate-acidic volcanics) intercalated with small amounts of clastics. These lithologies have mainly experienced granulite and amphibolite facies metamorphism, but locally have been retrogressed to greenschist facies. In comparison, strata of corresponding age in the east section of central Tianshan is very limited in extent, and composed mostly of migmatites and injection gneiss as well as hornblende gneiss.

The 4675 m thick Koksu Group forms the lower sequence of the Early Proterozoic stratigraphy. The Group is exemplified by the sequence exposed in the Koksu area to the south of Xingdi Village, which is composed of alternating layers of hornblende-plagioclase gneiss, hornblende-plagioclase leptynite, amphibolite with biotite-plagioclase gneiss, biotite-plagioclase leptynite intercalated with schist, and calc-schist. Amphibolitic layers increase in abundance upwards in the Group. The rocks are severely migmatized (especially adjacent to granite bodies) and deformed. The protoliths of the Koksu Group were intermediate-basic volcanics and greywackes, and metamorphism was of the low-amphibolite facies of medium-pressure type.

The 5000 m thick Xingditag Group forms the upper sequence of the Early Proterozoic strata within the Kuruktag and central Tianshan regions. It consists chiefly of a series of low-grade metamorphosed, fine-grained clastics and carbonates, including quartz schist, meta-arenite, metasiltite, quartzite and marble intercalated locally with gneiss, crystalline schist and migmatite. Their original rock sequence was mainly intermediate-basic (and intermediate to acidic) volcanics-and greywacke-flysch formations. The metamorphic grade is greenschist low-amphibolite facies. This sequence is covered unconformably by the Yangjiblag Group of the Changcheng System. The zircon U-Pb ages from the migmatites (migmatitic granites), granodiorites and quartz diorites of this group are 1912, 1920 and 1800 Ma respectively (Bureau of Geology and Mineral Resources of Xinjiang Uygur Autonomous Region 1993).

Migmatization is widespread in the Xingxingxia area, within the eastern segment of central Tianshan. Injected migmatite and migmatitic gneiss are common, and locally, migmatitic granites exist (He Gaopin and Li Yingyi 1988). Within the central segment of central Tianshan, a 368 km^2 blue quartz-bearing diorite-granodiorite-plagioclase body intruded into the Early Proterozoic Xingditag Group. This intrusion is unconformably overlain by the Sinian System to the east of Kuler, which yields U-Pb zircon dates of 1800 and 1900 Ma (Bureau of Geology and Mineral Resources of Xinjiang Uygur Autonomous Region 1993). According to the explanatory text of the 1: 2 000 000 Geological Map of Xinjiang (1985), the Early Proterozoic intrusives of this region are dominated by gneissic acidic rocks such as gneissose granite, two mica granite and smaller amounts of gneissic granodiorite and diorite. These lithologies are severely migmatised and injected by successive intrusions. Compositionally, they are mainly silica-oversaturated alkaline rocks.

The Beishan Group form the Beishan Mountains and is over 5400 m thick. It consists of a series of cordierite-bearing medium and high-grade metamorphosed rocks including biotite-plagioclase gneiss, hornblende-plagioclase gneiss and mica-quartz schist intercalated with large amounts of amphibolite: greenschist and lesser quantities of marble. The protolith sequence was an intermediate-basic-wacke formation. Granites and migmatites are well-developed, and progressive metamorphic zones are observed in the Beishan Group. Metamorphic grade is dominated regionally by low pressure amphibolite facies. The Early Proterozoic metamorphic rocks from the Tianshan-Beishan section, such as hornblende-plagioclase gneiss, amphibolite, biotite-plagioclase gneiss and biotite-hornblende leptynite were originally formed from basalt, andesite dacite and rhyolite protoliths (Zhang Zhide 1990). In combination with the wacke-dominated volcano-sediments, they make up an intermediate-basic volcanic and wacke flysch formation.

Structural patterns are generally dominated by closed, linear, and overturned folds which are accompanied by thrusts and crystalline schistosity. The axes of the folds all extend in an E-W direction, showing that the folds were formed principally by N-S compression (Wang Yunshan and Chen Jiniang 1989). These patterns are especially well developed within the Kuruktag area. The E-W trending metamorphic facies belts indicate that the temperature conditions decreased northwards. A good example of this is seen in the Halik Mountain of southern Tianshan and Kuruktag area.

In summary, the lithological, metamorphic and structural data suggest that the Tianshan-Beishan mobile belt was an island arc system adjacent to the active continental margin north of the Tarim Protoplatform.

Dacandaban-Qilian section. The Early Proterozoic rocks in this section are distributed in the area 36°-40° N and 92°-104° E. Their general trend is NW-SE, but to the west of 90° E, they turn to ENE-WSW. The Early Proterozoic rocks exposed in Dacandaban and Jun Ul Mountains on the northern periphery of Qaidam Basin and, eastwards, in Dulan area and Ngola Mountain mostly belong to the Dacandaban Group and to a lesser extent the Gokeshan Group. In contrast, those situated at the northeastern border of Qaidam Basin belong to the Shaliuhe Group.

The lower part of the 3614-9000 m thick Dacandaban Group is chiefly composed of garnet-bearing sillimanite biotite-plagioclase gneiss, biotite leptynite, biotite hornblende plagioclase gneiss, amphibolite, cordierite-bearing sillimanite garnet plagioclase gneiss, staurolite-biotite plagioclase gneiss, migmatites and migmatitic gneiss intercalated with diopside-bearing forsterite marble, wollastonite marble and kyanite-bearing two mica-quartz schist. The upper part of the sequence is composed of biotite-quartz schist, biotite leptynite, sillimanite-bearing-biotite-quartz schist and amphibolite intercalated with several layers of quartzite, marble and wollastonite-bearing marble. These lithologies are generally metamorphosed into amphibolite facies, although near Dulan they reach high-amphibolite facies and granulite facies. Geochronological studies of the Dacandaban Group yield a range of ages: the hornblende-plagioclase gneiss yields a zircon U-Pb age of 2205 Ma, and the K-feldspars from the pinkish red pegmatite dykes cut into this Group produced two Rb-Sr isochrons of 1568 and 1463 Ma (Chen Jiniang and Wang Yunshan 1985).

The Shaliuhe Group is dominated by mica schists such as kyanite-bearing two mica quartz schist, garnet-bearing staurolite and two mica leptynite, with lesser amounts of metavolcanics and carbonates. Their protoliths are considered to have been pelitic, psammitic-intermediate-basic volcanic and carbonate formations. Low amphibolite facies of a medium-pressure facies series is extensive in the field, and high greenschist facies (Barrovian garnet zone) occurs locally. However, progressive metamorphic zones are very limited.

The Early Proterozoic rocks from Qilian Mountains to the east of Dacandaban and Jun Ul Mountains, extend as a narrow NW-SE trending belt within Gansu Province. In contrast, within Qinghai Province they occur as large lenticles in the eastern and western ends of the Qilian Mountains. (The central part of the Qilian mountains are covered by Mesozoic and Cenozoic strata). The corresponding rocks exposed in Yemananshan Mountains are the Yemananshan Group, which makes up the framework of the central Qilian Mountains. The sequence is termed the Huangyuan Group in their eastern extension in the Datonghe area of Qinghai Province, whereas in the western extension along Beidahe River and the scattered outcrops in Qilian County the succession is known as the Beidahe Group. The corresponding rocks in the Maxian Mountains of the central part of Gansu Province are called the Maxianshan Group, and those in the Longshan Mountains of the southeastern part of Gansu Province are termed the Niutouhe Group. The lithologies, metamorphic and structural characteristics of these Groups are described below:

The Yemananshan Group extends in a ENE direction from the major peak of Annanba and the mountain pass of Dangjin in the west, but change to a NW trend eastwards to the Qiaotouzi area of northern Gansu. In total the sequence is between 5591 and 10818 m thick and can be divided into three parts. The lower part is comprised chiefly of biotite-plagioclase (monzonitic) gneiss, garnet-bearing sillimanite oligoclase gneiss and diopside-plagioclase gneiss intercalated with hornblende-plagioclase gneiss, garnet-bearing sillimanite-staurolite-two mica schist, staurolite-two mica quartz schist, diopside hornblende (tremolite) biotite schist, biotite plagioclase hornblende schist, (garnet-bearing) tremolite biotite quartz schist, tremolite plagioclase leptynite, (scapolite) diopside marble and metaconglomerate. This complex sequence also shows a great deal of lateral variation. In addition, there are various migmatitic gneisses, leptynites and biotite quartz schists exposed. However, the rocks nearer the top of the lower part of the stratigraphy show rhythmic layering and do not show much lateral variation.

The middle part of the succession is mainly composed of metacarbonate rocks such as dolomite and (muscovite-graphite-bearing) marbles intercalated with garnet-bearing (staurolite) mica quartz schist, biotite plagioclase gneiss, diopside-plagioclase leptynite, calcite-chlorite schist, amphibolite, muscovite (actinolite) schist and garnet-bearing zoisite hornblendite. The marble layers are very thick and often show no lateral variation.

Finally, the upper part of the succession is composed of biotite (muscovite, monzonitic) quartz schist, (carbonic) chlorite quartz schist, calcite mica quartz (muscovite or chlorite) schist, hornblende schist, marble and crystalline carbonate rock intercalated with quartzite, actinolite schist and metabasic volcanic rocks.

The 3796 m thick Huangyuan Group is subdivided into the Liujiatai and Dongchagou Formations. The Liujiatai Formation is chiefly composed of biotite-andesine (monzonitic) gneiss, hornblende-K-feldspar gneiss, migmatites, amphibolite, dolomite marble and wollastonite-bearing forsterite-diopside marble. The overlying Dongchagou Formation consists of garnet-bearing staurolite-biotite (garnet-bearing two mica) schist, quartz schist and amphibolite intercalated with staurolite-bearing sillimanite-biotite (andesine) gneiss, (andradite or forsterite-bearing) diopside marble.

The protoliths of these two formations are dominated by pelitic psammites, with some carbonates as well as small amounts of intermediate-basic (intermediate-acidic) volcanics, constructing a pelitic and psammitic clastic formation with carbonated and intermediate-basic volcanic interlayers. They experienced high-greenschist, low-amphibolite and high-amphibolite facies metamorphism of the medium-pressure facies series, of which low-amphibolite facies is more extensive. Progressive metamorphic zones are locally developed, which are concordant to WNW regional tectonic structures.

The 546-10239 m thick Beidahe Group includes the following rock assemblages: its lower horizons are composed of biotite (garnet-bearing) plagioclase gneiss, garnet-bearing biotite-oligoclase gneiss, plagioclase-horblende gneiss intercalated with two mica-quartz schist. Further up, there are biotite (sericite)-quartz schist, muscovite schist (with or without garnet), hornblende plagioclase schist intercalated with plagioclase-hornblende schist, banded marble and biotite-plagioclase gneiss. Towards the top of this sequence, the marble and schist layers alternate rhythmically. The middle part of the Group is composed of quartzite and dolomitic or coarse-grained or banded marbles interlayered with (chlorite) muscovite schist and sericite schist. These lithologies are overlain in turn by plagioclase-biotite-quartz schist, two mica-quartz schist, hornblende schist, actinolite schist, epidote-zoisite-actinolite schist, quartzite, leptite, marble, crystalline carbonates, phyllites and meta-arenites. The protoliths of the Beidahe Group include predominantly psammitic pelites, with more minor pelites, psammites, carbonate and intermediate-basic volcanics. These lithologies make up a carbonate rock-intermediate-basic volcanic-flysch formation which changes upwards into volcanic-flysch formation.

The Group is subject to medium-pressure amphibolite facies metamorphism. According to the studies on the Early Proterozoic metamorphic rocks in the Heidaban and Diaodaban areas of northern Qilian Mountains (Chi Hongxing 1988), two types of recrystallization and deformation occurred in these areas: for example, the two mica leptynite was firstly metamorphosed into low-amphibolite facies and subsequently into a high-amphibolite (sillimanite-bearing) facies; the early-phase schistosity S_1 has been strongly folded. The second-phase schistosity S_2 is the axial plane foliation of the second metamorphic event. Locally, streaky, banded and scattered porphyritic migmatites related to regional migmatization can be encountered.

The Maxianshan Group is also strongly affected by migmatisation: the lower part of the Group is composed of augen biotite-plagioclase (monzonitic) migmatite, migmatitic granites intercalated with migmatitic plagioclase-gneiss and amphibolite. These lithologies give way upwards to dolomite, biotite K-feldspar (monzonitic, oligoclase) migmatite intercalated with hornblende schist, migmatitic biotite gneiss and migmatitic granites as well as dolomitic marbles. The middle part of the Group is composed of banded biotite-monzonitic migmatite, migmatitic biotite K-feldspar gneiss, hornblende-andesine gneiss intercalated with augen biotite-plagioclase migmatite, migmatitic biotite-plagioclase gneiss, hornblende-plagioclase schist, garnet-bearing muscovite-quartz schist and dolomite. This sequence is followed by biotite-quartz schists, gneiss, (biotite) hornblende schists, and garnet-bearing muscovite-quartz schist and biotite-chlorite schists which make up the upper part of the Group. The protoliths are littoral-hypabyssal flysch and acidic, basic volcanics-carbonate rock formations. They were subjected to regional metamorphism ranging from high greenschist to low amphibolite facies. Rb-Sr whole rock dating of the Group yielded an age of 1062 Ma, whereas an altered basic dyke intruded into the Group yielded an age of 1079 Ma (Gansu Bureau of Geology and Mineral Resources, 1990).

The 7863-10423 m thick Niutouhe Group can be divided into three sequences. The lower sequence is composed of dolomitic, muscovitic or banded marbles intercalated with biotite (hornblende-plagioclase) gneiss, biotite-quartz (biotite-hornblende) schist, mica schist garnet-bearing chlorite-plagioclase schist, sericite (garnet-bearing two mica quartz, garnet-bearing mica quartz) schist and marble. Distinct migmatization occurs, and abundant garnet-rich gneiss is found in the east of the region. In contrast, in the west of the region, schists are common and are intercalated with iron-bearing quartzite. The middle part of the Group is composed of hornblende schist, hornblende gneiss interlayered with migmatitic biotite gneiss, chlorite-sericite-quartz schist, calcite-chlorite schist, biotite-plagioclase schist, garnet-bearing mica quartz schist intercalated with marble and metamorphosed pyrite. The upper sequence is composed of biotite-quartz schist, biotite gneiss intercalated with plagioclase gneiss, meta-arenite, arenarious slate, calcite-chlorite-epidote schist interlayered with crystalline carbonate, rhyolitic prophyrite, andesitic porphyrite and metatuffstone.

The original protoliths of the lower and middle successions are likely to be a volcanic-carbonate-flysch formation which was metamorphosed to low amphibolite and high-greenschist facies. The upper sequence consisted of a volcanic-flysch formation which was metamorphosed into low-greenschist facies.

In summary, the Dacandaban-Qilian section was subjected to a medium-pressure facies metamorphism, mostly corresponding to high-greenschist/low-amphibolite facies. It has been estimated that the P-T conditions of metamorphism are: P>0.35 GPa and T=570-660 °C. High-amphibolite to granulite facies metamorphic grade were formed in very few areas. Progressive metamorphic zones are well developed. These principally consist of staurolite- and kyanite- (and to a lesser extent hornblende- and sillimanite-) bearing zones. Locally, a continuous change from garnet zone to sillimanite zone occurs: this is exemplified by Huangyuan-Datong area, where the Huangyuan Group displays a south to north progression through the garnet, kyanite and sillimanite zones. These zones are distributed roughly in concordance with the WNW regional structural trends.

The common protoliths of the Dacandaban-Qilian section are intermediate-basic volcanic sequences and flysch formations. If the Tianshan-Beishan section in the mobile belt is considered to have formed in an island arc system, the Dacandaban-Qilian section, a constituent of the same belt, would have formed in an active continental margin near to the island arc. The whole mobile belt may be assumed to have been an accretional zone on the northern periphery of the Tarim Protoplatform. Therefore, a type of Wilson Cycle may have been formed between the Tarim and North China Protoplatforms.

3.3.2 Kunlun Mobile Belt

Western Kunlun-Altun section. The Western Kunlun and Altun Mountains are situated on the southern periphery of the Tarim Protoplatform. They are located between 36°-40° N, 74°-78° E and 36°-40° N, 88°-93° E respectively. The western Kunlun area is referred as the western and southern periphery areas of Xinjiang, including the Alai Mountains at the junction between the former Soviet Union and the western margin of the Tarim Protoplatform, the Kunlun Mountains and some localities of the Karakorum Mountains.

Both the Western Kunlun and Altun sections have extensive exposures of Precambrian rocks. The widespread Early Proteozoic rocks in this region can be grouped into two series: the lower series is termed the Bulungkoal Group where it is exposed in the major peak area of Kongur Mountain of western Kunlun. However, in the axial areas of western Kunlun and in the Tieklik Mountain it is termed the Karakax Group. In both these areas the lower series is unconformably overlain by upper series strata, termed the Kungur Group and Elinkat Group in the respective areas. The Elinkat Group is in turn covered unconformably by the Salajastak Group which belongs to the middle Proterozoic Changcheng System.

The 8000-14,700 m thick lower series Bulungkoal and Karakax Group comprise various gneisses, quartzites, crystalline schists and marble rocks. Metamorphic minerals such as sillimanite, staurolite, kyanite and garnet occur extensively. The protoliths are normal flysch sediments such as feldspar-quartz sandstone, pelites and carbonate, intercalated with small amounts of volcanic and pyroclastic rocks. The rocks were subjected to greenschist-amphibolite facies metamorphism and in most areas to migmatization. Progressive metamorphic zones and, in some places, ductile shear zones marked by mylonites are also present.

The approximately 8000 m thick upper series Kongur Group is chiefly composed of various quartz schists, marble, crystalline limestone, amphibolite and leptynite intercalated locally with gneiss. In comparison the Elinkat Group is about 10,000 m thick, and consists of various greenschists intercalated with small amounts of quartz marble overlain mainly by schists, schistose conglomerate, marble, slate and metafeldspathic arenite. Both Groups originally consisted of flysch material and were metamorphosed progressively to greenschist-amphibolite facies. The Elinkat Group is dominated by low amphibolite facies in the west, and by greenschist facies in the east. The granophyre (keratophyre) injected into this Group at the upper reaches of Ulugustan River has been dated as 1764 Ma by means of Rb-Sr method (from the Explanatory Text for the Geological Map of Xinjiang, 1:2 000 000; 1985). The protolith of the Early Proterozoic volcanic rocks of this area include basalt, andesite, dacite and rhyolite. These were metamorphosed into hornblende-plagioclase schist, plagioclase-hornblende schist, hornblende-plagioclase gneiss and biotite-hornblende leptynite. In addition, there are felsite, spilite-keratophyre, quartz dacite and pyroclastic rocks (Zhang Zhide 1990).

The Proterozoic granitic rocks from Kunlun Mountains are mainly distributed on the northern slopes of the western Kunlun Mountains. Their intrusive forms are very complex and they show extensively developed gneissosity. The main rock types include gneissose granite, biotite-plagioclase granite and granodiorite. Chemically these intrusions are of calc-alkaline and peralkaline-subalkaline affinity, showing features of high SiO_2, K_2O, and low Na_2O contents. They probably belong to an in situ type of magmatism related to melting and metasomatism (Zhang Zhide 1990).

The oldest rocks in the Altun Mountains are the Late Archean Milan Group which outcrop in the northern slope of central Altun and as an E-W trending belt of small faulted blocks. In the region of Ruojiang County, Xinjiang, the sequence reaches 3287 m in thickness. It has been assumed that these blocks might have been captured by successive deformations in the Early Proterozoic mobile belt. They are dominated by leptynite and granulite in the upper and lower parts of the sequence respectively. Lithologically they include rocks such as amphibolite, biotite-hornblende leptynite, biotite-bearing monzonitic leptynite, pyroxene (hornblende)-bearing K-feldspar leptynite, biotite-pyroxene-plagioclase leptynite, leptite, biotite-diopside-plagioclase leptynite, banded hornblende-microcline migmatite and biotite-bearing K-feldspar leptynite. All of them show obviously banded or streaky structures. Additionally, there are hypersthene-plagioclase granulite, hornblende-hypersthene granulite and hypersthene-bearing granulite intercalated in the sequence. The amounts of granulite tend to decrease, sometimes to the point of absence, from west to east, and leptynite is similarly replaced by gneiss. Banded migmatites also occur, being mainly developed in the northern part of the upper sequence. The protoliths of this group were volcanics of basaltic, dacitic and rhyolitic compositions and a greywacke formation (Zhang Zhide 1990), which experienced high-amphibolite granulite facies metamorphism.

The amphibolites from the correspondent group within Qinghai Province have yielded a zircon U-Pb age of 2462.5 Ma ((from the Explanatory Text for the Geological Map of Xinjiang, 1:2 000 000; 1985), which has been postulated to be younger than the formation age of this Group. Therefore, it may be reasonable for the geologists from Xinjiang to assume this Group to be Late Archean in age.

The only Early Proterozoic rocks exposed in this area belong to the upper series of the Altun Group. It is mainly comprised of mica (quartz) schist and biotite (plagioclase) gneiss intercalated with marble, quartzite and biotite-hornblende schist. In the lower part of the sequence it has been interlayered with hornblende-plagioclase leptynite, hornblende-plagioclase gneiss, biotite-plagioclase leptynite, amphibolite and marble. The gneiss and schist contain garnet, staurolite, kyanite or andalusite. The protoliths are mostly flysch sediments, whereas the interlayered volcanics are similar to those from Kunlun Mountains. The metamorphic grade is predominantly amphibolite and to a lesser degree greenschist facies. In general metamorphism is therefore a medium-low pressure, intermediate type. Progressive metamorphic zones occur and deformation is very strong, with accompanying migmatites and granites. The Group is usually in fault contact with strata belonging to neighbouring groups.

Eastern Kunlun Section (Southern Periphery of the Qaidam Basin). The eastern Kunlun section is located between 35°-37° N and 90°-100° E on the southern periphery of the Qaidam Basin. The Early Proterozoic rocks in this area chiefly belong to the 2614-3553 m Jinshuikou Group. Their distribution starts in the west in the Kumkuli Basin, trending WNW-ESE to E-W for a 1000 km through Golmud, Nomhon, Xiangride to the Xinghai hot spring area. The Group is intruded by five great migmatitic granite belts and usually occurs as discontinuous belts or remnants in the migmatitic granites, or as nappes of different size overlying the Permian and Triassic sequences. The main rock types in the lower horizons of the sequence are biotite-plagioclase gneiss, andalusite-bearing biotite schist, garnet-bearing sillimanite biotite plagioclase gneiss, garnet-bearing staurolite leptynite intercalated with amphibolite, diopside marble and forsterite marble. Within the upper horizons of the sequence, the main rock types are biotite-quartz schist, biotite leptynite and biotite-plagioclase gneiss intercalated with quartzite, marble, and locally with hyperthene-hornblende gneiss and two-pyroxene granulite. Migmatization is extensive and stronger towards the bottom of the sequence. The protoliths are mainly feldspathic pelitic arenites and more competent argillites, dolomitic argillites, carbonate rocks and intermediate-basic volcanics as well as tuffs. These lithologies construct pelitic-arenitic clastic formations and carbonate rock formations with smaller amounts of intermediate-basic volcanic interlayers. Metamorphic grade is chiefly amphibolite facies, but locally reaches granulite facies. (Low pressure metamorphism is common, but medium-pressure metamorphism does occur in the east of the area.) In some localities, progressive metamorphic zones can be found, and retrogression of metamorphosed lithologies to greenschist facies also occurs.

Migmatic granites occur as a belt on the southern periphery of Qaidam Basin and intrude the strata from Early Proterozoic to Triassic in age, constructing the eastern Kunlun granite belt. The main rocks include granite and granodiorite which have porphyritic and granitic textures. These intrusive rocks are associated with the Early Proterozoic metamorphic rocks. They are commonly concordant with their country rocks, and contacts between granite and country rock are mostly gradational. The relics in the migmatitic granites show streaky banding and have similar orientations to the country rock foliation.

A zircon U-Pb age of 2469 Ma has been obtained from the migmatitic granite at Xiangher of Dongxia, Huangyuan (Comprehensive Studies on the Metamorphic Rocks of the North of Qinghai-Tibet Plateau, the volume for examination 1984). Additionally, K-feldspars from the pegmatite dykes injected into Dacandaban Group at Dololx, at the northern periphery of Qaidam Basin, have yielded a Rb-Sr isochron age of 1563 Ma (Wang Yunshan and Chen Jiniang 1984).

The lithologies described above, together with their geochemistry, structural and metamorphic characteristics can be considered the products of an island arc environment of formation. However, in the Altun Mountains at the eastern extension of the mobile belt, only sediment-dominated flysch formation is preserved. However, the volcanic interlayers have the same features as those from western Kunlun Mountains. Further eastwards at the southern periphery of Qaidam Massif only small amounts of intermediate-basic volcanic rocks are exposed. Such a variety of rock assemblages in a mobile belt may have resulted from different environments of formation within the same mobile belt. Alternatively, their differences could result from suffering varying effects of orogeny and thus subsequently differing state of erosion. Consequently, the southern periphery of the Tarim Protoplatform was probably a marginal accretionary zone within an island arc environment during Early Proterozoic time.

3.4 Formation and Accretion of the Yangtze Protoplatform

3.4.1 Emergence of Yangtze Protoplatform

On the Yangtze Protoplatform, old metamorphic rocks are restricted to sporadic exposures along the southwestern and northern borders. Based on geophysical data, it was deduced that there are boundaries within the Archean gneisses underneath the Sichuan Basin and that the crystalline basement of the Yangtze Protoplatform was mainly consolidated in Early Proterozoic time. This conclusion is largely based on the absence of old age values reported for the few outcrops of metamorphic rocks.

A metamorphic complex covers the southwestern margin of the Yangtzi Protoplatform in an area striding 28°30'–30°31' N and 101°20'-102' 40° E. Severely cut by later structural deformation, it crops out as a N-S oriented stripe but exhibits an E-W structural trend. Recent studies show that the complex is composed of an Archean greenstone belt (represented by the Kangding Group) and large amounts of granitoids. The former consists of amphibolite, granulite, leptynite and schist and has protoliths of a submarine tholeiite-calc-alkaline series volcanic suite and pelite-sandy flysh formations (Lu Jiemin 1986). The latter are largely hornblende plagioclase gneiss and granitic gneiss derived from tonalite-trondhjemite-granodiorite and some K-rich granite (Lu Jiemin 1986; Zhai Mingguo 1986). The complex has passed through high-greenschist, amphibolite to granulite facies regional dynamic-thermal metamorphism and demonstrates a distribution of facies zones parallel to structural orientation (Lu Jiemin 1986; Dong Shenbao 1986). The gneiss yielded a palingenitic zircon U-Pb concordia age of 2478 Ma (Wu Maode 1988), and a whole-rock isochron age of 2.4 Ga (Yuan Haihua et al. 1985). In addition the hornblende plagioclase gneiss and two-pyroxene granulite have a Pb-Pb isochron age of 2957±304 Ma (1σ) (Yuan Haihua 1985). The structures and the allocation of metamorphic facies zones jointly indicate that the complex is more likely to be a boundary marking later tectonic deformation than a Late Archean accreted continental margin. Similar metamorphic rocks occur in the Cangshan-Ailao Mts. to the west, with a lithology nearly identical to that of the Kangding complex. However this metamorphic sequence was controlled by later NW-SE trending structures generated by subsequent tectonic deformation. These rocks also strongly indicate a possibly widespread Archean basement is present on the western margin of the Yangtze Protoplatform.

In the southwestern part of the Yangtze Protoplatform, the middle sector of the borehole-geophysical profile across the Sichuan Basin (Fig. 3.26) revealed a set of gneisses at a depth of 6-10 km, which are overlain by the neritic-shelf facies Late Proterozoic Banxi Group (Ou Qingxian 1986). The gneisses are very similar to those of the Kangding complex and suggest the presence of an Archean terrane beneath the Sichuan Basin. Furthermore, The Sibao orogeny possibly involved the crystalline basement in tectonic deformation. If this factor is taken into account, then another inference can be reached i.e. - that under the Late Proterozoic Banxi Group there still exists an Archean basement.

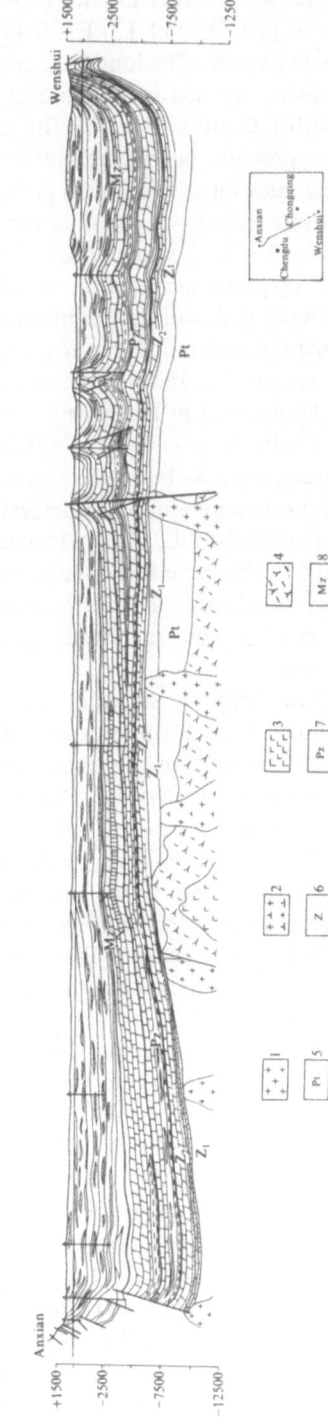

Fig. 3.26. The integrated geological interpretation profile of Anxian-Wenshui in Sichuan Basin (modified after Ou Qingxian 1986, unpublished): 1. Granite; 2. granodiorite; 3. ultrabasic rock; 4. gneiss; 5. Late Proterozoic Banxi Group; 6. Sinian; 7. Paleozoic; 8. Mesozoic

Situated in the inner side of the central northern margin of the Yangtze Protoplatform, in an area between 110 53°-111 17° E; 30 45°-31' 20° N, is a dome structure surrounded by the Sinian System. The long-axis of the dome trends N-S. A suite of metamorphic complexes is exposed in the core of the dome, which was previously termed the Konglin Group. Part of the complex consists of metasedimentary rocks including graphite-bearing actinolite (garnet) biotite plagio-clase gneiss, schist and biotite leptynite, intercalated with granulite marble and quartz-ite. These lithologies indicate a protolith of very mature terrigenous clastics rocks. Lu Liangzhao et al. (1988) defined it as a khondalite series. Another part of the sequence is made up of biotite leptynite, amphibolite, epidote-amphibolite schist and chlorite biotite schist, reflecting protoliths of intermediate-acid and basic volcanics. The complex has gone through metamorphism of greenschist amphibo-lite to granulite facies, and was regarded as Early Proterozoic in age because of the lack of geochronological data. However, Liu Huanliang (1987) obtained a whole-rock isochron age of 2988±141 Ma from the gneiss and a zircon U-Pb concordia age of 2850±15 Ma from the fine-grained biotite leptynite, which convincingly date the complex as Archean in age. The prominent regional structural styles are exhibited as a series of NE plunging isoclinal closed folds. This orientation is perpendicular to the border of the North China and the Yangtze Protoplatforms, indicating that at the time the Archean complex was developed, the northern margin of the Yangtze Protoplatform was not a boundary of continental crustal accretion, but was a rifting zone in the primitive Archean crust.

A comprehensive geophysical study by seismic, aeromagnetic gravity and magnetotellurgic sounding methods of the northern Anhui-Shanghai profile has re-vealed the layering and the block structure of the crust and mantle in the eastern Yangtze Protoplatform. It was found that a widespread Precambrian crystalline base-ment of medium to high metamorphic grade capped the lower crust (Fig. 3.27; Chen Lusheng et al. 1987). These intermediate- to high-grade metamorphic rocks are strickingly similar in lithology to the Archean Dabie and Jiaonan Groups.

In summary, it is therefore quite reasonable to envisage that an Archean crystalline basement extensively existing beneath the Yangtze Platform, and that the Yangtze Protoplatform became a craton as early as the end of the Archean.

Based on the seismic sounding data, Zhu Jieshou (1985) set up a velocity model which shows that the crust in the various tectonic provinces are all either three lay-ered or multi-layered. The South China Plate has the highest velocity (6.3-6.4 km/s) and distinct thickness changes, varying gently from 30 km in the southeast coastal area to 45 km in the north west. The average velocity of the North China Plate is 6.2-6.3 km/s., which is lower than that of the South China Plate, and the velocity layers are indistinct inside the crust. In some regions low-velocity layers occur in 35-46 km. However the lateral variation of the thickness and velocity distribution are some-what dramatic and the crust models of neighbouring regions are hard to correlate. The average density disparity (ß) between crust and upper mantle obtained by the statistical relation between P-wave velocity and the density of rocks is 0.45 g/cm^3 for the South China and 0.54 g/cm^3 for the North. These geophysical features imply that although both the North China and the Yangtze Protoplatforms have Archean

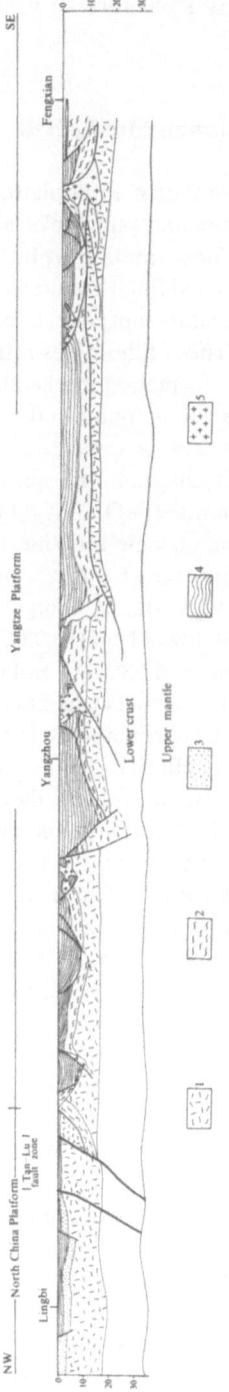

Fig. 3.27. The integrated and interpreted section from Fengxian to Lingbi in the lower reaches of Yangtze River (simplified after Chen Husheng, 1987): 1. Hypometamorphic rocks; 2. Epi-metamorphic rocks; 3. Middle Proterozoic strata; 4. Phanerozoic strata; 5. granite

basements, they are unlikely to have been formed by the splitting of a unified craton.

3.4.2 Southern Sichuan-Central Yunnan Mobile Belt

On the southwestern margin of the Yangtze Protoplatform, 100 km south of the Kangdian greenstone belt, a suite of metamorphic rocks are exhumed in a window in the surrounding Triassic cover. These metamorphic rocks were termed the Dahongshan Group by Bai Jin (1981) and Shen Yuanren (1982). The Group is composed of amphibolite, chlorite schist, albite leptynite, mica schist, quartzite, carbonaceous phyllite and impure marble. These lithologies reflect protoliths formed in a eugeosynclinal volcanic-sedimentary formation marked by spilite and keratophyre. An analogous formation of the same age is found in the southern Kangdian greenstone belt.

On the (Na_2O+K_2O) versus to SiO_2 diagram, the spilite projects onto the alkalic olivine basalt area. However, on both the $<FeO> / MgO$ to SiO_2 and $<FeO> / MgO$ versus $<FeO>$ diagrams it falls in the oceanic tholeiite area (Wu Xiche and Duan Jinsun 1982 and unpublished data). In contrast, the keratophyre has a chemical character of slightly low SiO_2 (54.7%), high alkalic content $(Na_2O+K_2O=7.09\%)$ and total Fe content $(Fe_2O_3+FeO=16.53\%)$, lower MgO (0.93%) and higher TiO_2 (1.64%), with $Na_2O / K_2O=10.4$ and $Fe_2O_3+FeO=16.53\%$ (Li Bohui 1980).

Hu Aiqin et al. (1986) reported a U-Pb concordia zircon age of 1665+13.6/-10.9 Ma as well as a hornblende Sm-Nd isochron age of 1657±82 Ma, and considered that the two values denote the upper age limit of the Dahongshan Group. Taking the low REE content as proof of a depleted mantle origin, they defined the Sm-Nd model age of 1.9-2.0 Ga as the age of mantle differentiation and further delimited the diagenetic age of the Dahongshan Group between 1.6-1.9 (or ~2.0) Ga. However, Wu Maode (1988) obtained a whole-rock Rb-Sr isochron age of 1720 Ma and zircon U-Pb ages of 1900 Ma and 2506 Ma from the lower part of the Group. Therefore the Dahongshan Group is generally estimated to be of Early Proterozoic, rather than Middle Proterozoic age (Dong Shenbao et al. 1986; Cai Xuelin et al. 1986; Tang Jiafu 1988).

Multiple-stages of tectono-thermal events have affected the Dahongshan Group. Over 30 K-Ar and Rb-Sr age data concentrated on four crest values which indicated four major events: the Jinning movement at 800-900 Ma; the Kunyang uplift at ~1.1 Ga; the Chengjiang movement at 700 Ma and the Caledonian movement at 500 Ma (Qian Jinhe and Shen Yuanren 1984; unpublished).

A successive change in metamorphic grade from subgreenschist to high-amphibolite facies is seen in the rocks. Syenite and sillimanite associations appear in the meta-argillites accompanied by vigorous migmatization. The metamorphism was of regional intermediate pressure dynamic-thermal type and was completed with the Lüliang orogeny at the end of Early Proterozoic time. The resulting crystalline basement was then covered by Middle Proterozoic sediments. Evidence of the extensive nature of the crystalline basement is given by the increase in the depth of the Moho from the southeast (41 km) to northwest (61 km). The Early Proterozoic mobile belt marked by the Dahongshan Group also sits on a N-S trending gradient zone where the Moho plunges abruptly (Chen Yuankun and Wu Shanglong 1982).

The most remarkable structures within the Dahongshan Group are the broad folds trending ENE to E-W. Chen Zhiliang and Chen Shiyu (1987) correlated this struc-ture style with the Middle Proterozoic Kunyang and Huili Groups. However, these broad open folds are the superimposed products of later deformation, whereas the early syn-metamorphic folds are tight recumbent ones (Photo. 3.2), with hinge point-ing at 340°-350° (Bai Jin and Zhang Xueqi 1981). As ductile deformation is very common, it is hard to exclude the possibility of a ductile shear zone parallel to the axial plane of the folds.

The rock assemblage, regional geophysical anomaly and early structural style display a primitive N-S orientation for the mobile belt, which is approximately par-allel to the southwest boundary of the Yangtze Protoplatform. Thus it is regarded as an accretion boundary.

3.5 Petrotectonic Assemblage

A series of Proterozoic volcano-sedimentary rocks were deposited inside and on the margin of the strongly eroded Archean greenstone belts and granulite-gneiss ter-ranes. They are in such a sharp contrast to the Archean lithologies in both petrogen-

Photo 3.2. The recumbent fold in the Dahongyu Group marbles

esis and structural styles that it is evident that the evolution of the crust entered a new stage. These rock series are promising materials in the study of the Archean-Proterozoic boundary and are of epoch-making significance in continental evolution.

Tectonic differentiation in the Early Proterozoic resulted in the observed diversity in petrotectonic assemblages. The presence of new rock types and mineral deposits also mark the beginning of a new era. The Early Proterozoic of China consists of three types of rock assemblages similar to those found elsewhere in the world (Condie 1982; Windley 1984). However, these rock assemblages (described below) reveal a slightly different tectonic environment: a stable environment which resulted in the development of highly mature rock assemblages.

Feldspar quartz sandstone shale carbonate rocks assemblage: The poorly-matured flysch formation in the Shanxi-Henan rift zone is an example of this type of assemblage. It contains continental tholeiite and has a thickness of 10 000 m. The carbonate rocks in it are multi-stratified magnesian rocks rich in stromatolites, totalling 4000 m in thickness.

Similar rock assemblages are found in the western sector of the Yanshan-Yinshan mobile belt and the upper metamorphic series of the Jiao-Liao collision belt. In the latter case quartzite is rare and carbonate rocks are prominently of magnesium-rich type. This assemblage yields an extremely large high quality magnesite deposit which is rare elsewhere in the world and constitutes an enigmatic formation in the Proterozoic of China. Although recrystallized, the magnesite beds still retain their oblique beddings, ripples, pesolitic structures, and have syngenetic breccia which indicate a high-energy environment. Mudcracks, hail prints and nodules are found in the dolomitic marble making up the floor of the deposited bed. A stromatolite found inside as well as around this deposit indicates an intertidal-flat environment of formation (Zhang Qiusheng 1988). The source of such huge deposits is still a riddle. However, since the erosion source area, composed of the Archean complex, could not have supplied so much magnesium, the formation of the deposit must be connected with vigorous volcanic activity.

Calc-alkaline volcano-sedimentary series: Represented by the Qinglong greenstone belt in the Qing-Luan collision belt, this series is composed of calc-alkaline volcanics such as pillow basalt, andesite and rhyolite, and turbidite series of wacke-argillaceous rocks, totalling 5000 m in thickness. The formation contains BIF and small copper deposits. This assemblage is thought to indicate an island arc environment of formation (Condie 1982). It also occurs in the eastern sector of the Yinshan-Yanshan mobile belt and on the periphery of the Tarim Basin, where it contains carbonate rocks which are largely destroyed by tectonic activity and denudation. Despite being formed in a greenstone belt, the BIF within the assemblage are different from other world examples in their association with turbidites, and contain small-scale Lake Superior type features.

In a global context, BIF are mainly formed in the Early Proterozoic (Goldish 1973; Windly 1984). However, those in China were produced mainly in the Archean

Era. Large-scale iron deposits are mainly found in the Archean greenstone belt and granulite-gneiss regions, such as the Anshan iron ore field and the Qian'an iron ore field of the Middle Archean and the Huoqiu iron deposit and the Wutai iron deposit of the Late Archean. The Early Proterozoic Lüliang Group yields a Lake Superior-type banded iron deposit, however, they have much smaller reserves when compared to the deposits of Archean age.

Recently Zhai Mingguo and Windley (1990) made a comprehensive study of the Archean and Early Proterozoic BIF of North China. They pointed out three marked characteristics: 1) Most BIF are Archean in age, their formation reaching a peak during the late Archean; 2) BIF in China occur in both high-grade regions and greenstone belts; the former being more significant; 3) Oxide-iron formation is predominant in nearly all Archean BIF; silicate- and carbonate-iron formations are rare. The few carbonate iron formations occur in Lower Proterozoic sediments. The presence of such large amounts of BIF within high-grade regions is unusual, but there is no difference in chemical composition between BIF in high-grade regions and in greenstone belts. However, there are geochemical and petrological differences between their associated supracrustal sequences. The regional geology, rock association and geochemistry of associated metavolcanics and orthogneisses suggests that the high-grade type of BIF formed in arcs or back-arc basins, which were intruded by voluminous tonalites and granodiorites. These were later converted by deformation and high-grade metamorphism into orthogneisses. Such processes of accretion and crustal growth are different in kind but not in principle from those operating today.

Bimodal volcano-sedimentary rock assemblage: This assemblage mainly consists of basic and acid volcanic rocks, but also contains sedimentary formations in the upper horizons. It occurs on the southern margin of the North China Protoplatform, in the Jiao-Liao mobile belt, and on the southwestern margin of the Yangtze Protoplatform. Lithologically the sequence constitutes the spilite-keratophyre series of the Dahongshan Group. The basic volcanic rocks usually have geochemical features of ocean crustal (or ocean island) tholeiite, while the sedimentary rocks are largely turbidites made up of arenaceous, argillaceous and carbonate rocks.

A number of large-scale boron deposits are found in the Jiao-Liao mobile belt, which are scarce throughout the world and of great significance. Based on the major mineral components the boron deposits are divisible into three types: ascharite type, ludwigite-magnetite type and ludwigite-ascharite-magnetite type. They take stratified shapes, preserved in magnesian marble and are syngenetic with submarine volcanic rocks. Volcanic exhalation supplied a large quantity of ore-forming materials which were then subject to high temperature-pressure metamorphism of amphibolite grade (Zhang Qiusheng et al. 1988).

Red beds: Abundant magnesium-rich carbonate rocks are found in association with red beds. In the Shanxi-Henan rift zone, large amounts of purple to violet coloured feldspar quartzite, phyllite and slate occur, marking a change in atmosphere from an oxygen-poor to oxygen-rich state. This evidence coincides well with the oxygen-rich environment reflected by the presence of stromatolites.

Fine clastic rocks: Another difference between the Archean and Early Proterozoic is the REE geochemistry of the Early Proterozoic fine clastic rocks. In complete contrast to those of the Late Archean, the REE patterns contain a negative Eu anomaly (Nance and Taylor, 1976, 1977; McLennan et al. 1979). REE content increases, with a trend towards increasing REE abundance and Eu anomaly value. The REE patterns of the fine clastics are identical to that of potassic granite. This is due to the extensive area of such basement material underlying - and acting as source regions for - Early Proterozoic sediments (Taylor and McLennan 1981).

3. 6 Metamorphism

Except for a few Late Archean greenstone belts (e.g. the Wutai and Dongwufenzi greenstone belts) that have preserved progressive metamorphic zones from greenschist to low-amphibolite facies, most Archean metamorphic rocks are in amphibolite facies to granulite facies, having reached temperatures above 700 °C, pressures greater than 0.6 GPa, a geothermal gradient of 26-28°C/km, and having acquired a "planar" distribution (Dong Shenbao et al. 1986). The mechanism of metamorphism is still unexplained since the metamorphosed Archean terrane was exposed when Proterozoic materials began to be deposited. Therefore it can be inferred that at least 20 km thick top layer of the Archean crust had been denuded. Even if any progressive metamorphic zones did occur, they may have been removed.

The Early Proterozoic rock formations have also gone through extensive metamorphism during tectonic deformation processes. Compared with the Archean they have the following features:

1. With time, the metamorphic type changes from medium-high temperature regional metamorphism to a low temperature, dynamic metamorphism with a descending temperature and shrinking extent, indicating an irreversible decline of heat flow in geological history (Zhang Shouguang 1989). Regional dynamic-thermal metamorphism is predominant, reaching low-greenschist to low amphibolite facies and covering both medium and low pressure facies series.

2. The metamorphic belts were developed in intercratonic collison zones and marginal accretion zones as well as intracraton aulacogens, and therefore are usually arranged in linear fashion in accordance with the tectonic belts.

3. Difference in tectonic belts caused the variation in metamorphic degree and type. In collision and continental margin mobile belts, metamorphism is largely of the regional dynamo-thermal type in low amphibolite facies, occasionally reaching high amphibolite facies (e.g. in the Jiao-Liao belt), reflecting higher heatflow in the tectonic zone, thinner crust or environmental mobility. Meanwhile in the intracratonic aulacogens (e.g. in the Shanxi-Henan Rifted Province), metamorphism is usually of the regional low-temperature dynamic type mainly in low greenschist facies, revealing a thick crust and low heatflow.

4. Distinct metamorphic zonation occurs even in the low-greenschist facies producing minerals of different metamorphic grade. In collision zones and marginal mobile belts it generally demonstrates a reduction of metamorphic degree from the fringe to centre of the mobile belt, showing migrating towards rising along the centre of the belt. Although there is no perfect paleogeothermal trough recovered by paleogeothermal distribution in some mobile belts, they are still a useful criterium to reconstruct a compressional-shearing environment.

3. 7 Tectonics

With the coexistence of mobile belts and rigid crustal blocks, the tectonic styles of the Early Proterozoic mobile belts show remarkable difference with those of the Archean:-

1. The most distinct feature of Early Proterozoic structures are their linear distribution. Lineations are seen not only in the inter-cratonic collision belts (e.g. the Jiao-Liao collision belt and the Qing-Luan collision belt), but also in continental margin accretion belts (e.g. in the Yinshan-Yanshan and Qinling-Dabie mobile belts), as well as within intracratonic rift zones (e.g. the Shanxi-Henan Rifted Province). It shows that the post-Early Proterozoic tectonics did not cause ductile deformation and structural superimposition on the Archean craton.

2. The Early Proterozoic mobile belts that lie between the Archean cratons run across the structural trend of the cratons. This emphasises the rigidity of the Archean cratons. Alternatively, the reversed fan-shape fold systems in the intracratonic troughs also indicate that the deformation of the Early Proterozoic troughs did not change the rigidity of the Archean basement.

3. The diversity in tectonic environment determined the variation of regional structural styles. For instance, with the exception of the orogenic root zones, the collision and the accretion-zones are characterized by recumbent folds and accompanying nappe ductile shearing-zones. This reflects a deep-level horizontal shortening and shear-compression mechanism (Fig 3.28). In the trough, however reversed folds and accompanying thrust-faults are predominant, revealing a shallow-level lateral compression.

4. Apart from the accretion zones that frequently evolved into later structural deformations, most mobile belts have undergone an episode of deformation during the Lüliang Movement. Observed superimposition, such as in the Jiao-Liao mobile belt, is represented by first-phase N-S recumbent folds and by second-phase coaxial-level vertical folds. These folds are the product of different stages of the same process. Although they have gone through a series of later deformations, and an later E-W trending fold system has been created by a sinistral shearing on the west side of the belt, the linear structural zones are still a remarkable feature.

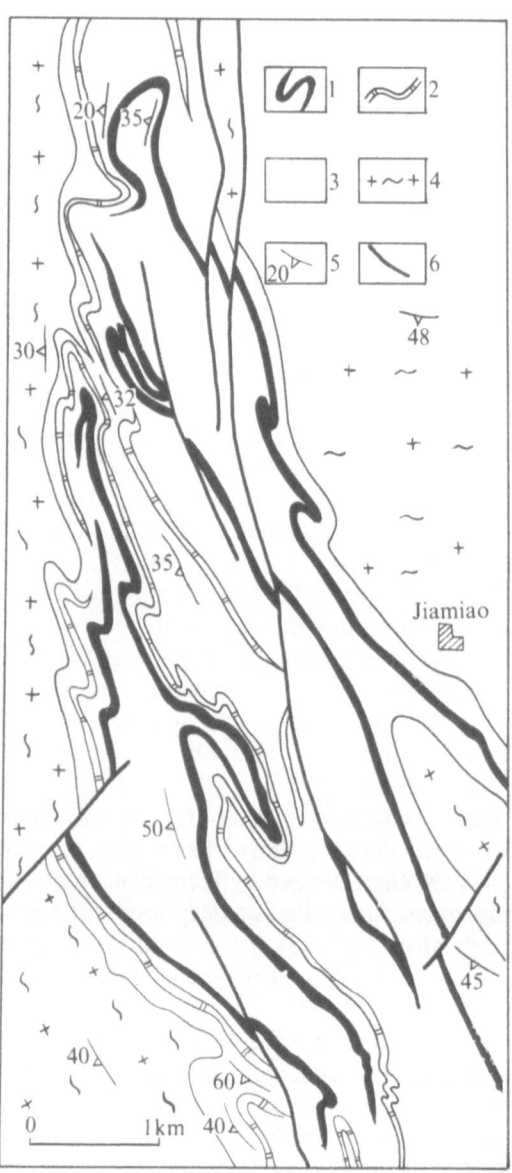

Fig. 3.28. The plane configuration of the polyphase deformed supracrustal rocks in the Dabie Complex of the Jia Miao area (by Suo Shutian): 1. Magnetite quartzite, magnetite amphibolite; 2. Marble, calc-silicate rocks; 3. Biotite plagioclase gneiss, plagioclase amphibolite; 4. Old granite, granodiorite; 5. Foliation; 6. Fault.

3.8 Post-orogenic Granites and Basic Dyke Swarms

Intrusives in orogenic belts can indicate the crustal mobility during and prior to the intrusion by their composition, attitude as well as deformation. The postorogenic intrusives have homogeneous constituents and are discordant to the country rock. In sharp contrast to the highly deformed wall rock, they show little deformation. Where these intrusions occur, these aspects clearly imply an advanced state of consolidation within the fold belt at the time of their emplacement.

Synorogenic granites also occur in the North China Protoplatform, but with remarkably different characteristics from their post-orogenic counterparts: in the Liao-Ji (Liaoning-Jilin) mobile belt a number of rock bodies of various sizes are found composed of biotite plagioclase granite of medium-coarse-grained texture and containing oligoclase-andesite (60-65%), quartz (25-30%), biotite (3-5%) and microcline (2-3%). Geochemically, their average composition is SiO_2 =71.45%, TiO_2 =0.08%, Al_2O_3 =16.13%, Fe_2O_3 =0.32, FeO =0.67%, MgO =0.82%, CaO =0.81%, Na_2O =6.46%, and K_2O =1.58%, with Na_2O+K_2O=8.04% and K_2O/Na_2O=0.24. The plagioclase granite has low Rb and a K/Rb ratio of 1007-1321 in accordance with the high Na/K ratio (6.1-4.5; Zhang Qiusheng 1988). The Na-riched granites are seen to cut through the older strata but, owing to the involvement of folding with the strata, appear in conformity with the wall rock (Mu Kemin et al. 1989). Zhang Qiusheng et al. (1988) reported a couple of U-Pb concordia ages of 2073 Ma and 2066 Ma for the synorogenic granite in the Liao-Ji mobile belt.

The postorogenic granites are seen widely in the North China Protoplatform and they are commonly characterized by their intrusive granitic texture, discordant orientation of structures, K-rich microcline and having concentrations of $K_2O > Na_2O$. Within the Liao-Ji mobile belt, the Early Proterozoic post-tectonic granites intruded in the Liaohe Group are represented by structurally massive, medium to coarse grained biotite granite. Mineralogically, it is composed of microcline (40%), oligoclase (30%), quartz (20%) and biotite (10%), and geochemical analysis shows a composition of SiO_2 =75%, TiO_2=0.1%, Al_2O_3 =12.23%, Fe_2O_3 =0.31%, FeO =0.94%, MgO =0.22%, CaO =0.8%, Na_2O=3.76% and K_2O =5.28, with K_2O/Na_2O=1.40 (Zhang Qiusheng et al. 1988; Mu Kemin et al. 1989).

In the Wutai Mts. region of the Hebei-Shanxi rift zone, the Lüliangian granitoids are of post-tectonic type and have yielded a biotite K-Ar age of 1810±29 Ma. Characterized by high alkaline and K_2O / Na_2O (K_2O+Na_2O=7.3-10%, K_2O / Na_2O=1.54-6.15), the granitoids are made up of hornblende adamellite and biotite granite. The hornblende adamellite has a mineral composition of microcline perthite (35-40%), plagioclase (15-20% An_{10}), quartz (20%), hornblende and biotite (3-15%), and shows a mean chemical composition of SiO_2=69.50%, TiO_2=0.91%, Al_2O_3 =13.32%, Fe_2O_3 =2.59%, FeO =3.30%, MgO =1.58%, CaO =2.25%, Na_2O =2.87%, K_2O =4.98%, with K_2O/Na_2O=1.73. In comparison, the biotite granite shows a mineral combination of microcline perthite (30-35%), quartz (25-30%), oligoclase (20%), orthoclase (10-15%) and biotite (10%), with biotite and minor mica assuming indistinct orientation of polysynthetic crystal. Chemical analysis give concentrations of

SiO_2=73.75%, Al_2O_3=12.38%, TiO_2=0.21%, Fe_2O_3=1.45%, FeO=0.89%, MgO =0.15%, CaO =1.16%, Na_2O =3.25%, K_2O =5.02% and Na_2O/K_2O=1.54. Evidently, the various Lüliang intrusive granitoids have homogeneous compositions, high alkaline concentrations and a larger K_2O content than Na_2O, which generally reveals that they originated by remelting sialic crust in relatively shallow position during the forming of the platform (Bai Jin et al. 1986). In the Lüliang Mountains, the Early Proterozoic granites are mainy distributed in the Guandi Mount, the middle sector of the Mountains. They are usually termed the Guandishan granites, and have a K-Ar isochron age of 2083 Ma and a U-Pb concordia age of 2053 Ma (Wu Tieshan et al. 1984). These bodies are concentrated in an area as large as 2000 km^2 and compositionally are biotite granites. Three stages in development of these bodies can be discerned: In stage A a medium-coarse-grained biotite granite is formed; followed in stage B by a coarse-grained porphyroid biotite granite and in stage C by a fine-medium-grained biotite granite (Wu Tieshan 1984). The main mineral composition includes microcline (30-45%), oligoclase (20-30%), quartz (30-35%) and biotite (5-10%). The chemical composition is SiO_2=73%, TiO_2=0.22%, Al_2O_3=13.62%, CaO=0.99%, Fe_2O_3=0.65%, FeO =1.71% and MgO =0.43%, with Na_2O+K_2O=8.24% and K_2O / Na_2O=1.49. Despite the fact that the Guandishan granite is divisible into fine- and coarse-grained, it is otherwise universally homogeneous and massive and no orientation or other deformation is found.

Distinct diversity in composition, texture and structure are found between syntectonic and post-tectonic granites. Syntectonic granites have a genetic correlation with volcanic rocks and usually make up one endmember of a compositional series with basic rock types. For example, the syntectonic granite in the Liao-Ji mobile belt is marked by rich Na_2O (with K_2O/Na_2O<1), reflecting a crustally derived magma differentiated at depth. As the granite has experienced the same tectonic deformation as the wall rocks, it has a pronounced structural orientation. This is especially the case for the schistosity and gneiss schistosity at the edge of the intrusive body. Meanwhile, the post-tectonic granite shows a remarkable homogeneity in composition. Those in the Liao-Ji mobile belt and the Hebei-Shanxi rift belt have the same composition, and are characterized by a high alkaline content (K_2O+Na_2O> 8%, K_2O/Na_2O value around 1.5), reflecting derivation from a low degree of crustal melting. The abundant microclinic and metasomatic texture also reflect a composition similar to that of the lowest melting point of granitic crust. This type of granite has a simple shape, a distinct contact with country rocks and no orientation texture. This is in sharp contrast to the country rocks which bear highly developed deformation and foliation, indicating that it was formed in a cratonized stable environment. This proves that the orogenic belt had been consolidated even before the intrusion of the magma.

Early Proterozoic postorogenic magmatism is the product of the tectono-thermal event at the end of Lüliang Movement. A group of isotopic ages ranging from 1.8 to 2.0 have been obtained from rocks of this period. Some basic intrusives and dyke swarms, as well as pegmatite dykes, were intruded before and after granitic magmatism (e. g. in the Songshan and Wutai Mts. respectively; Ma Xingyuan et al. 1981b; Bai Jin et al. 1986). The basic dykes cut across the fold belt and are seen unconformably overlain by an assemblage of conglomerate-quartzitic sandstone-shale and carbonate belonging to the middle Proterozoic Changcheng System in the Wutai Mountains. However, basic dyke intrusion continued to occur, reaching a maximum in - and then ceasing entirely within - mid-Proterozoic times.

3.9 Stromatolites and Microfossils

According to reports from other countries, the oldest stromatolites in the world are found in rock sequences of 3.4-3.5 Ga in age (Walter et al. 1980). However Archean stromatolites are generally very limited in distribution (Cloud and Semikhatov 1969; Martin et al. 1980; Hofmann et al. 1985) and relatively simple in form. Up to the present, the earliest stromatolites in China were found in Early Proterozoic strata and chiefly in the North China Protoplatform where most stromatolites are in the Early Proterozoic Hutuo Group. These stromatolites, together with those in Gantaohe Group from Taihang Mountains, make up the typical Early Proterozoic stromatolite series in China.

Early Proterozoic strata contain not only some stromatolites such as *Conophyton, Jacutophyton, Stratifera, Collumnaefacta, Collumnacollenia Omachtenia, Kussiella, Gymnosolen, Jurusania, Boxonia, Minjaria* and *Tungussia* etc. which are common examples in Mid-Late Proterozoic time, but also some new ones including *Pilbaria, Externia, Kussoidella, Confunda, Vertexa, Tibia* and *Kanpuria.* (Zhu Shixing et al. 1987) which are not found in the Mid-Late Proterozoic sequence, but are widespread in many Early Proterozoic sequences overseas.

Early Proterozoic stromatolites are uncomplicated in their micro-structure and textures: for example stromatolites in the Hutuo Group show mainly banded and thread-like features. Mid to Late Proterozoic stromatolites, however, tend to vary in time from having banded, linear to coagulated microstructures. This variance seems to be related with the increasing complexity of components with time. As to the micro-textures, the stromatolites from Hutuo Group are commonly characterized by macula, plaque and minor filiform types. Some microfossils are made of spherical cells of 0.6-1 μm in diameter. These fossils make up the linear microstructures in stromatolites, showing curved chains composed of five to six cells. These cells and chains without sheaths are some kind of fossilised primitive bacteria, which suggests that Early Proterozoic stromatolites were formed by the activity of some primitive and simple prokaryotes (Zhu Shixing et al. 1987).

Early Proterozoic stromatolites are complicated in form and show cyclic changes in characteristics, which can be used to subdivide them into different stromatolite assemblages. Whether representing types, sorts, order of occurrence or in microstructures, these assemblages cannot be compared with those of Mid to Late Proterozoic series. Studies on the Wutai Mountains and their adjacent areas indicate that the Early Proterozoic stromatolite series can be subdivided into five assemblages (Table 3.2; Zhu Shixing et al. 1987). Early Proterozoic stromatolite series occur

Assemblages	Representative Stromatolites
V	*Niushania miushancensis* *Mistassiaia niushanecsis* *Omachienia denticulata*
IV	*Nordia tianpengnaocensis* (f. nov.) *Pilbaria of. inceria formis* *Microsiylus granularis* (f. nov.) *Pseudogymnosolen beidaxingense* (f.nov.) *Conophylon balios* (f.nov.) *Tungussia striolata* (f.nov.) *Eucapsiphora longoienuia*
III	*Alcheringa majuscala* (f.nov.) *Pilbaria perplexa* *Liaoheelia fuscieulata* *Jacuiophyion bulbosum* (f.nov.) *Asperia minuta* (f.nov.) *Tibiu fanancensis* (f.nov.) *Gruneria strumata* (f.nov.) *Collumnae facia composita* (f.nov.) *Carnegia daguanshancensis* (f.nov.)
II	*Gynnosolen simpplex* (f.nov.) *Kussicilla plana* *Dfulmekella djulmekensis* *Nanlouella bulbosa* (f.nov) *Collumnacollenia rantamaa* *Conistratifera regularis* (gr. eg.f.nov.)
I	*Kussoidella planicolumnaris* (f.nov.) *Djuimekelia tuanshanziensis* *Kanpuria bulbosaa* (f.nov.) *Conistraiifera irregularis* (gr.et.f.nov.)

Table 3.2. The assemblage subdivision of early Proterozoic stromatolites in China (revised from Zhu Shixing et al. 1987).

beneath later series. Some mutual sorts exist in both series and this may suggest heredity and cyclicity from one to another. However, between the two series, differences of the mutual types in the order of occurrence and specific members together with their relatively simple microstructures of the lower series stromatolites are specific characteristics of Early Proterozoic stromatolites (Zhu Shixing et al. 1987).

Stromatolites have both organic and sedimentary features. Studies have revealed that stromatolites are formed by the activities of algae, predominated by blue algae and bacteria. Indeed, bacterium fossils such as *Primaevifilum septatum Schopt*, *Siphonophycus antiguns Schopf, Warrawoonella radia Schopf* are found even in the stromatolites dated at 3.4-3.5 Ga. Consequently, the occurrence and development of stromatolites reflects the evolutionary features of early life forms. In general, the flourishing and decline of stromatolites reflects the evolutionary development of algae. The widespread stromatolites in the Early Proterozoic Hutuo Group and in its corresponding strata suggest a flourishing period for the prokaryotes such as blue algae. The considerable large existence of blue algae which can produce free-oxygen through photosynthesis may indicate that free-oxygen content in the atmosphere had exceeded the Pasteur level, that is, over 1% of oxygen content of the present atmosphere (Cloud 1976). The Early Proterozoic sediments represented by the Hutuo Group may be important products in the turning point of the crustal evolution from a reducing to oxidation environment. Thus, they may mark the initiation of a new environment of formation in crustal evolutional history, with the initiation of a new geological period characterized by the extensive occurrence of algae and other life forms in an oxidation environment.

4 Mid-Late Proterozoic (Pre-Sinian) Crust

Huang Xueguang
Tianjin Institute of Geology and Mineral Resources
Tianjin, China, 300170.

4.1 Distribution and General Features

General consolidation of the Chinese Protoplatform and the current tectonic framework of China's continental crust was achieved following the close of the Lüliangian orogeny at the end of Early Proterozoic times. Thus, Mid-Proterozoic times initiated the extant era of rigid continental plates.

The Mid-Late Proterozoic (Pre-Sinian) strata are broadly distributed in the major part of China (Chen Jinbiao et al. 1981, Wang Hongzhen et al. 1984) and are divisible into stable and a mobile types of sedimentary environment. The stable environments of deposition are predominantly situated on protoplatforms, median massifs and deposited cover sequences on the basement. In contrast, mobile environments are prevalent in the inter-continental and continental margins and as inliers in the Phanerozoic fold belts, including amongst others the Qinling, Qilian, Tianshan and Kunlun foldbelts (Fig. 4.1).

The abortive attempt at extension in the protoplatforms gives rise to aulacogen-style taphrogenic-orogenic rifting with consequent block movements leading to continent-wide subsidence of the protoplatform. There are two major NNE-NE trending aulacogens in the North China Protoplatform. These are situated (1) within the Yanshan to Taihangshan regions; (2) within and to the north of the Xiongershan to Zhongtiaoshan regions. They are separated from each other by the Shanxi arch. A third west-trending aulacogen is termed the Mid-Proterozoic Zhartai aulacogen in the Yinshan Mts., Inner Mongolia. All these aulacogens developed at triple junctions related to domal uplifts and are a key indication of the existence of stable continents and plate tectonics (Fig. 4.2). Coeval with this rifting event is the development of the anorogenic belt in the Yanshan range with the emplacement of anorthosite, rapakivi granites and extrusive associates, which occurred between 1.6-1.4 Ga ago. The intrusion of basic dyke swarms in the protoplatforms also implies the existence of widespread dilatant areas in the continental crust.

The Late Proterozoic (pre-Sinian) times were characterised by convergence between the protoplatforms. For instance the North China and Yangtze Protoplatforms experienced a Wilson cycle characterized by oceanic crust subduction and continental collision, followed by subsequent resplitting.

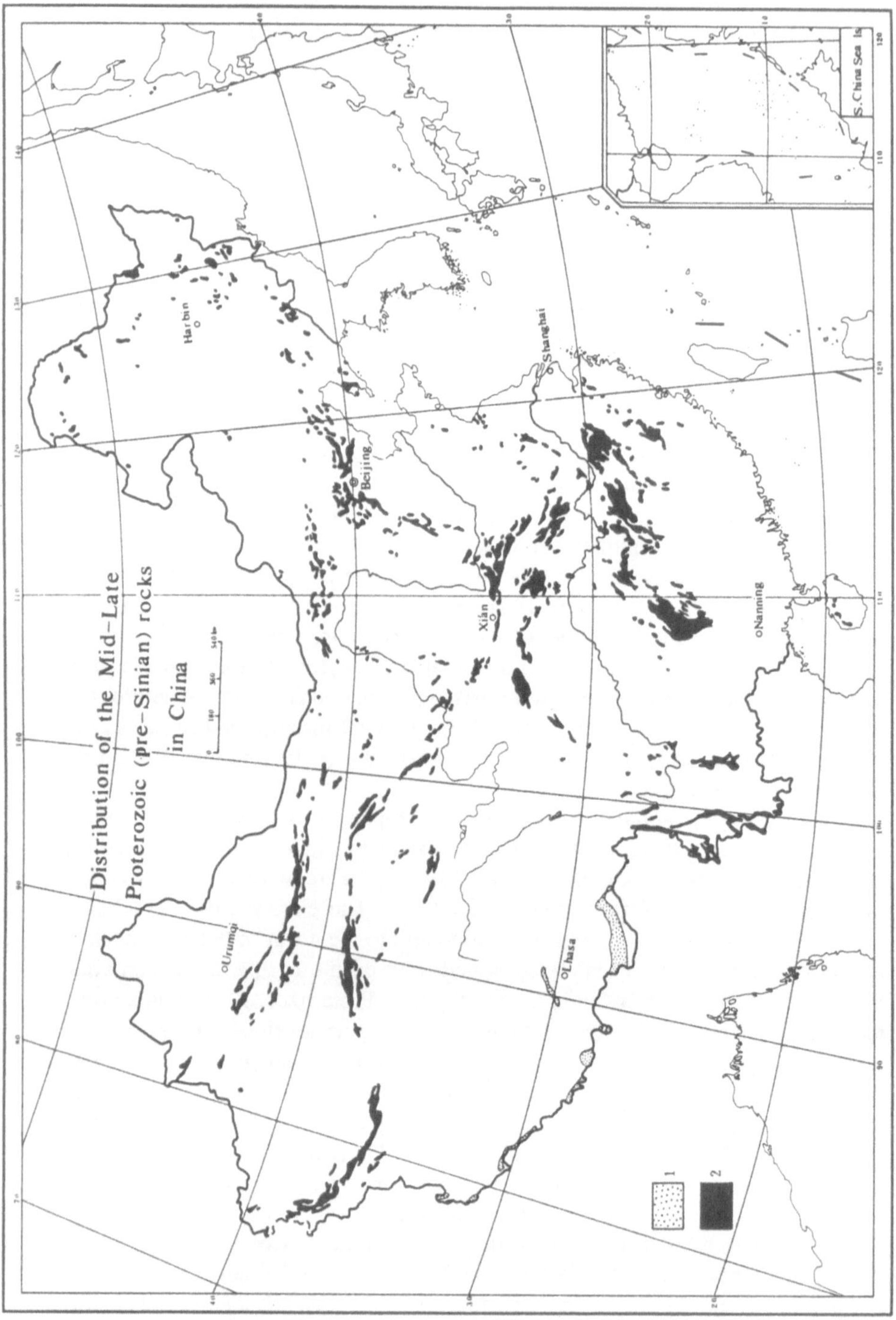

Fig. 4.1. Distribution of the Mid-Late Proterozoic (pre-Sinian) rocks in China 1. Pre-Sinian rocks; 2. Mid-Late Proterozoic rocks

4.2 Aulacogens and Depression Belts in the North China Protoplatform

At the end of Early Proterozoic times, the North China Protoplatform was consolidated by the Lüliang-Zhongyue orogeny. From this time until the Mesozoic Indosinian-Yanshanian movement, most of the North China Protoplatform did not undergo any extensive orogenesis and appears to act as a rigid block. The Yanshan and Xiong'er-Hangao aulacogens formed on the North China Protoplatform were transformed into a stable paraplatform sedimentary cover during the mid-late stage of Middle Proterozoic times. The sedimentary cover was also deposited in the depressed zone on the east and west margins of the protoplatform. In addition, a marginal-sea trough developed on the north and south margins of the North China Protoplatform (Fig. 4.1). Table 4.1 shows the features and correlation of its sequences.

4.2.1 The Yanshan Aulacogen

The Yanshan aulacogen is one of the two great graben-like deep troughs formed within the North China Protoplatform during early Middle Proterozoic times. Some of its characteristics are similar to the "Aulacogen" described by Shatsky (1955), and Salop et al. (1969). The Yanshan aulacogen is a generally NE-trending, large-scale trough-like basin whose subsiding centre is distributed along the belt of the Jixian, Tianjin and Chaoyang, and western Liaoning regions (Fig. 4.3). Its northwest end lies west of Zhanjiakou, and it penetrates into the Taihang Mt. area to the southwest, in plan appearing as a triangular aulacogen (Fig. 4.1). There is a 10 000 m thick sequence comprising perfect and well-preserved strata of Mid-Late Proterozoic age in the aulacogen, the most complete of which is the Jixian section. The criterion of division and correlation of the Mid-Late Proterozoic in North China were constructed in the 1930's by Gao Zhenxi et al. (1934). Many geologists have made multidisciplinary studies on the section (see Wang Yuelun et al. 1979, 1980: Wang Yuelun 1984; Chen Jinbiao et al. 1980). The Changcheng System, the Jixian System and the Qingbaikou System of the Jixian section are commonly used in China as chronostratigraphic units of Mid-Late Proterozoic age (All China Commission on Stratigraphy, 1983). In terms of crustal evolution, the boundary between the Dahongyu and the Gaoyuzhuang Formations is the most singnificant.

The evolution of the Yanshan aulacogen can be divided into three stages as follows:

Initiation stage. The Yanshan aulacogen was formed at about 1800 Ma and was mainly controlled by synsedimentary extentional faults, producing a generally NE-extending large, trough-like basin in which the Changzhougou Formation was deposited. The bottom of the sequence is fluviatile facies sediments mainly comprising coarse-grained conglomerate-feldspathic wacke and fine-grained conglomerate-

Table 4.1. Correlation of Middle and Late Proterozoic rock assemblages in North China Platform

Unit		The Interior of North China Protoplatform		North margin of North China Protoplatform		South margin of North China Protoplatform		Depression in the east margin of North China Protoplatform		Depression in the west margin of the N. China Protoplatform (Yinchuan region)
		Yanshan aulacogen	Xiong'er-Hango aulacogen	Chartai aulacogen	Bayan Obo geocline	East Qinling	North Huaiyang	South section	North section	
Upper Proterozoic	Sinian	Qingbaikou System sandstone rock - shale carbonate rock assemblage	Dongpo Fm. Luoquan Fm. Donglia Fm. Huanglianduo Fm.					Kangan Group Suxian Group Xuhuai Group sandstone-shale-carbonate rock assemblage	Jinxian Group Wuxingshan Group sandstone-shale-carbonate rock assemblage	Zhengmuxian Formation
	Qingbaikou		Luoyu Group sandstone-shale-carbonate rock assemblage			Taowan Group	Foziling Group bimodal volcanics flysch assemblage	Bagongshan Group — clastics-shale-carbonate-rock assemblage	Xihe Group Yongning Group	Wangquankou Group sandstone-shale carbonate rock assemblage
Middle Proterozoic	Jixian	Gaouzhuang Fm. of Changcheng System to Jixian System carbonate rock-shale-quartz sandstone assemblage	Ruyang Group sandstone-shale-carbonate rock assemblage	Shagan Group clastics shale and carbonate rock assemblage		Kuanping Group sandstone basalt carbonate rock silicalite assemblage				Huangqikou Group sandstone-shale -carbonate rock assemblage
	Changcheng	Changcheng Fm. to Dhongyu Fm. wacke - quartzite shale - carbonate bimodal volcanics assemblage	Xiong'er Group bimodal volcanics (potassic basalt- trachyte- rhyolitic) assemblage	Chartai Group clastics-carbonate-black shale assemblage	Bayan Obo Group sandstone, shale carbonate, flysch assemblage					

quartz wacke containing 5-20% or more feldspar. The sediments are immature, poorly sorted and rounded. These characteristics demonstrate that the sediments experienced quick transportation and fast deposition. Oblique bedding dips synthetically (Photo 4.1), and reflects the features of aqueous transportation. The statistical dips of oblique bedding and pebbles show that the river flowed from ENE to WSW (Fig 4.4). Due to ongoing crustal extension, the southeast wall of the basin constantly subsided along syndepositional faults. Hence the basin developed into a fault trough with its cross-section steep in the northwest and gentle in the southeast. The sedimentary section of the basin is therefore wedge-shaped (Fig. 4.5a).

From the middle of this stage, marine transgression occurred in this area resulting in the deposition of a littoral facies around Jixian where the sequence is thicker and consists of quartz sandstone, quartzitic sandstone and quartzite with occasional silty shale.

Development stage. The aulacogen continued to develop between 1970-1550 Ma during sedimentation of the Chuanlinggou, Tuanshanzi and Dahongyu Formations. Under the control of syndepositional fault, the southeast wall continued to subside (Fig. 4.5b,c). The Chuanlinggou Formation is composed of silty shale and black shale, whereas the Tuanshanzi Formation consists of pelitic-arencaceous ferruginous dolomite. They were deposited in a closed to semi-closed low-energy reducing environment. The Dahongyu Formation is a rhythmic sedimentary sequence consisting of quartz sandstone, silty shale, and stromatolite-bearing dolomite, with volcanic rocks. The generation of the Yanshan aulacogen is related to mantle upwelling.

Photo 4.1. The cross-bedding in Changzhougou Fm. sandstone in Jixian, Tianjin

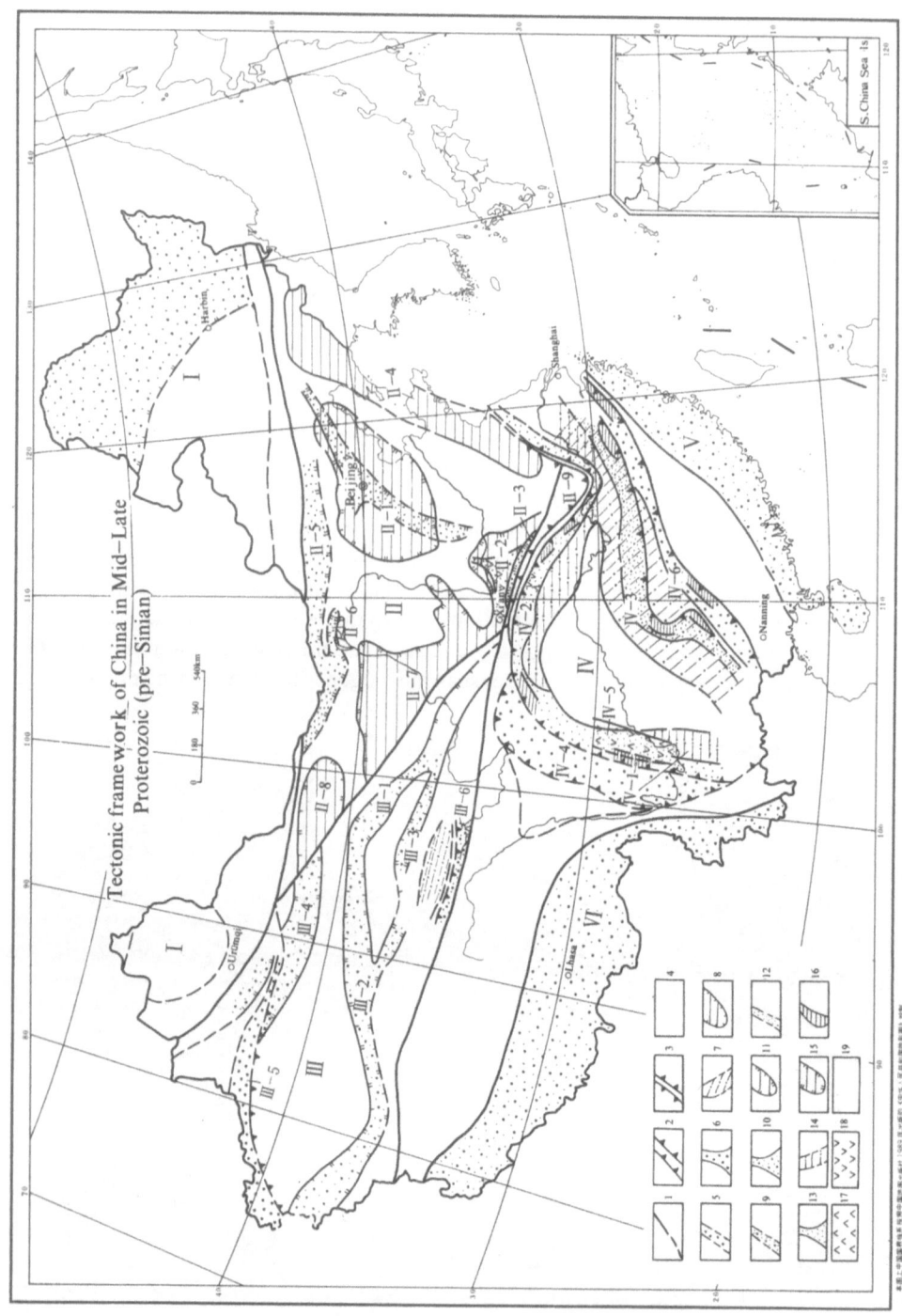

Tectonic framework of China in Mid–Late Proterozoic (pre–Sinian)

Fig. 4.2. Tectonic framework of China in Mid-Late Proterozoic (pre-Sinian) times: 1. Boundaries of tectonic provinces; 2. collision and subduction zone; 3. suture; 4. oceanic crust; early Late Proterozoic (Qingbaikou Period): 5. mobile belt; 6. aulacogen; 7. back-arc basin, 8. cover of platform in downwarping.

Mid-Late Proterozoic: 9. mobile belt, 10. aulacogen, 11. cover of platform in downwarping belt.

Middle Proterozoic: 12. mobile belt, 13. aulacogen, 14. back-arc basin, 15. cover of platform in downwarping belt; 16. complex in trench (ophiolite): 17. acidic volcanics; 18. basic volcanics.

19. Archean-Early Proterozoic shield. I. South margin of Siberia-Mongolia Plate; II. North China Protoplatform II-1. Yanshan aulacogen, II-2. Xiong'er-Hangao aulacogen, II-3. northern Qinling-northern Huaiyang ocean trough, II-4. Jiaoliao-Xuhuai depressed belt, II-5. Bayan Obo geocline, II-6. Chartai aulacogen, II-7. depressed belt in western North China Protoplatform, II-8. Beishan depressed belt, II-9. Qinling-Dabie block.

III. Tarim Protoplatform: III-1. Qilian Mt. aulacogen, III-2. western Kunlun mobile belt; III-3. aulacogen on the southern margin of Qaidam block, III-4. Tianshan aulacogen; III-5. Late Proterozoic southern Tianshan mobile belt; III-6. Mid-Late Proterozoic eastern Kunlun mobile belt.

4. Yangzi Protoplatform: N.-1. Middle Proterozoic trench-arc-basin system on the west margin, N.-2. Mid-Late Proterozoic trench-arc-basin system on the north margin, N.-3. Middle Proterozoic trench-arc-basin system on the southeast margin, N.-4. Late Proterozoic mobile belt on the west margin, N.-5. Late Proterozoic rift in western sichuan, N.-6. Late Proterozoic mobile belt on the southeastern margin; V. Cathaysia; Mid-Late Proterozoic mobile belt; VI. Mid-Late Proterozoic mobile belt on the northern margin of Gondwans

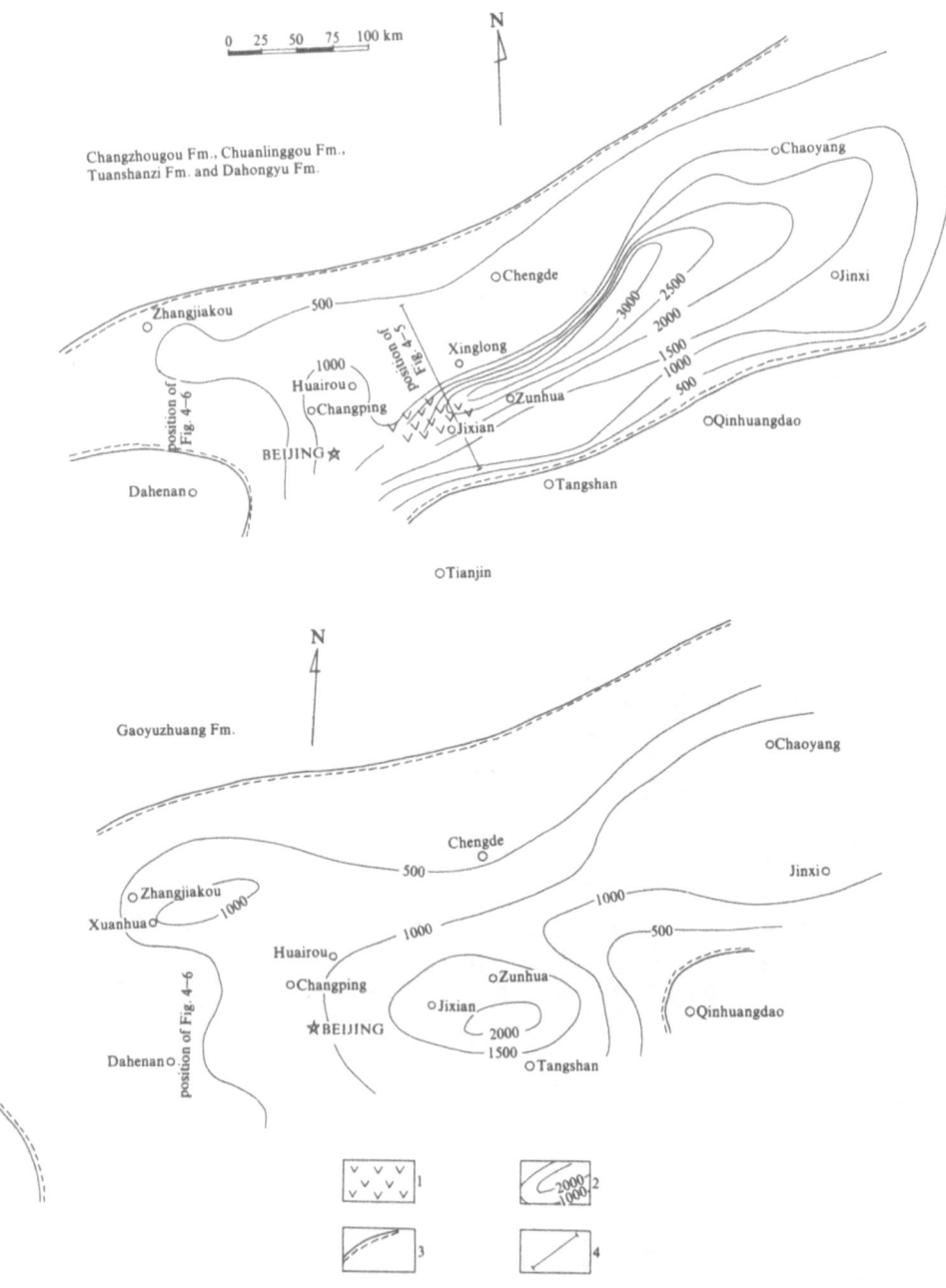

Fig. 4.3. Isopach of Changcheng System in Yanshan Mt. region, Hebei Province: 1. Volcanic rock; 2. isopach; 3. margin of transgression; 4. position of cross section

As a result, volcanism is the dominant characteristic during the development stage of the aulacogen. This magmatic activity can be divided into early and late stages which differ considerably in nature and scale (Fig. 4.5c).

In the early (Chuanlinggou) stage, the chemistry of magmas ranged from calc-alkaline to basic, with both eruptive and intrusive magmatism occurring. This resulted in the formation of breccia (agglomerate) pipe, thin-bedded lavas with amygdaloidal structure and dykes of plagioclasic porphyrite. The breccia (agglomerate) pipes, 200-300 m in length and 100-150 m in width, are complexes consisting of agglomerate, volcanic breccia, tuffaceous sand conglomerate, tuff and basalt. The lavas are basaltic rocks made up mainly of microlites of plagioclase. The rocks are significantly rich in sodium (Na_2O=3.50-2.52%) and poor in potassium (K_2O=0.15-0.85%), and are distributed over small areas.

In the late (Tuanshanzi-Dahongyu) stage, which correlates with the further development of rifting, mantle upwelling resulted in the eruption of large quantities of magmas which include eruptive breccia (agglomerate), lava and tuff, and intrusive crossed-veins of potassium diabase, lamprophyre and syenite porphyry. The volcanic rocks can be divided into four kinds, i.e. lava, volcanic breccia, tuff and tuffaceous rocks. Moreover, a lot of volcanic necks developed. The lavas are poorly-differentiated potassic basic rock predominantly comprised of potash minerals which include orthoclase (~60%) and minor amount of sanidine, olivine and augite. Petrochemically, the volcanics are poor in SiO_2 (36.14-44.06%), rich in K_2O (8.25-119.3%) and low in Na_2O (0.14-0.24%), commonly rich in Cr, Ni, Co, Ti, with low FeO/MgO ratios (2.44-1.43). This geochemical data show that the primitive magma is basic in composition. Meanwhile the rocks contain abnormal amounts of potassium and high concentrations of Be, Ba, Zr. and Sr, indicating crustal contamination. The volcanism in the late stage began with the explosive and eruptive phases resulting in the extrusion of potassium-rich basic rocks components and ended with small-scale intrusive phase of acid rocks marked by syenite, which exhibits the bimodal nature of magmatic activity during rifting (Ren Fugen 1986, a, b).

Late stage magmatic activity occurred about 1.6-1.7 Ga ago. The dominantly eruptive rocks described above were generated in connection with the anorthosite, gabbroic complex and the Miyun rapakivi granite which occur in the northwest wall of the syndepositional faults. This large-scale magmatic activity occurred in the anorogenic tensile environment when the syndepositional faults reached the upper mantle. The continental tholeiitic magma is associated with deep faulting in upper mantle and/or lower crust. It underwent a slow and complete differentiation, during which it was contaminated by upper crustal material.

Magmatism is related to the development of the northeast branch of the triangular aulacogen. Because the northwest branch of the aulacogen was segregated by the Changping-Huairou submarine uplift, the Changzhougou Formation comprises mainly alternate quartz sandstone and silty shale with fishbone cross-bedding and lenticular bedding which exhibits typical features of tidal sedimentation. The sediments are thin (generally 100-200 m) and show planar distribution (Fig. 4.3). In addition to these features the sediments of the Chuanlinggou, Tuanshanzi, Dahongyu and Gaoyuzhuang Formations overlapped from the basal levels upwards to the west

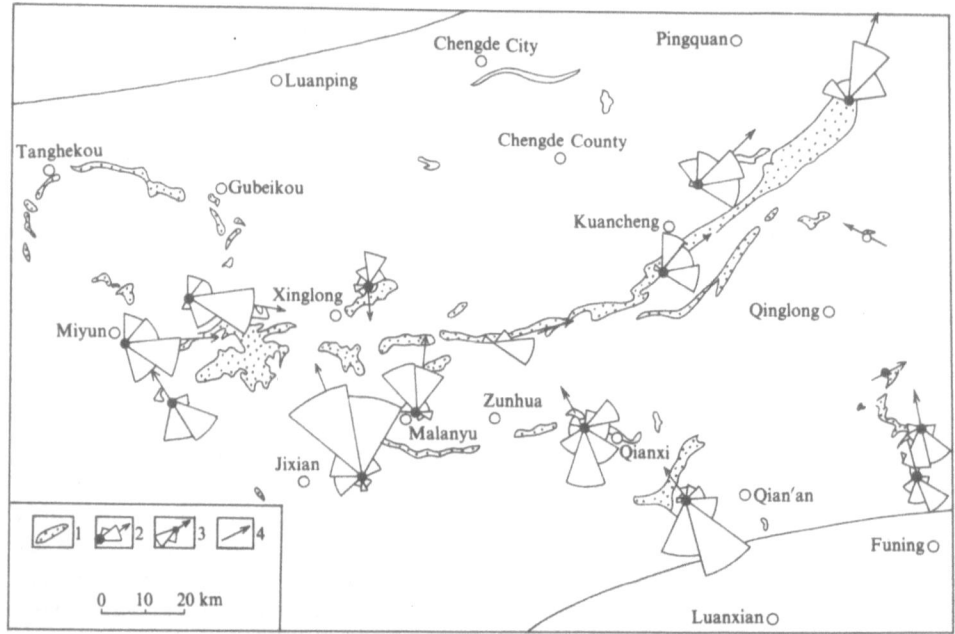

Fig. 4.4. Paleocurrent trends from directional data of the early Changzhougou age in Yanshan Mts. (after Wang Changyao 1987). 1. outcrop of Changzhougou Formation; 2. trends indicated by oblique bedding; 3. trends indicated by pebbles; 4. average trends surveyed less than 10 times.

and south (Fig. 4.6). Kidney-form and oolitic hematites (actually ferruginous stromatolite and oncolites) are present at the bottom of Chuanlinggou Formation, the first red bed found in the Mid to Late Proterozoic in this area, and they demonstrate that the density of oxygen in the atmosphere had risen markedly by these times. On the contrary, in the south branch of the aulacogen, the base of the rifted basin tilted slightly to the north. During the mid to late Middle Proterozoic, this area became high land. Due to marine regression, the deposition of new layers overlapped to the north and probably did not extend into the Xiong'er-Hangao aulacogen.

Waning stage. The end of the Dahongyu age also marked the end of the major period of volcanic activity. From the beginning of Gaoyuzhuang times onwards, extensive marine transgression occurred and reached a climax during this stage. The marine transgression reached halfway towards the Wutai Mt. to the west (Photo 4.2) and towards the present western Bohai Bay to Shanhaiguan-West Shandong oldland to the east. Furthermore, data from petroleum exploration drilling demonstrate the presence of the Gaoyuzhuang Formation in the northern part of North China Plain. The paleo-marine environment formed during this time was similar in nature to that of an epicontinental sea (Huang Xueguang 1985). The stratigraphic sequence mainly consists of carbonate rocks which are distributed in approximately planar fashion and show evidence of tidal rhythms. Nevertheless, to a certain extent the crust was still tectonically active. First, the sediments are controlled by the downthrow of the

NW

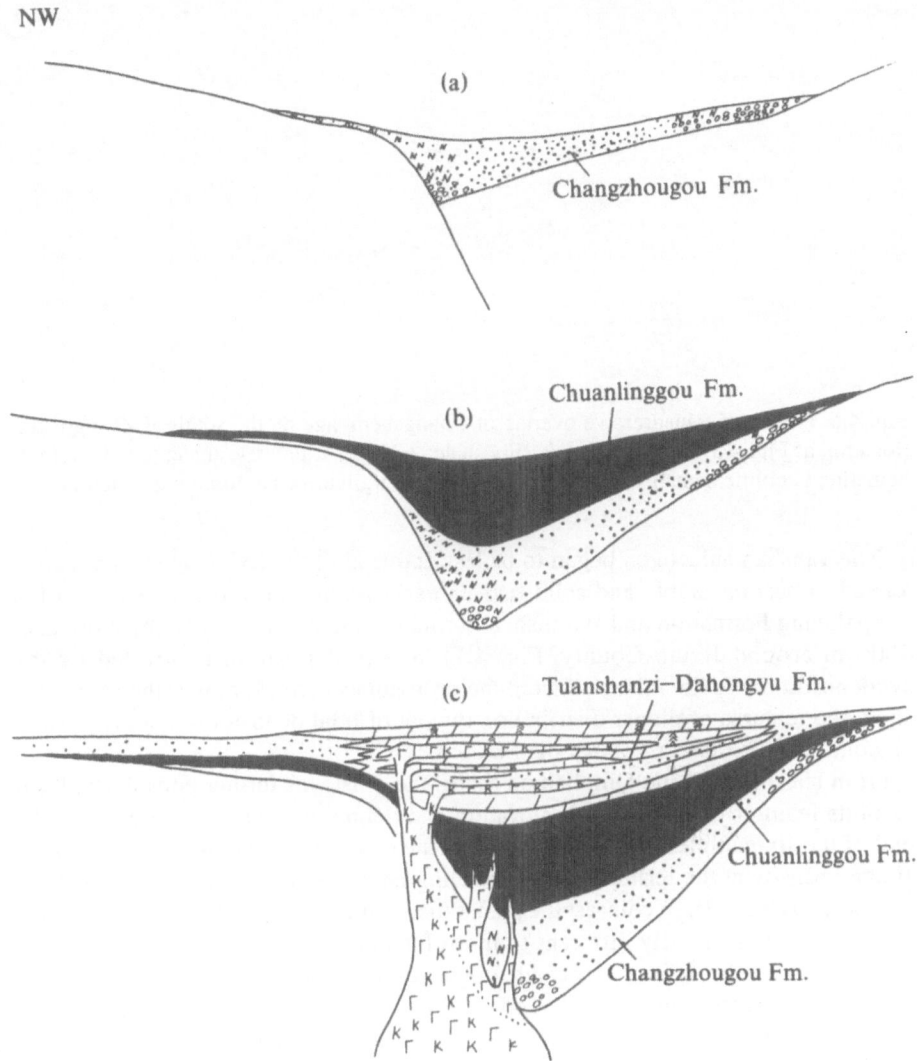

Fig. 4.5. Schematic cross-section showing the development of the Yanshan aulacogen.

southeast wall of the syndepositional fault around Jixian County, where they are markedly thicker than those in other areas (Fig. 4.3). In comparison deep-water facies tumulose limestones occurred extensively in the middle part of the basin; secondly, uneven elevation and subsidence took place. Examples of this include the Qinling uplift after Dahongyu times, and the Luanxian uplift at the end of Gaoyuzhuang times, which sometimes resulted in a local disconformity on the margins of the basin, such as the Xingcheng movement around Jinxi. However, continuous deposition took place over the greater part of the basin.

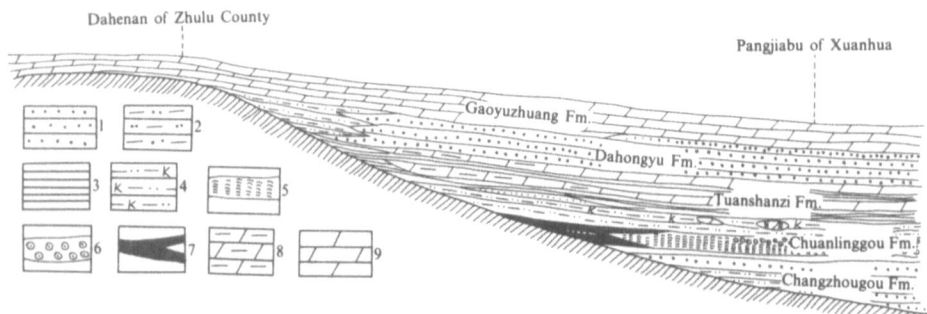

Fig. 4.6. Profile of transgressive overlap of Changcheng age on the south of Zhangjiakou (location as Fig. 4.3) 1. Sandstone; 2. silty shale; 3. shale; 4. potassic shale; 5. kidney-form hematite; 6. oolitic hematite; 7. massive hematite; 8. argillaceous dolomite; 9. dolomite

The Yanshan aulacogen began to be less active after the Jixian age and the crust tended to become stable and solidified. In early Jixian times, the thickness of the Yangzhuang Formation and Wumishan Formation increased notably (by more than 4000 m around Jixian County, Fig. 4.1), due to deposition controlled by the syndepositional faults inherited from the protoaulacogen. Moreover the sediments show the features of planar distribution typical of tidal deposition in a clear-water epicontinental sea and of planar distribution.

From late Jixian to the Qingbaikou times, the crust was further consolidated and exhibits features of both overall elevation and depression. The Qinyu uplift at the end of the Jixian times, the Yuxian uplift at the end of the Xiamaling times and the Jixian orogeny at the termination of the Qingbaikou times are seen everywhere in the basin (Table 4.2). In contrast the planar distribution, small but constant thickness of deposits are markedly different to those features controlled by synsedimentary faults. Although the sedimentary centre migrated from time to time, little facies change is found and the rock units classified on the Jixian section are consistently correlatable over a large area, especially for the Qingbaikou and the Cambrian Systems. Despite a long discontinuity between them, they occur together everywhere and constitute a substantial platform sedimentary cover.

4.2.2 Xiong'er-Hangao Aulacogen and Qinling-Dabie Block

The Xiong'er-Hangao aulacogen lies astride the juncture of Henan, Shaanxi and Shanxi provinces. The different divisions of Middle and Upper Proterozoic in the three Provinces are shown in Table 4.3. However in the southern North China Protoplatform, Middle Proterozoic rifting had different characteristics. It began with intense volcanism resulting in formation of the Xiong'er-Hangao triple arm aulacogen. The 160 km long north branch of this structure was inserted into the North China

Protoplatform. The aulacogen was limited by faulting on the east and west sides and remoulded by later deformation on the south border. However it still shows the characteristic features of a triple junction (Fig. 4.1, 4.2 and 4.7).

The Lower Middle Proterozoic Xiong'er Group (or Xiyanghe Group) is unconformable with units of the underlying Archean metamorphic complex (such as the Taihua Group) and Lower Proterozoic low-grade metamorphic rocks. It is a suite of tremendously thick unmetamorphosed or slightly metamorphosed eruptive rocks with Rb-Sr isochron ages of 1710±73.6 Ma, 1439±35 Ma, 1459±48 Ma, 1454±36 Ma (Hu Shouxi et al. 1988), 1675 Ma (Guan Baode et al. 1980), and U-Pb ages on zircon of 1545 Ma (Dong Shenbao et al. 1986). Thus the Group was inferred to have a lower age limit at 1700 Ma and an upper age limit at 1400 Ma, roughly corresponding to the Dahongyu Formation volcanics in the Yanshan aulacogen. However, a more recently obtained U-Pb zircon age for the Xiyanghe Group, a correlative of the Xiong'er Group, is 1820 Ma (Sun Dazhong, et al. 1988). Hence the Xiong'er Group could be older than the current age data suggests. The Ruyang Group and Luoyu Group (or the Gaoshanhe Formation-Shibeigou Formation), which are above the Xiong'er Group, have a rock assemblage comprising sandstone, siltstone, pelite and carbonate rocks. A Rb-Sr isochron age of 1267 Ma has been reported for the Yunmengshan Formation at the base of the Ruyang Group (Guan Baode et al. 1980). Meanwhile the K-Ar age for the glauconite from the Beidajian Formation in the same Group ranges between 1115-1149 Ma (Working Group, 1987). Recently, the Cuizhuang Formation of the Luoyu Group has been dated as 1138, 1149 and 1159

Photo 4.2. The Gaoyuzhuang Fm. overlapping onto Archaean gneiss

Overlying: Early Cambrian Fujunshan Fm., of Canglangpu stage		
Qingbaikou System (371 m thick)	Jixian Movement 800 Ma	
	Jing'eryu Fm., sandy conglomerate, glauconitic sandstone and shale, limestone	203 m
	Weixian Uplift	
	Xiamaling Fm. sheetlike siltstone	168 m
	Qinyu Uplift 1.05 Ga	
Jixian System (4000 m thick)	Tieling Fm. sandshale, dolomite and stromatolitic limestone	333 m
	Hongshuizhuang Fm., shale and silty shale	131 m
	Wumishan Fm., chert-dolomite, dolomitic, bituminous dolomite	3398 m
	Yangzhuang Fm., red and white argillaceous microlitic dolomite containing silt	707 m
	Luanxian Uplift 1.4 Ga	
Changcheng System (4151 m thick)	Gaoyuzhuang Fm., quartzite, sandstone and volcanic rocks chert-dolomite	1588 m
	Dahongyu Fm., quartzite, sandstone and volcanic rocks chert dolomite	408 m
	Qinglong Uplift	
	Tuanshanzi Fm., pelitic-arenaceous ferruginous dolomite, fine-grained sandstone and shale	518 m
	Chuanlinggou Fm., silty shale and shale	889 m
	Changzhougou Fm., conglomerate, sandstone, cuartzite	854 m
	1.8 Ga	
Underlying: Archean Qianxi Group metamorphic complexes.		

Table 4.2. Stratotype of the Mid-Late Proterozoic of Jixian Section in the Yanshan aulacogen

Ma by the K-Ar method using glauconites (Guan Baode et al. 1980; Ma Guoqian et al. 1980) and 1125 Ma by the Rb-Sr isochron method using illite (Li Yunjun 1988). The Sanjiaotang Formation of the Luoyu Group has glauconite ages ranging 1012-1089 Ma. Thus we can date the boundary of the Cuizhuang Formation and the Sanjiaotang Formation at 1050 Ma, comparable with that between the Jixian System and the Qingbaikou System in the Yanshan aulacogen. The Ma'anshan Formation has a K-Ar age of 1168 Ma on glauconite which is close to the average glauconite K-Ar age of the Beidajian Formation at 1157±30 Ma (Guan Baode et al. 1980). Many Chinese geologists now believe that the upmost tillite of the Luoquan Formation may be of Late Sinian age.

The Xiong'er Group has a maximum thickness of more than 7100 m, comprising lavas, subordinate pyroclastic rocks and minor sedimentary rocks. In the north part of the aulacogen, alluvial sediments at the bottom of the sedimentary sequence consist of <100 m thick bedded clastic rocks. The lower part of this deposit consists mainly of yellow and yellow-green pebbled feldspathic quartz sandstone, whereas the upper part of the sequence is comprised of red and purple mudstone, shale and sandy shale. Volcanic rocks in the Xiong'er Group have the features of central-vent and crack-vent eruption. In the early stages of volcanism, eruption was mainly continental facies type with a centre around Xiaoshan Mt. and Xiong'er Mt. areas. A

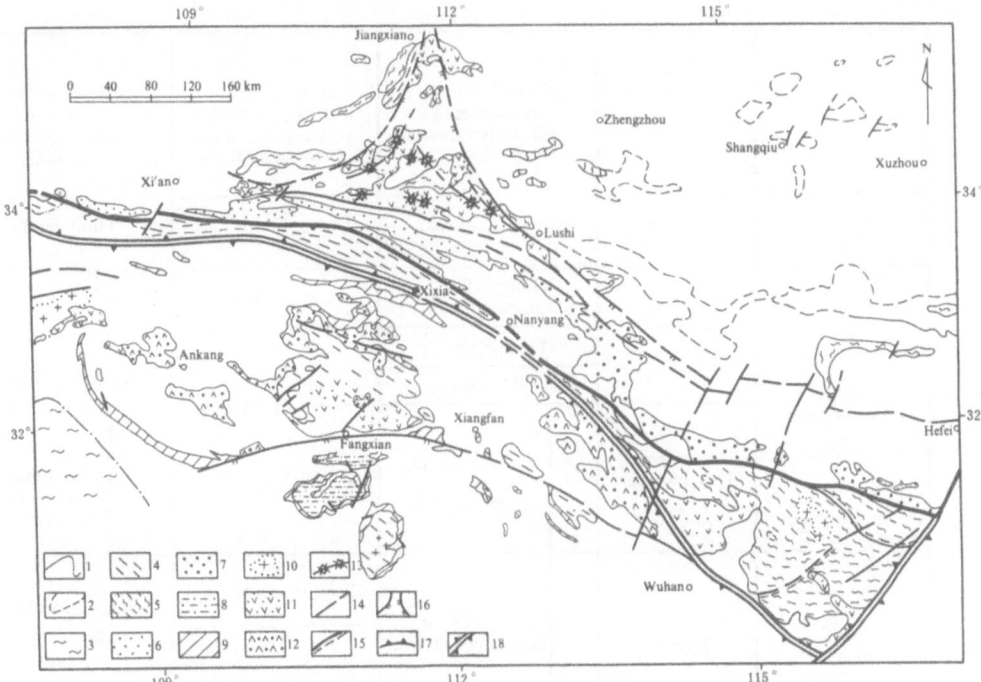

Fig. 4.7. Tectonic sketch map of the Qinling-Dabie region: 1. Boundaries of exposure; 2. Concealed strata; 3. Archean; 4. Lower Proterozoic; 5. undifferentiated Archean-Lower Proterozoic; 6. Middle Proterozoic mobile belts; 7. Late Proterozoic mobile belts; 8. Mid-Late Proterozoic basin sediments; 9. Sinian system; 10. Granites and migmatitic granites; 11. Middle Proterozoic volcanic rocks; 12. Late Proterozoic volcanic rocks; 13. Volcanic fissure and central eruption rents; 14. Faults; 15. Aulacogen; 16. Late Proterozoic suture; 17. Phanerozoic suture; 18. Phanerozoic suture (redrawn after Ma Xingyuan et al. 1987)

small quantity of subaqueous sediments were deposited in the south and east of the aulacogen. In the middle stage, the sea trough and bay to the south increased in size. Continental and subaqueous volcanic eruptions occurred alternately. Vesicles in the lava decrease but pillow structure and jasper bands increase upwards. In comparison eruption in the east proceeded mainly under water, whereas by the late stage continental-facies eruptions were still occurring around Xiaoshan Mt. The south sea-trough continued to expand and was filled by an increasing degree of pyroclastic and sedimentary rocks.

The volcanic sequence of the Xiong'er Group is comprised of mainly potassic basalt-trachyte-rhyolite, subordinate trachy basalt-trachyte belonging to alkaline volcanic rocks, and partly subalkaline volcanic rocks. The volcanic rocks of the Xiyanghe Group are of bimodal type, with SiO_2 ranging discontinuously between 59-67% (Tang Min et al. 1988). Their petrochemistry exhibits high concentrations of potassium and iron as well as low calcium and aluminium contents, of which high potassium is the predominant feature. The Na_2O/K_2O ratios of potassic basalt are

Erathem	System	Zhongtiaoshan-Leshan area (Guan Baode et al. 1980)		Luonan area of Shannxi, Lushi area of west Henan (Li Qinzhong 1980)	Songshan area of west Henan (Ma Xingyan et al. 1981)	
Lower Paleozoic	Cambrian	Lower Cambrian		Lower Cambrian	Dongyao Fm.	
Upper • Proterozoic	Sinian	Dongpo Fm. Luoquan Fm.		Luoquan Fm.		
		Dongja Fm.			Hongling Fm.	
		Huanglianduo Fm.				
	Qingbaikou		Luoyukou Fm. Sanjiaotang Fm. Cuizhuang Fm.	Shibeijou Fm.		Heyao m. • Luotuopan Fm. Puyu Fm. • Ma'anshan Fm.
Middle • Proterozoic	Jixian		Beidajian Fm. Baicaoping Fm. Yunmengshan Fm.	Fengjiawan Fm.	Bingmagou Fm.	
		Xiaobeigou Fm.		Gaoshanhe Fm.		
	Changcheng	Xiyanghe Group		Xiong'er Group		
Lower • Proterozoic				Tietonggou Fm.	Songshan Group	
Archean		Taihua Group		Taihua Group	Dengfeng Group	

Table 4.3. Comparison of the Middle and Upper Proterozoic in Xiong'er-Hangao aulacogen

less than two and decrease with the increase of SiO_2 contents. This change is gradual in the volcanic rocks with SiO_2 content of less than 68%, but rapid in those with SiO_2 over 68%. Rocks containing $SiO_2 < 62\%$ also have much lower Na_2O/K_2O ratios. This demonstrates that the volcanic sequences are not comagmatic differentiates and their acid components may come from the anatexis of continental crust. The $^{87}Sr/^{86}Sr$ initial values are between 0.7056-0.7141 which also show that upper-mantle-derived magma has mixed with the material of partial anatexis of the lower crust (Sun Shu et al. 1985). The total abundances of REE are low but LREE values are relatively high with poorly-fractionated patterns. The La/Yb ratio range from 8 to

18, Eu/Eu*=0.6-0.9 and Ce/Ce*=0.77-0.82. As volcanism progressed, the values of REE, SiO_2, K_2O, LREE/HREE, Ce/Ce* increased and those of Al_2O_3, FeO, MgO, Eu/Eu*, Eu/Sm decreased (LiZenghui et al. 1988). The geological and geochemical data demonstrate that the volcanism of the Xiong'er Group is related to paleo-rifting (Sun Shu et al. 1985; Li Zenghui et al. 1988; Ouyang Jianping et al. 1988; Tang Min et al. 1988): the aulacogen is filled with a large amount of lava and fluviatite sediments with the environmental characteristics of a down-warped trough overlying an active mantle. Hence, it is thought that the Xiong'er-Hangao aulacogen is a three-branch rift system produced by crustal uplift above a mantle hot spot. At the end of the eruption of the Xiong'er Group, the aulacogen ceased developing. Rift activity gradually died away and the rift was later buried by cover sediments.

The Ruyang Group ranges in thickness between 1500-5000 m and its corresponding strata are unconformable with the Xiong'er Group in many places. Generally there was hiatuses in deposition between the two groups. Their distribution was also obviously controlled by basin margin faults. Compared with the Xiong'er Group, deposition of the Ruyang Group expanded slightly eastward and westward. In the northern part of the aulacogen, terrigenous clastic rock assemblages were deposited in a delta-tidal flat environment. In other areas littoral facies quartz sandstone, tidal flat facies, neritic facies and clastic rock assemblages occur. These assemblages are dominanty comprised of sandstone with 6-8% shale and are derived from the east, north and west (Guan Baode et al. 1980; Sun Shu et al. 1985). Littoral facies terrigenous clastic rocks and neritic facies carbonate rock assemblages also occur in the southwest part of the sea basin, where abundant and various kinds of algal stromatolites are found in the carbonate rock (Li Qingzhong et al. 1980; 1985). The cross-section of the aulacogen is asymmetric and the depositional centre shifted westward with time. Both the high maturity of the terrigenous clastic rocks and the frequently tidal nature of deposition demonstrate that the depositional rate was low in the aulacogen (Sun Shu et al. 1985).

Following the formation of the Ruyang Group, the Luoyo Group and its corresponding strata were deposited as the sea basin shrank southwards and expanded eastwards. The eastern margin of the basin has terrestrial facies and littoral facies clastic rock assemblages. In comparison the western part has littoral neritic shale, carbonate rock and sandstone assemblages, whereas the southwestern part contains relatively deep-water facies black carbonaceous slate and silliceous rock assemblages. Deposits have a planar distribution and a thickness < 500 m. The deposition centre of the Group lies perpendicular to the axial trend of the aulacogen (Sun Shu et al. 1985). All these features are characteristic of a quasi-cover and demonstrated the waning of the aulacogen.

The southern margin of the North China Protoplatform is a complicated mobile belt which remained active for a long time. An intensely compressed and faulted Early Proterozoic-Archean metamorphic rock suite occurs on the north of the Qinling-Dabie block. The suite is divided into several groups formerly regarded as of Middle to Late Proterozoic age, with the Kuanping Group, the Taowan Group and the Erlangping Group situated in the east of the belt. However geologists have disputed the ages of each Group. Recent studies have clarified some of these issues. At present, it is commonly accepted that the Kuanping Group and the Sujiahe Group are of the Mid-Late Proterozoic age and the Taowan and Foziling Groups are of Sinian age. The Kuanping and Sujiahe Groups are the infill in the E-W directed marine trough bounded by the North China Protoplatform and the Qinling-Dabie block.

The Kuanping Group has a thickness of over 8000 m and extends from Baoji, Shaanxi eastwards to western Henan, in a narrow zone 800 km long and 10-50 km wide. It forms a relatively complete volcano-sedimentary cycle and is limited from north to south by deep fractures. The lower part of the sequence consists of greenschist intercalated with quartzite, quartz marble and sericite quartz-schist. The protoliths are felsic clastic rocks with basic volcanics. In contrast the middle part mainly consists of mica quartz-schist with amphibolite, quartz marble and less felsic granulite, and the upper part of the sequence comprises amphibolite, biotite marble and mica plagioclase quartz-schist. The protoliths of the middle members are basic volcanics with carbonate rocks and calcic siliceous rocks and the upper members are felsic rocks with basic volcanics. The U-Pb isochron age of zircon (origin unidentified) from quartz schist is 1730 Ma and the Rb-Sr isochron age of zircon from metamorphosed basic volcanics is 1411 ± 30 Ma (Zhang Qiusheng 1980; Gao Hongxue, et al. 1988; 1989).

The volcanic rock series of the Kuanping Group contains rhyolitic, dacitic, subalkaline andesitic pyroclastic rocks and lavas erupted in submarine environments. Petrochemically the Group is a bimodal volcanic rock series comprising the basic components of tholeiites and acidic calc-alkaline series rocks. The basic lavas are characterized by low SiO_2 (45-49%) and K_2O (less than 0.4%) but high TiO_2 (more than 1%) contents. The low and small range of values <FeO>/MgO (1.2-2) demonstrate that the volcanics experienced weak differentiation and possess the characteristics of an undifferentiated magma (Zhang Qiusheng 1980; Sun Shu et al. 1985). The CaO contents vary largely, from 5.6 to 12.48%. Because CaO has an evident negative correlation with Zr, Y, La, Ti and P, the change of CaO contents can be interpreted in terms of CaO-rich minerals, such as plagioclase and clinopyroxene taking part in shallow-level magmatic fractionation. The basaltic rocks have a low contents of ÝREE and exhibit LREE depletion. This indicates that the basaltic magma is derived from a depleted source and that the magmatism occurred under low-pressure conditions. There exists an obvious linear relationship between Ti, P, Y, La, Zr and other elements, with the order of element uncompatibility being P=La>Zr, Y=>Ti. These relationships also demonstrate that plagioclase and clinopyroxene (but not garnet) are the main crystallising minerals during solidification (Wan Yusheng et al. 1988). Hence, the Kuanping Group was formed in the marine trough of a continental margins, under control of an extensional plate boundary.

The Kuanping Group has undergone 3 stages of metamorphism and deformation:- in Late Proterozoic, Late Caledonian-Early Variscan and Indo-Sinian-Yanshanian orogenies respectively. However, it is in the first stage that the fundamental features of the Kuanping Group metamorphic rocks were formed, which are demonstrated by 3 metamorphic facies zones: chlorite-biotite zone (subgreenschist facies), almandine zone (high greenchist facies) and staurolite zone (low amphibolite facies), formed by regional dynamo-thermal metamorphism. During the

Caledonian orogeny, ductile shearing caused retrogressive metamorphism of subgreenschist and locally high greenschist facies in the previously high-grade metamorphosed rocks (Cong Rixiang et al. 1990). The Late Proterozoic formation in the Kuanping Group created tight, reversed and recumbent folds and penetrative crystalline schistosity. Finally during late Caledonian-Early Variscan a series of ductile shear and large refolding reordered the structural pattern into a number of tectonic slices which were then juxtaposed as a tectonic complex (Gao Hongxue et al. 1988; Zhang Shouguang 1990; Zhao Ziran et al. 1990).

The age of the Taowan Group has long been disputed. The Regional Geological Party of Shaanxi Province (19??) first defined it as Early to Middle Proterozoic age. However, later isotopic dating and paleontological studies suggested it to be of Sinian (Geng Shufang 1989), Cambrian or even Odovician age (Gao Hongxue et al. 1988; Zhang Weiji et al. 1989). Recently a Rb-Sr isochron age of 569±66 Ma (Geng Shufang, 1989) has been obtained in the mica schist on top of the Taowan Group and tiny shell fossils and collophane are found in the Miaowan Formation (Wang Zhendong et al., 1989). Hence the Taowan Group is regarded as Sinian-Early Cambrian in age.

The Erlangping Group was previously classified as Mid-Late Proterozoic in age. Meanwhile, recently geologists have found Crinoidea, Syringopora, Tabulata, Tetracoralla, *Paroductus* and fragments of Brachiopods as well as some stems of higher plants which lived during the Ordovician-Triassic period in the strata corresponding to the Erlangping Group, the Yunjiashan Group and the Zimugou Formation. In the middle of the Erlangping Group, Radiolaria and Porifera have been found which are younger than those existing in the Cambrian period. The lower part of the Erlangping Group has the Rb-Sr ages of 517±84 Ma, 573±65 Ma (leptynite) and 681±39 Ma (volcanic rocks). Therefore, the Erlangping Group should be regarded as Early Paleozoic in age (Yao Zongren 1987; Hu Shouxi et al. 1988).

Various interpretations are given on the evolution of the south margin of the North China Protoplatform. Wang Hongzhen (1981) and Wang Hongzhen et al. (1982) discussed the evolutionary history of the continental margin between the North China Platform and the Yangtze Platform in the light of tectonic events. They suggested that the margin experienced a long-term development, and converged, folded and uplifted in the Indo-sinian period. During this time, the marine environment waned and the margin of the North China Protoplatform and the Yangtze oldland experienced more than one crustal consumption zone, produced by the collision between the continent and island arc-type crust. The Kuanping Group was deposited in faulted basins amongst the islands built by the Qinling Group, situated at the south margin of the North China Protoplatform in Middle Proterozoic time. Meanwhile the early accretional crustal consumption zone was formed in the continental margin area on the south margin of the North China Protoplatform. Wang Qingchen et al. (1989) suggested that the metamorphic complex is actually mélange produced by tectonization of rocks formed in various environments. As such it is a petrotectonic assemblage, rather than a lithostratigraphic unit, and hence should not be called a "Formation" or "Group". Clearly the fossils found in the complex can only be used to estimate the age of the rock containing the fossils, not to infer the age of adjacent rocks. They also hold that the frequent discoveries of Paleozoic fossils in the metamorphic complex have very important significance and indicate that these metamorphic sediments had been formed in the Paleo-Tethys ocean. (This ocean separated the North China Platform from the Yangtze Plate during most of Paleozoic times and began to be subducted by Devonian times, although the two continents did not collide until Triassic times.) This, of course, does not exclude the existence of the Precambrian basement but reveals the complexity involved in the study of Precambrian structure.

In the eastern section, north Huaiyang area, the low grade metamorphic rocks of the Sujiahe and Foziling Groups occur adjacently and have corresponding ages and horizons. The two Groups constitute relatively young sediments deposited in the sea trough and overlie the Late Archean Dabie and Luzhenguan Groups. The Henan Regional Reconnaissance Party obtained a number of U-Pb ages ranging between 693-604 Ma on zircons in the lower part of the Sujiahe Group (Yang Sennan et al. 1983). In comparison, microflora fossils within the mid-upper part of the Foziling Group suggests that it approximately corresponds to the early Late Proterozoic Qingbaikou Group in the Yanshan aulacogen. However, no microflora fossils and isotope ages of the volcanics in the lower part of the Foziling Group have been obtained, so the possibility that a part of Foziling Group is of Middle Proterozoic age cannot be excluded (Yang Sennan et al. 1983).

Substantial progress has been made in the study of the upper limit of the Fozling Group and its stratigraphic correlation with other sequences in recent years. The upper age limit of the Foziling Group is probably Early Paleozoic (Ma Baolin and Zhang Zhaozhong 1988): The Foziling Group is now known not to correlate with the Permo-Carboniferous Meishan Group (Jin Fuquan et al. 1987). Similarly, the Xinyang Group, has been found to contain fossils of Azonomoleter luber, Liophotriletes naimoua, Stenozonotriletes naumoua, Poridlecotriletes noumoua in carbonaceous schist. In addition, faunal fossils of marine facies such as Polychaeta, calcareous algae and bivalves occur in crystalline carbonate rocks (Hu Shouxi et al. 1988). There are spores of continental origin, some Scolecodont, Acritarchs and minor Chitinozoa fossils in sandy slate in the mid-upper part of the Xinyang Group. All of these fossils indicate that Xinyang Group was formed in the Devonian (Hu Shouxi et al 1988; Gao Lianda et al 1988). Prior to this work, the Xinyang Group was assigned to the Late Proterozoic and regarded as corresponding to the Foziling Group. Therefore, in the east section of the south margin only the Foziling and the Sujiahe Groups are of the Late Proterozoic age and the presently-published data delimited them as belonging mainly to the Sinian System.

Ma Xingyuan et al. (1987, 1989) envisaged that the development of the northern Qinling-northern Huaiyang aulacogen split the Qinling-Dabie block (II-9 in Fig. 4.1) off from the North China Protoplatform and formed a tectonic belt similar to an island arc. The detached block went through a complicated evolution until it was accreted onto the Yangtze Protoplatform, which in turn collided with the North China Platform during Phanerozoic times.

Strata Division	Formation / Group	South Section (west of Tan-Lu Fault)[1] — Xuhai Group	North section (east of Tan-Lu Fault; south Liaoning)[2] — Wuhangshan Group
Upper Proterozoic — Qingbaikou System (Sinian) / Bangongshan Group (Lower Series)	Liulaobei Fm (Xihe Group)	Marlstone and shale with limestone. (Acritarchs, Phaeopha and microflore fossils are contained within the shale)	Shale, siltsone and shale with limestone containing abundant Acritrachs and microflora fossils
	Wushan Fm (Xihe Group)	Glauconite-bearing pebblyd quartz sandstone and quartz sandstone	Pebble-bearing sandstone and quartz sandstone with glauconite quartz sandstone.
	Caodian Fm (Yongning Group)	Purple quartzitic conglomerate and sandy conglomerate	Conglomerate, pebble-bearing sandstone and feldspathic wacke
Underlying Strata	Fengyang Group (Pt_1) / Anshan Group (Ar)		

Table 4.4. Comparison of strata division and depositional characteristics of the Qingbaikou System in south and north sections of the Jianlian-Xuhuai depression belt . [1]Regional Geological Surveying Party, Geological and Mineral Resources Bureau of Anhui Province, 1985); [2]Compilation Group of Regional Stratigraphic Table of Liaoing Province, 1978.

4.2.3 The Jiaoliao-Xuhuai Depression Belt (North China Protoplatform)

The NNE-trending Jiaoliao-Xuhuai depression belt developed in the Late Proterozoic (from the Qingbaikou to Sinian times) in the eastern North China Protoplatform (II-4 in Fig. 4.1). In the south section of the depression belt on the west of the Tan-Lu Fault, during the early Qingbaikou stage some nearly east-trending foredeeps were filled by a scattered sequence of molasse. This consists of red ferruginous sandstone and conglomerate, represented by the Caodian Formation of the Bagongshan Group.

Following the deposition of the Caodian Formation a marine transgression occurred in the area, forming a littoral facies glauconite-bearing quartz sandstone of high compositional and textural maturity (the Wushan Formation). The advancing marine transgression and deepening sea water then resulted in the creation of a low-energy hydrostatic area where glauconite-bearing, fine-grained continental-derived clastics and carbonate rocks (mainly carbonaceous shale and pelitic limestone) were deposited forming the Liulaobie Formation. The shale contains Acritarchs, macroscopic Phaeophyta and microflora fossils, and the pelitic limestone bears lenticular stromatolite and doncolites. The shale gives a Rb-Sr isochron age of 840±72 Ma (Regional Geological Surveying Party, Bureau of Geology and Mineral Resources of Anhui Province 1985). Deposition is continuous from the formation of the Liulaobei Formation to the overlying Sinian System. The north section of the depression belt included the Jiaodong area to the east of the Tan-Lu Fault, south Liaodong peninsula and south Jilin Province. The southern part of the Liaodong peninsula contains typical successions, within which the Yongning and Xihe Groups corresponds to the lower and upper parts of the Qingbaikou Group respectively. Table 4.4 shows their characteristics. The lowest part of the Xihe Group in south Jilin has a glauconite K-Ar age of 818 Ma. The Dushan Formation in Korea, corresponding to the Xihe Group in China, has a K-Ar age of 853 Ma (Research Group of Upper Precambrian Shenyang Institute of Geology and Mineral Resources 1986). Hence, the south and north sections of the depression belt are very similar and completely comparable in strata sequence, rock assemblages, contact with overlying and underlying strata, fossils, paleomagnetism, and isotope ages (Chang Shaoquan, 1980; Xu Xuesi 1982; Project Co-operation Group 1984; Regional Geological Surveying Party, Bureau of Geology and Mineral Resources of Anhui Province 1985; Upper Precambrian Research Group, Shenyang Institute of Geology and Mineral Resources 1986). The strata of the two sections are classified as cover sediments and were formed in tidal environment within an epicontinental sea.

4.2.4 The Chartai Aulacogen and Bayan Obo Geocline

By Middle Proterozoic times the north western margin of the North China Protoplatform was to some extent still tectonically mobile. Middle Proterozoic sequences are preserved in the Chartai Group and the Bayan Obo Group (II-5,II-6 of Fig. 4.1). In the south, the 500 km-long Chartai Group extends eastwest in Langshan-Se'erteng Mts. Area, whereas the 800 km-long Bayan Obo Group, situated from north of the Chartai Group, stretches eastward to Huade (where it is known as the Huade Group) along the margin of the North China. These two groups are separated by an uplifted sequence of metamorphic rocks of Late Archean age (Fig. 4.8).

The Chartai Group mainly comprises metamorphic conglomerate, quartz sandstone, quartzite, dolomitic slate, stromatolite-bearing crystalline limestone and dolomite, carbonaceous slate and intermediate to basic volcanics. In ascending order, it is divided into the Shujigou, Zenglongchang, Agulugou and Liuhongwan Formations which have the depositional features shown in Table 4.5. The table shows that the lower part of the Shujigou Formation is an assemblage of pebbly feldspathic quartz sandstones and sandy conglomerate which have low compositional and textural maturity and represent rapidly deposited strata formed during the extensive sagging period of the faulted basin. The pebble content in conglomerate is variable, reflecting the composition of the underlying basement. The upper part of the Shujigou Formation consists of highly mature quartzite and quartz sandstone together with (alkaline) basalt, basaltic trachyandesite, andesite and rhyolitic dacite volcanic rocks. The Zenglongchang Formation was deposited on top of the volcanics and is composed of silty mudstone and stromatolite-bearing limestone which indicate that the rifting and subsidence declined and the depositional environment changed into a relatively stable littoral-tidal belt. The Agulugou Formation is comprised of black carbonaceous slate with carbonaceous microlitic limestone and large-sized copper, lead, and zinc sulphide deposits. It represents a closed and extensively chemically reduced environment, typical of a confined sea produced just before the closure of a rifted basin.

The Dongwufenzi Group, beneath the Chartai Group, underwent a metamorphism in the Early Proterozoic (Ma Xingyuan 1989), and yielded a zircon U-Pb age of 2025 Ma from amphibole-plagioclase gneiss. However, the First Geological Party of Inner Mongolia obtained the U-Pb metamorphic ages of 2531±0.3 Ma for biotite-plagioclase gneiss and 2581±0.4 Ma for amphibole-plagioclase gneiss (The Working Group on the Geological Time Scale of China, 1987). The age suggests that the Dongwufenzi Group is of Upper Archean age.

The lithology, formation, stromatolite assemblage and depositional environment of the Chartai Group are all comparable with the major part of the Changcheng System (from the Changzhougou Formation to the Dahongyu Formation) in the Yanshan aulacogen. The Chartai Group has been dated by the U-Pb zircon method as 1516 Ma and 1612 Ma (Dong Shenbao et al., 1986). This demonstrates that the Chartai Group comprises Middle Proterozoic aulacogen sediments deposited on the northern margin of the North China Platform which were inserted into the basement of the protoplatform from the west eastwards (Fig. 4.7). In contrast to the Yanshan aulacogen, the Chartai Group experienced deformation and greenschist facies metamorphism. This Group thus became part of the basement of the North China Plat-

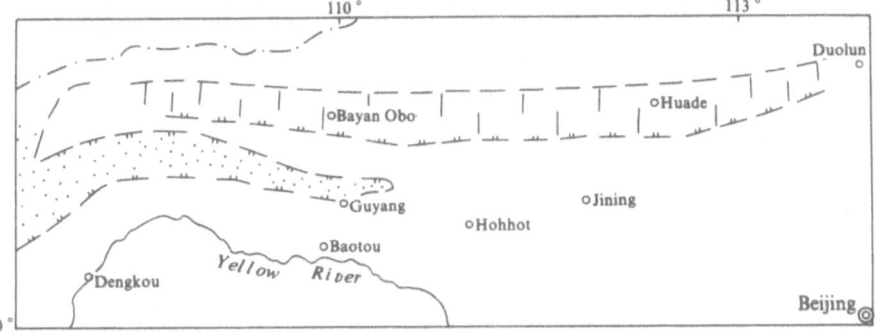

Fig. 4.8. The distribution of the Middle Proterozoic Chartai aulacogen and the Mid-Late Proterozoic Bayan Obo geocline (after Ma Xingyuan 1989).

Table 4.5: The sedimentary features of the Chartai Group (modified after Wang Ji et al. 1987)

Erathem	Group	Fm.	Member	Thickness	Lithology	Primary sedimentary structure
Middle Proterozoic	Shinagan				stromatolite-bearing quartz sandstone and slate with limestone	
	Chartai	Argulugou	2	>400 m	grey-black and grey stromatolite-bearing carbonaceous microcrystalline limestone and carbonaceous shale	Flat laminated bedding
			1	418 m	grey-black carbonaceous slate and phylitic slate with silty shale	Thin-bedded flat laminated bedding with a thickness of 1-5 mm
		Zenglongchang	2	180 m	green-grey crystalline dolomitic limestone and banded siliceous limestone containing stromatolite	medium to thick bedded
			1	113 m	grey metamorphic siltstone with carbonaceous dolomitic slate	medium to thin bedded horizontal bedding, lenticular bedding and wavy bedding
		Shuijigou	2	725 m	grey fine-grained quartzite and quartz sandstone with feldspathic quartzite intercalated with grey-green metamorphic andesite at the bottom	medium to small sized convergent and trough cross bedding, small-sized symmetric and asymmetric ripples; amygdaloidal structure in volcanics
			1	300 m	grey and dark grey pebbled feldspathic quartz sandstone, and iron-ore bearing conglomerate, sandy conglomerate and metamorphic tuff at the bottom	medium to large-sized convergent, planar, wedge-shaped and trough cross bedding with a thickness of 30-50 cm and asymmetric ripples on the layers
Upper Archean	Dongwufenzi				hornblende plagioclase gneiss, amphibolite, biotite plagioclase gneiss, biotite leptynite and magnetic complex	

form. The crustal movement that resulted in its inversion occurred 1400 Ma ago and is called the Se'erteng orogeny (Ma Xingyuan et al. 1989a). Subsequent marine transgression resulted in the deposition of the Shinagan Group. The Group is less than 2000 m thick and consists of an assemblage of quartz sandstone and slate intercalated with limestone. According to stromatolite-based studies and comparison of strata, the Shinagan Group approximately corresponds to the Gaoyuzhuang Formation and the Tieling Formation within the late Middle Proterozoic Jixian System. However, it is much thinner than the Jixian section sequences.

The Bayan Obo Group is well known for its extensive REE and iron deposits. It was divided into nine rock assemblages (H_1-H_9) and assigned to the Proterozoic in 1950s by No. 214 Geological Party of Ministry of Geology and Mineral Resources. It was classified into nine formations in an unpublished regional reconnaissance report in 1966. It overlies the metamorphic Late Archean Sanheming Group, (a correlative of the Dongwufenzi Group; Dong Shenbao et al. 1986) which consists partly of biotite plagioclase gneiss and amphibole plagioclase gneiss yielding metamorphic ages of 2500 Ma. The Group was considered to range in age from the Cambrian (Dulahala Formation) to the Silurian (Huhe'ailigeng Formation). Subsequent studies based on stromatolites and isotopes confirmed its age: isotopic dates range from Pb-Pb lead ages between 1350-1650 Ma, U-Pb isochron method ages of 1500 Ma and an Rb-Sr age of 750 Ma (Dong Shenbao et al. 1986). Thus the Bayan Obo Group is of Middle Proterozoic age and corresponds approximately to the Chartai Group. The age of 750 Ma indicates that part of the Group maybe of Upper Proterozoic age. The upper part of the Group consists of the Ardeng, Alehuduge and Huhe'ailigeng Formations in ascending order. They are not in contact with the lower part of the Bayan Obo Group but are located on the north of geosyncline-platform boundary. It is reasonable to suppose that they are a part of the same geosynclinal sedimentary succession. Part of these three formations may either be of Lower Paleozoic or Upper Proterozoic age.

In contrast to the Chartai Group, the Bayan Obo Group has a thickness of more than 10,000 metres and is 90% composed of clastic rocks and clay, together with minor carbonate rocks and volcanics (see Table 4.6 for details). There are slump structures formed by underwater landslides (Li Jiliang et al., 1981) and flysch deposits, which show that the Bayan Obo Group was deposited on the continental slope during a process of continuous sagging. Laterally the lithology and thickness of the strata change considerably especially in the lower part of the sequence, where it shows great changes over a short distance. This demonstrates that the paleogeographic environment was very complex. The tectonic setting of the Bayan Obo Group is demonstrably different from that of the Chartai Group: it developed on the continental slope along the craton margin (Fig. 4.8). The basin was not a symmetrical trough, but a geosyncline with one side below sea-level (Ma Xingyuan 1989). However, the Bayan Obo and Chartai Groups may be metamorphosed and folded by the same tectonic movement, for they are both metamorphosed to greenschist facies during regional low-temperature dynamic-metamorphism (Dong Shenbao et al. 1986), and constitute the basement which was consolidated in the North China Platform.

Erathem	Group	Formation	Thickness	Lithology	Primary sedimentary structures
Lower Proterozoic	Bayan Obo	Huhe'ailigeng	565 m	quartzite, slate, marl and siliceous limestone	medium-bedded horizontal bedding
Middle Proterozoic	Bayan Obo	Alehuduge	980 m	varicoloured phyllite and slate with minor quartzites	medium to thick bedded horizontal bedding
Middle Proterozoic	Bayan Obo	Ayadeng	395 m	stromatolite bearing siliceous limestone, crystalline limestone and dolomitic limestone	thick-bedded and massive
Middle Proterozoic	Bayan Obo	Huji'ertu	2448 m	slate, limestone, marl, quartzite and fine-grained sandstone with uralite epidotite	Upper part: thin flat bedding, bed thickness 2-5 mm. Lower part: convergent and trough cross bedding, mud cracks and symmetric and asymmetric ripples
Middle Proterozoic	Bayan Obo	Baiyinbaolage	2166 m	quartzite, quartz sandstone and sandy, silty and argillaceous slate with sandy limestone	major horizontal bedding with minor trough cross bedding, small sized flute cast and symmetric and asymmetric ripples
Middle Proterozoic	Bayan Obo	Bilute	2501 m	carbonaceous slate, siliceous slate and fine grained feldspathic quartz sandstone	horizontal laminated bedding, graded bedding, small-sized flute cast and symmetric and asymmetric ripples
Middle Proterozoic	Bayan Obo	Halahuoqiante	866 m	siliceous and sandy limestone, marl, calcareous slate, quartz sandstone and pebbled feldspathic quartz sandstone	small-sized herringbone and trough cross-bedding, convergent oblique bedding, stiring bedding, wavy bedding and horizontal bedding (i.e. flysch)
Middle Proterozoic	Bayan Obo	Jianshan	743 m	carbonaceous, siliceous and sandy slate, potassic slate, dolomite, quartzite and lens of limestone	horizontal bedding with a thickness of 1-2 cm and flyschoid formation
Middle Proterozoic	Bayan Obo	Dulahala	686 m	quartzite, fine grained quartz conglomerate, sandy conglomerate, pebbled coarse-grained feldspathic quartz sandstone	small-sized planar and trough oblique bedding and unperfect graded bedding
Upper Archean	Sanheming (Dongwufenzi) Group			biotite plagioclase-gneiss and amphibole-plagioclase gneiss	

Table 4.6. Division and sedimentary features of the Bayan Obo Group

The Bayan Obo aulacogen has a unique style of magmatic activity. Its volcanism can be divided into two stages. The early stage (Jianshan stage) volcanics are represented by alkali-ultrabasic rocks (i.e. carbonatite, and K-rich alkali-acid volcanics). There are also alkaline dykes consisting of ultrabasic rocks, gabbro, alkaline diabase, aegirinite, alkaline syenite, and carbonatite which are narrowly distributed within secondary faulted basins. The 300 m thick K-rich alkaline volcanics were derived from trachytic lava and tuff. The late stage E-W trending Huji'ertu stage rocks, represented by epidote amphibolite derived from an intermediate-basic volcanic protolith, were derived from fissure-type effusive eruptions. Finally, the formation of these alkaline magmatic rocks was followed during the post-geosynclinal stage by intrusion into the central part of the aulacogen of a layered basic-ultrabasic assemblage, comprising mostly gabbro, troctolite, diorite, peridotite (Wang Ji et al. 1987).

Petrochemically the alkaline dykes have high K, Na and higher Li_2O, Nb_2O_5 and BaO than normal igneous rocks. The REE pattern of the gabbro is a smooth and gentle curve without Eu depletion, which indicates that gabbro was the product of the fast emplacement of deep, primitive, undifferentiated magma. The carbonatite rocks have a dolomitic composition and appear to be of an enigmatic volcano-sedimentary origin. Some characteristics seem to suggest derivation from deep sources: for example, its occurrence was controlled by E-W trending structures; it contains lava flow structures, such as oriented columnar apatites, ferruginous schlieren and xenoliths of wall rock. The dolomite also contains a lot of alkali silicate, phosphatic and ferruginous minerals, and lesser amounts of barite, aegirine-augite, zircon, monazite, bastnaesite and spinel. Petrochemically it is higher in SiO_2, FeO, P_2O_5 and K_2O than ordinary carbonate rocks and rich in characteristic elements of carbonatite, such as Nb (Ta), Ce, Ti, Fe, Sr, Th, Ba and Zr. In addition, Sr, Nd, S, C and O isotope systematics indicate that it is derived from a deep source. Analysis of the dolomite and its apatite have $^{87}Sr/^{86}Sr$ values of 0.7030, 0.7041 and 0.7036, $d^{34}S$ values of -4 to +4 (from 131 samples) which form a tower-shaped graph distribution, ^{13}C values of -6.57‰-+0.36‰ and ^{18}O values of 8.28-19.39‰. Eschynite has an average $^{147}Sm/^{144}Nd$ value of 0.10475 and $^{143}Nd/^{144}Nd$ of 0.5115075, both less than the average values for the Earth and similar to that of continental basalt and alkaline basalt (Wang Ji et al. 1987; Chen Hui et al. 1987). However, the dolomite also has some sedimentary features. For example, it contains some terrigenous clasts, with sedimentary bedding and siliceous bands and concretions; it has microfossils and nannofossils of Thallophyta, paleospore and algae (Institute of Geochemistry, Academia Sinica 1988; Yang Ziyuan 1989). In summary, this layered (or lenticular) ore-bearing dolomite is neither simply sedimentary, nor purely volcano-sedimentary in origin. It is perhaps a kind of hot brine with mantle-derived chemistry (i.e. rich in Mg, Fe, Na, K, Ba, Sr, Nb, TR, F, P and CO_2). The brine may have risen up along a deep fracture, entered the faulted sea basin, then mixed with sea water of the basin and was eventually deposited to form the dolomite (Chen Hui et al. 1987) and a gigantic Nb-REE-Fe deposit. It was the intensive faulting in the Bayan Obo aulacogen that provided a favourable tectonic environment for alkaline magma activity.

4.2.5 Depression Belts in the Western North China Protoplatform

In the mountainous region around the Yinchuan City in the western North China Protoplatform, the Middle Proterozoic Huangqikou Group and disconformably overlying Wangquankou Group outcrop sporadically (Liao Huarui, 1989). They unconformably overlie the Lower Proterozoic Zhaochigou Group due to the absence of the Qingbaikou Group and are covered by the Upper Sinian Zhengmuguan Formation. The Huangqikou Group corresponds to the Changcheng System and largely consists of quartzite, quartzitic sandstone sandwiched with slate and dolomite. The Wangquankou Group is equivalent to the Jixian System, and consists of a pile of thick dolomite with cherty bands, siliceous limestone with quartzite, shale and stromatolite-bearing dolomite. The bottom of the Group is made up of glauconite sandstone which yields an age of 1291 Ma, equivalent to the Middle Jixian System (Yang Zhende et al. 1988). Seismic survey and deep drilling reveal an over 3000 m thick basement of the Middle Proterozoic age under the widespread cover of Mesozoic and Cenozoic age. The Proterozoic sequence mainly comprises quartzite, andesite, siltstone, shale and dolomite which sits unconformably on the Archean gneiss. This Proterozoic strata can be correlated with the sequence exposed around Yinchuan and hence is of Changcheng-Jixian period age (Zhang Kang et al. 1981). The two combine to make up the depression belt in the western North China Protoplatform (II-7 in Fig. 4.1).

The Beishan depression stretches E-W and is bounded on its north and south sides by an old massif. Inside the depression the Middle Proterozoic Baihu Group, which forms part of the Changcheng system. The Group consists of a sequence of medium to low grade metamorphic facies rocks derived from littoral and neritic protoliths, with an exposed thickness of 3269 m. The sequence constitutes a complete cycle, consisting of terrigenous clastics, clay and carbonate rocks, overlain by intercalated basic-intermediate volcanics and tuff. The late Middle Proterozoic Jixian System Pingtoushan Group is composed of dolomite overlain by limestone bearing abundant stromatolites, with a thickness of 3075 m. The Late Proterozoic Qingbaikou System Dahuoluoshan Group is also dominated by carbonate rock, intercalated with minor clastic and siliceous rocks totalling 1575 m in thickness. Both the Pingtoushan and the Dahuoluoshan Groups are unmetamorphosed and preserve intact sedimentary textures. Various typical carbonate intraclastic textures, large amounts of stromatolites, tidal bedding, "l" shaped crossbedding as well as birdseye and crack structures collectively identify a stable sedimentary environment strongly influenced by tidal activity (Zhao Xiangsheng et al., 1984).

In summary, depressions occurred in the western North China Protoplatform in Mid-Late Proterozoic time. Those developing earlier had strong topographic contrast and consisted of coarse clastics, with some intermediate-basic eruptives in the intensely downwarped sectors. Later ones have a decreased intensity and are composed of fine-grained clastics and stromatolite-bearing carbonate rock. The rock assemblage in the depressions is made up of quartzite, shale and carbonate rock types, which make up a stable sedimentary cover.

4.2.6 Gravity Tectonics in the Wulashan Group

Gravity tectonics implies that the process of rock deformation occurs primarily under the influence of gravity. The gravity tectonics in the Late Proterozoic Wufoshan Group, Songshan region, is a typical example (see Ma Xingyuan et al., 1981b). The complete sedimentary sequence of the Wufoshan Group (see Table 4.3) begins with a basal conglomerate and is followed respectively by the quartz sandstone of the Ma'anshan Formation (Wm) and thick-bedded limestone of the Hongling Formation (Wh). The cycle includes three sub-cycles: (a) the Ma'anshan Formation to the Puyu Formation (Wp); (b) from the Luotuopan Formation (Wl) to the Heyao Formation (Wh) and (c) from the bottom to the top of the Hongling Formation respectively. A hiatus in deposition occurs between each subcycle. Moreover there are obvious rhythmic and a variety of synsedimentary structures (such as flexure and convolution structures) within each cycle. All these features indicate the mobility of the crust during sedimentation. In fact, the Wufoshan Group was deposited during the remobilization of the basement. The complicated gravity tectonics in the Wufoshan Group was induced by the tilt of the basement block in the southern part of this region, resulting in the transport of the whole rock mass along prominent gliding structures. For example, the plan distribution of the Wufoshan gravity gliding structure presents a southwards convex Wufoshan main arc and small arcs in the east and west end of the structure. It has a length of about 23 km and a width of 5 km, consists of an underlying system, a main lubricating layer, a slip surface and a gliding system (Fig. 4.9). Slip breccias and sandstone dykes also occur.

The underlying system of the glide structure comprises basement rocks, including the Dengfeng Group (Ard) and the Songshan Group (Pt$_2$'s),and the Ma'anshan Formation of the Wufoshan Group. The Songshan Group has a series of inclined and overturned linear folds which possess NNE-SSW or nearly N-S trending axial lines and west-dipping axial planes. The Ma'anshan Formation, sitting unconformably on the basement rocks, makes up a large flexural belt which is generally a north-dipping, smooth and undulating regional monocline_(Fig. 4.9). Deposition of the Wufoshan Group was controlled by a series of E-W trending normal faults. These faults also cut through the strata of the Ma'anshan Formation in its late stage and controlled the development of the gravity gliding structure. In addition, the approximately N-S-trending axial gentle warping and cross-fault were superimposed on the downwarping fold zone. These features are the result of the inhomogeneity of basement flexure and tear during the gliding process.

The main lubricating layer of the glide system is the Puyu Formation. Sometimes it is a part of the gliding system and the boundary plane between it and underlying system is the main slip plane. In some other regions it was combined with the underlying system to form a uniform body and its ceiling is the main slip plane. In still more cases, both its upper and lower boundary planes are slip planes of which the upper one is more important. The presence and absence of the Puyu Formation is determined by the separation and convergence of the upper and lower slip planes. In general, the formation was intensively destroyed at the top of the arc and well preserved in the two wings of the glide outcrop. The synsedimentary slump structures (such as flexure and convolution structures) are well-developed and formed a large-sized dome in the sliding process (Fig. 4.9). This resulted from the sliding and lifting of a hard layer of the underlying Ma'anshan Formation. Due to its use as a tectonic deroofing zone, the Puyu Formation has a lot of small-sized structures and cleavage owing to the movement on the upper and lower main slip planes and along its inner shearing plane. In places it was scraped and squeezed untill it disappeared or formed isolated lumps.

Generally the upper slip plane of the Puyu Formation is more important and extends over the whole area, whereas the lower slip plane is not developed at the reflex arc on the two wings of main arc. It has obscure fracture trace and is evident only when the upper lubricating layer converges with it or cuts down to the Ma'anshan Formation. The main lubricating layer shows a bend like a sine curve in plan view. The main slip plane, generally with a northward inclination, is steep in the upper and gentle in the lower part of the Puyu Formation (Fig. 4.9). The main slip plane developed mainly along the bedding, i.e. the lower main slip plane occurred along the ceiling of the Ma'anshan Formation and the upper main slip plane along the bottom of the Luotuopan Formation. The two walls of each slip plane rubbed each other in the sliding process to produce excavation and scraping; for example, the upper main slip plane often dug and cut the bedding of the main lubricating layer, the Puyu Formation, whereas the lower main slip plane scraped the top of the strata of the Ma'anshan Formation. The structures in the two walls of the main slip plane are very discordant. The upper wall appears to have had a different slip process from the lower part, and there are diverse slipping directions in different places. Hence the associated structures in the two parts of the slip plane (such as cleavage, small fold, etc.) are not uniform. Because the slip plane is bent, the relative displacements and fault structure of two walls are not identical. In cross-section taking in the plane of the sliding direction, the fault is spoon-shaped: at the back of the slide the upper hangingwall slides down, whereas the front of slide constitutes a horizontal "reverse slide" (i.e. thrust fault). These slides are normal faulted, reverse faulted and thrust faulted respectively according to the relative displacements of the upper and the lower walls. Where an overturned fold in the back part of the slip plane, a thrust fault occur in the underlying system (Fig. 4.10). In addition to the main slip plane, there are also subordinate slip planes among the slipped masses, the blocks and the sheets. Some of them, together with main compression-shear mechanic features, were often generated in connection with the folding of the sliding system. Hence, the slide faults, differing from ordinary normal and reverse faults, have a special mobile style, showing a unique nature and constitute an unusual type of fault.

The slide system consists mainly of the Luotuopan, Heyao and Hongling Formations above the main slip plane. It is a tectonic unit including different orders of slip masses, slip blocks and slip sheets. In the Yanshi area, it comprises slip masses I, II, III and IV., together with several slip blocks and slip sheets which are the product of cutting by longitudinal faults parallel to the arcuate zone of the glide system. Transverse faults orthogonal to the arcuate system also occur. Most of them result from the tearing caused by the differential sliding during the development process of the

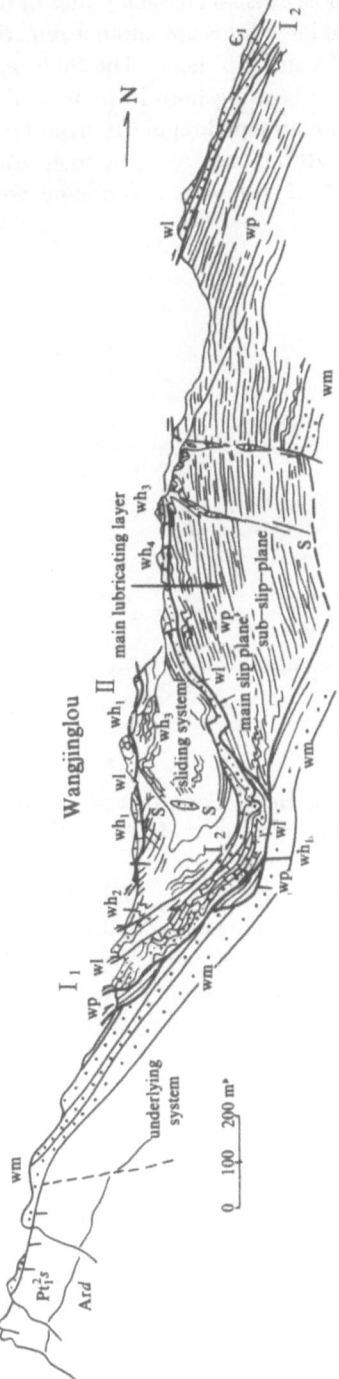

Fig. 4.9. Structures of the Wufoshan Group on the eastern slope of Wangjinglou hill in Yanshi, Henan (after Ma Xingyuan et al. 1981): The main slip plane is very clear. The main lubricating layer was squeezed leaving only a remnant at its root and formed a dome in north Wangjinglou. the Luotuopan Formation sandy conglomerate entered the fissures as sandflow to produce sandstone dykes and wedges. I_2-slip mass is a down-facing recumbent syncline. There are three klippes of II-slip mass on the ridge

glide system. This is one of the most evident characteristics of the sliding structures in this area. Meanwhile small and large folds are another remarkable feature developed in the slip masses, slip blocks and slip sheets. The folds on the arcuate zone of the sliding structure have obvious banding both in plane and in section. Various styles of folds in cross-sections are arranged regularly from the back (south) to the front (north) of the arcuate zone. At the back of the arcuate zone, there are sliding overturned folds, superimposed fold structures comprising down-facing folds re-

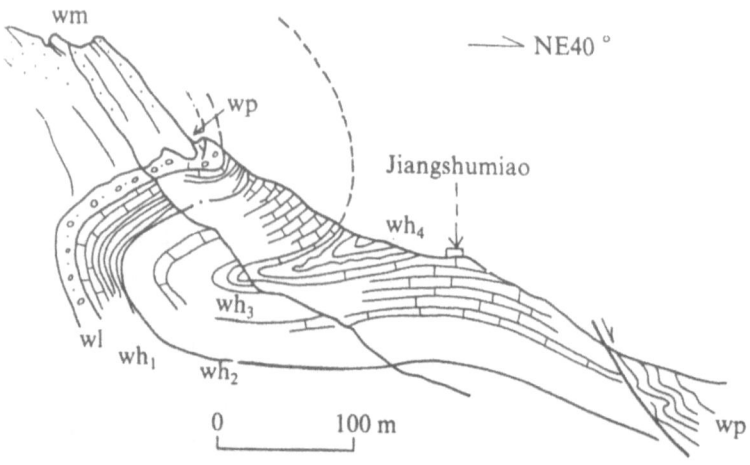

Fig. 4.10. The Recumbent folds in the Heyao Formation of the Wufoshan Group gliding system south of Jiangshumiao, Yanshi, Henan (after Ma Xingyuan et al. 1981b).

sulting from plastic flow, down-facing recumbent folds (Fig. 4.8) and down-facing overturned folds; in the middle part of the zone, continuous normal-inclined folds appear, whereas at the front of the arc, compression results in folds which are mainly inclined and overturned with horizontal to upright or even south-overturned hinges. However, the front part of the slide system has only a small outcrop because of the unconformably overlying Cambrian System.

Slip breccia is also a part of the Wufoshan sliding structure. It is generated in connection with the whole gliding structure. It consists of complicated rock masses and pebbles of different compositions, sizes and shapes, mixed with pelitic, calcareous and clastic materials. The brecciated rocks are predominantly derived from green-grey and yellow fine-grained feldspathic quartz sandstone and rose thick-bedded stromatolite-bearing limestone of the Hongling Formation of the Wufoshan Group. Of secondary importance are slaty pelitic limestones of the Heyao Formation, pebbled quartz sandstones of the Luotuopan Formation and yellow-green and purple shales of the Puyu Formation (Fig. 4.11). They have different shapes and sizes, with a maximum length of 500 m, thickness of more than 30 m and a minimum volume of 3 x 2 cm^2. The large-sized gravels possess prominent rectangular, thickly tabular and irregular shapes, and very well preserved bedding, sedimentary structures and stromatolites. The components of interstitial material and cements in the slip breccia are also intricate: some of them are clay and calcareous materials, and other tiny segments of limestone, quartz sandstone and shale. The slip breccia, with compositionally complex gravels and matrix, is dissimilar to general fault breccias. The gravels and matrix came from the overlying Hongling Formation and the underlying Heyao Formation. Hence, the slip breccia is different from mélange and flysch: - it is a kind of special cataclasite resulted from gliding structure.

Sandstone dykes often occur in, and constitute an important part of, the Wufoshan gliding structure. They are found in the underlying system, the lubricating system and the slide system, and predominantly in sandy shale of the Puyu Formation. In the Wufoshan arc, there are more sandstone dykes on the top of the main arc and its two sides, and fewer in the two wings of the main arc. Similarly, more occur in the back part when compared to the front part of the N-S-trending section, which exhibits fissures produced during formation of the gravity gliding structure. The sandstone dykes have intricate shapes, such as slaty, wedge, pyramidal and irregular shapes (Fig. 4.12), controlled by the shapes and attitudes of the fissures and the lithology of the wall rocks. The dykes are pebbled or pebble-free coarse-grained quartz sandstones with textures of sedimentary clastic rocks, massive structures with usually no bedding. Their petrography demonstrates that they are identical with pebbley coarse-grained quartz sandstones of the Luotuopan Formation of the Wufoshan Group, and they have a close relationship with the latter in a number of places. The formation of the sandstone dykes is closely related to the development of the Wufoshan gliding structure. The space occupied by sandstone dykes is provided by faults and joints resulted from the mobilization of gliding rock mass under the influence of gravity (Fig. 4.8). The make-up of the dykes is the result of the capture of a large amount of water in the pebbled sandstone of the Luotuopan Formation which lies on the impermeable Puyu Formation. Under the influence of movement of gliding structure, the pebbled sandstone was partially liquefied to form water-saturated sandflows. Because the sand mass is affected by contact deformity, the more intense the tectonic movement is, the stronger the activity of sand and the greater the pressures. This results in greater ease of formation for the various types of sandstone dykes. This process is also indicated by sandstone dykes developed on the top of the main arc and in small arcs on each side, and by undeveloped ones in the two wings of the main arc. Lastly we should point out that sandstone dykes produced by the Wufoshan gravity gliding structure demonstrate the presence of water and the "buoyancy effect" of pore fluid pressure that are a prerequisite for the development of the gentle plane of the sliding structure and the movement of the hanging wall rock mass.

In summary, the hanging wall and footwall of the main slip plane have remarkably disharmonic structures in the Wufoshan gravity gliding structure. There are simple structures in the footwall (underlying system) and very complex ones in the hanging wall (slide system). The hangingwall system is an Alpine-type structure comprising a series of secondary slip planes and inclined overturned folds. Large- and small-sized closed recumbent folds are developed and their overturned limbs do

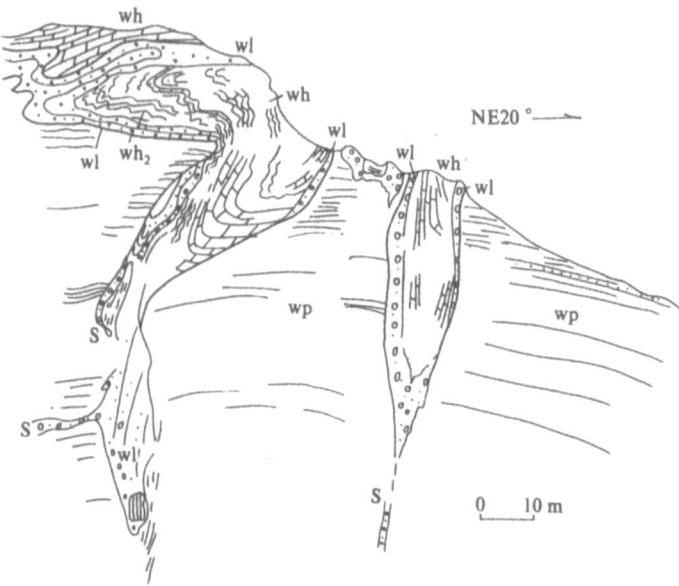

Fig. 4.11. Large-size rock and clastic wedges in the lubricating puyu Formation in Wangjinglu, Yanshi, Henan (after Ma Xingyuan et al. 1981b)

not become thinner. Such structures are characteristic of gravity glides. The occurrence of thick-bedded slip breccia, the development of hundreds of sandstone dykes as well as the disintegration, migration and modal expansion of slip mass, slip block and slip sheet all exhibit the significance of gravitational instability and the buoyancy effect of pore fluid pressure in the development of the Wufoshan arcuate tectonic zone. Gravity tectonics allows the cover succession to be deformed without any relationship to the basement. Also the intense structural deformation in some regions may be the result of slow epiorogenesis and uplifting and thus this movement does not conform to a known orogenic phase. Hence, a study of partial sliding structures resulting from gravity gliding provides a tectonic analysis which can well explain various structural phenomena and aid the evolutionary history of regional geological structure.

4.3 Tarim Protoplatform and Its Surrounding Mobile Belts

The Tarim Protoplatform is another old massif which developed in isolation in Mid-Late Proterozoic times. It covers the present Tarim and Qaidam Basins and the peripheral mountains where the Precambrian rocks are largely exposed. During the Mid-Late Proterozoic a series of taphrogenic troughs were formed on the protoplatform and a variety of sediments were deposited within them. On the margin a set of eugeoclinal type sediments were developed and turned into a Late Proterozoic collision belt (Fig.4.2).

In the surrounding mountains, including the Central Tianshan Mts., the Kuruktag region to the north, the Kunlun Mts. to the southwest, the Altun Mts. to the southeast and the Qilian Mts. to the east, there are widespread Mid-Late Proterozoic rocks developed in the taphrogenic troughs of all dictions, making up a remarkably complete sedimentary sequence ranging from the Changcheng through to the Qingbaikou System sucessions (Xinjiang Bureau of Geology and Mineral Resources, 1992; Institute of Geology of Gansu Province, Gansu Bureau of Geology and Mineral Resources, 1986). The taphrogenic trough in the Altun area, which was probably linked to the Qilian Mts. trough in the east, divided the crystalline basement of the Tarim Protoplatform into two massifs, i.e. - the Qaidam and the Tarim massifs. The Changcheng System is a thick transgressive sequence and the product of early taphrogenesis of the protoplatform. Two types of stable sedimentary deposits occur. One type is represented by the Bowam Group in the Kuruktag region, the Tianshuihai Group in the Kunlun Mts, the Tekes Group in the western section of the Middle Tianshan Mts. and the Bashikurgan Group in the Altun Mts. These constitute a series of neritic-littoral clastic-carbonate deposits. Lithologically they consist of meta-sandstone, graywacke, arkose, siltstone, quartzite and phyllite, with a small amount of basic, intermediate and acid volcanics and pyroclastics, reflecting weak volcanic activity. The other type of stable deposit is featured by well developed volcanics and volcanic sediments, consisting of spilite-keratophyre suite or intermediate-acid and intermediate-basic volcanics and pyroclastics. These types of rocks occur within the Sailajiazitag Group in the Kunlun Mts., the Xingxingxia Group in the Middle Tianshan Mts. and within the Julongguan and the Xinglong Groups in the Qilian Mts. U-Pb dating by Hu Aiqin et al. (1986) reported an age for the bottom of the Xingxingxia Group as 1900 Ma and an age limit for the top of the group of 1.4 Ga.

The Jixian and Changcheng Systems are commonly denoted by continuous sedimentation. A larger scale marine transgression occurred during the Jixianian, resulting in a more widespread sediments, such as the Kawalark and Liksu Groups in the central Tianshan Mountains, the Erqigan Group in the Kuruktag area, the Eramas and Langya Groups in the Kunlun Mountains, the Taladapan Group in the Altun Mountains and the Jingtieshan Group in the Qilian Mountains. During the continuous subsidence in the Jixian Period, huge thicknesses of neritic-littoral facies carbonate rocks were accumulated, partly deposited in a semi-closed bay. The rocks are mainly limestone, dolostone, marble, and sparse sandstone and siltstone, with scattered volcanics. The carbonate rocks bear large amount of stromatolites. Such features reflect a continuous rifting environment which gradually became more stable with time. Qingbaikouian times were characterised by a rather stable sedimentary environment which is best shown by the Palgantag Group in the Kuruktag area, where a set of littoral-neritic deposits were deposited in the Jixianian sedimentary basin. The lower part of the sucession consists of clastic rocks which give way upwards to carbonate rocks. The Qingbaikou System is roughly similar in other areas, but each local sequence has its own specialities. For example, the Kuxtay Group and Kaltax Group in the western Tianshan Mts. are dominated by carbonates with minor clastics; the Tianhu Group in the eastern Central Tianshan Mts. is a medium to high metamorphic sequence, ranging in age between 1000-660 Ma according to U-Pb age dating by Hu Aiqin et al. (1986). The ages are correspondent to those reported in "Precambrian in Northern Xinjiang and Its Ore Bearing Nature" (1990) and to the age scale set up in that region. The area covering the Altun Mts., the western Tianshan and the Qilian Mts. were uplifted into a land area at the termination of Jixianian times. Subsequent local subsidence resulted in the deposition of neritic sediments. Moreover, the Gongcha Group is largely composed of stable clastics and carbonate rocks with several interbeds of gypsum, suggesting a dying rift environment. Rifting finally ceased at the end of Qingbaikouan times. The Tarim orogeny (corresponding to the Jinning orogeny) created a widespread unconformity on top of the Qingbaikou System. Subsequently, light Late Proterozoic metamorphism resulted in the formation of low temperature dynamic greenschist facies rocks. K-Ar dating of Mid-Upper Proterozoic metabasalt yielded a whole rock age of 715 Ma, reflecting the age of Late Proterozoic metamorphic overprinting (Dong Shenbao et al., 1986). The extention of the metamorphic belts are coincident with the structural orientation.

The Mid-Upper Proterozoic rocks which outcrop on the south and north margins of the Qaidam massif suggests that rifting occurred mainly in the Middle Proterozoic Jixian Period. The Binggou Group of the Jixian System consists of dolomite and stromatolite-bearing siliceous dolomite, intercalated with limestone, chlorite-phyllite, chlorite-schist, siliceous slate and metasandstone. The protoliths are magnesian carbonate rocks with minor pelitic, taffaceous and calcaceous rocks. During the Jinningian they underwent a low-greenschist facies regional low-temperature dynamic metamorphism (Dong Shenbao et al. 1986).

In summary, although only fault contact between the Mid-Upper Proterozoic and its underlying strata can be seen in almost the whole Tarim Protoplatform, the Changcheng System Bowam Group unconformably covers the Lower Proterozoic Xingditag Group in the Kuruktag area. A basal conglomarate is present in some other areas. Moreover the presence of the Archean strata also demonstrates that the cratonic basement of the Tarim Protoplatform had been formed by the end of Early Proterozoic time. In the cratonic basement which is distributed in the Kuruktag and the Altun areas, diabase dykes with a length of over one hundred kilometres frequently occur. During the Mid-Late Proterozoic, the Changcheng System was controlled by rifting to form a thick flysch assemblage comprising pelitic material and volcanic clastics, and the overlying Jixian and Qingbaikou Systems were deposited by a quartzite-carbonate rock assemblage. In some areas interruption of sedimentation is apparent,with features denoting the rise and fall of sea level observed in the Mid-Upper Proterozoic formations. This is characteristic of cover rock deposited on continental crust. In contrast, there is a unconformity between the Mid-Upper Proterozoic and its overlying strata in the Tarim Protoplatform, illustrating the existence of two folded basements of Early Proterozoic and the Mid-Late Proterozoic age. In the Kunlun and the Altun Mountains, angular unconformities can be seen in the interior of the Mid-Upper Proterozoic, denoting the existence of multiple folded basements. So, compared with the North China Protoplatform, the Tarim Protoplatform shows a stronger degree of tectonic activity.

The sediments in the marginal mobile belt of the Tarim Protoplatform can be observed locally. The Arksu Group, distributed in the northwestern margin of the

Tarim Protoplatform, consists of sericite-chlorite-quartz schist, epidote-chlorite-quartz schist, tin-bedded quartzite, epidosite and glaucophane-schist. It lies underneath quartzite, epidosite and glaucophane-schist of Sinian age and, based on regional correlation, probably belongs to the Late Proterozoic. To the east, this suite of strata is widely distributed along the southern Tianshan Mountains. The protoliths are basic tuff, silicalcite and flysch formation, representing a thicker oceanic crust. Although accompanied upper mantle-derived rocks are not seen, it is a typical eugeosyncline formation, representing the extension of continental crust and the occurrence of an oceanic basin in Mid-Late Proterozoic times (Xiao Xuchang et al. 1990). In the middle Kunlun Mountains on the south margin of the Tarim Protoplatform, the newly established Late Precambrian Wanbagou Group comprises intermediate-basic volcanics which fall into the tholeiite area in the Na_2O+K_2O versus Na_2O+K_2O diagram. The Group may have formed in a back-arc basin type environment in Mid-Late Proterozoic time (Zhu ZhiZhi et al. 1985). A mobile belt with eugeosynclinal features may have existed in the northern and southern margins of the Tarim Protoplatform. Huang Jiqing et al. (1990) called the eugeosyncline on the north margin the South Tianshan Ocean or South Tianshan eugeosyncline. Similarly he termed the eugeosyncline on the south margin the Kunlun Ocean. He suggested that there are two eugeosynclinal belts across the whole area. At the end of Proterozoic times the South Tianshan Ocean and the Kunlun Ocean were subducted. The melange belt and the blueschist facies high-pressure metamorphic belt, distributed along the suture, recorded the nature and scale of the plate convergence. The blueschist belt has the isotopic ages of 729 Ma, 634 Ma and 539±25 Ma, which reflect the time of tectonic movement. Different opinions exist concerning the tectonic significance of the blueschist: for example Dong Shengbao (1986; unpublished) thought that the blueschist belonged to a glaucophane-schist belt within a craton environment.

4.4 Tectonic Regime of the Yangtze Protoplatform and Cathaysia

A mobile type tectonic environment existed in Mid-Late Proterozoic times in South China, represented by the development of a trench-arc-basin system around the Yangtze Protoplatform and characterized by the collision between the Yangtze Protoplatform and Cathysia (Fig. 4.1). These features reflect a complex tectonic evolution and form a specific tectonic pattern. But the general trend of evolution in this period is that the continental crust gradually increased in extent and tectonic activity became weaker and weaker. After the Jinning movement, much of the area entered a relatively stable stage of tectonic development and characterized by the conclusion of the third shield-forming stage of the Precambrian continental evolution (Table 4.7).

4.4.1 The Trench-Arc-Basin System around the Yangtze Protoplatform

Western Margin of the Yangtze Protoplatform: In the Middle-Late Proterozoic, two different types of mobile - the eugeocline belt marked by the Yanshan Group and the miogeocline belt marked by the Huili Group - developed on the Late Archean-Early Proterozoic crystalline basement at the western margin of the Yangtze Protoplatform in the central Yunnan-western Sichuan area (Fig. 4.12).

In the Yanbian Group, the eugeoclinal rock association is comparatively well preserved and recorded (Li Jiliang 1984; Zhang Yunxiang et al. 1988; Cong Bailin. 1988). According to Li Jiliang (1984), the ophiolite suite in the lower part of the Group consists (in ascending order) of cumulus complexes, diabase dykes, and pillow lavas with silicalcite and tuff. Disconformably overlying the ophiolite is a pre-flysch black sequence, followed by a relatively thick flysch (Fig. 4.13). The whole-rock Rb-Sr isochron age of the pillow lava and gabbro samples is 1106±58.5 Ma (Li Jiliang 1984); the isotopic age of the gabbro and ultrabasic bodies emplaced in the Yanbian Group is 1253 Ma and 1112 Ma respectively (Dong Shenbao et al. 1986). Therefore the Yanbian Group is considered to be a product of late Middle Proterozoic times. The Group is unconformably overlain by the molasse of the Sinian Lieguliu Formation.

The lower portion of this ophiolite suite, which is not completely preserved, is in fault contact with the Late Archean / Early Proterozoic crystalline basement. The lowermost portion comprises cumulus complexes of different sizes, which have an ultramafic core surrounded by ringed mafic rocks and consist mainly of harzburgite, lherzolite, plagioclase peridotite, troctolite, anorthosite, norite and gabbro. These rocks are gradational with each other and form a clear sequence, indicating roughly continuous crystallization. Sheet diabase occurs in the layered gabbro at the top of the cumulus complex as well as in the overlying basaltic lavas, suggesting multiple emplacements of diabase. The basaltic lava is massive at the base but pillow-like in the middle and upper parts of the sequence, with ratios of Al_2O_3/TiO_2 and CaO/TiO_2, and concentrations of Y and Zr and REE similar to those of the basaltic lavas of the Troodos and island-bay ophiolites as well as to some basalts of the modern mid-oceanic ridges (Li Jiliang, 1984). This sequence structures comprise alternating laminae of carbonaceous components derived from organic matter, hematite-limonite, clay and volcanic ash interbedded with siliceous laminae. Tuffaceous chert beds of turbidity current origin also locally occur and they are apparently related to the episodic generation of suspension currents.

The pre-flysch association of the Yanbian Group, tens of metres thick, consists of thin, dark grey tuffaceous slate which disconformably overlies andesitic-basaltic volcano-clastics at the top of pillow lavas. This indicates the characteristics of pre-flysch sediments formed in a slow, non-compensational, abyssal-type sedimentation environment with little source material. A small hiatus occurred between deposition of this sequence and the flysch material.

The upper part of the Yanbian Group is a flysch sequence about 5000 m in thickness, consisting of two divisions. The lower division is a volcano-clastic flysch characterized by volcanic and argillaceous components. The upper layer is composed of

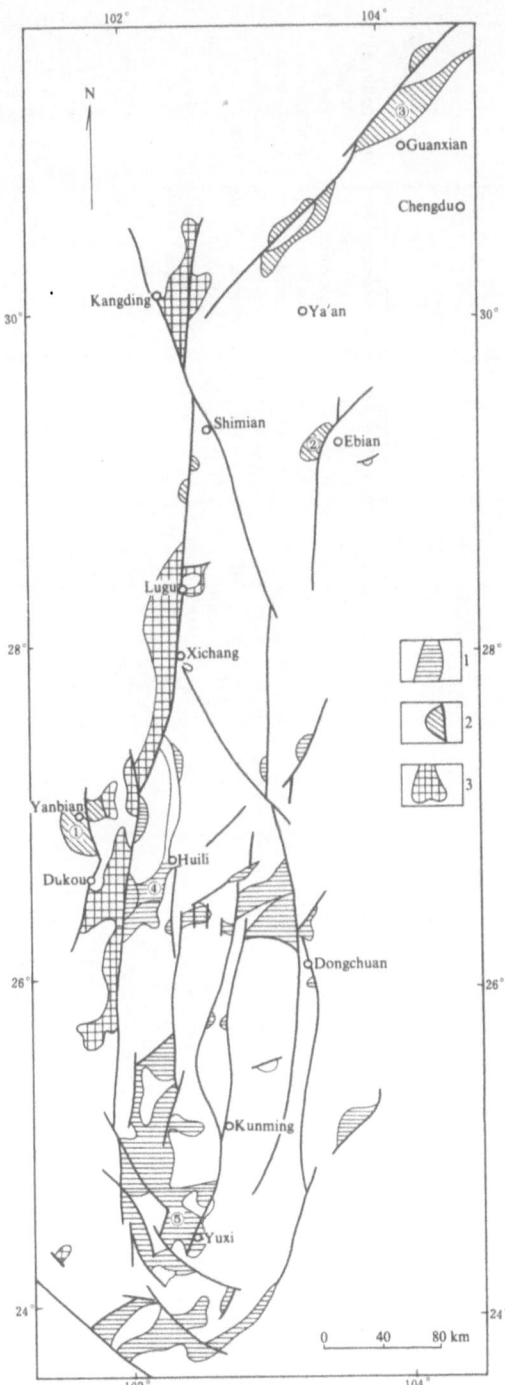

Fig. 4.12. Distribution of the pre-Sinian rock along the western margin of the Yangzi Protoplatform (Modified after Chen Zhiliang et al., 1987) 1. Pt$_2$ miogeoclinal sediments; √ Huili Group; ∞ Kunyag Group; 2. Pt$_2$ eugeoclinal; ¨ Yanbian Group, ≠ Ebian Group, ¬Huangshuihe Group; 3. Ar-Pt$_1$ crystalline basement

Stratigraphic unit			West margin of Yangtze platform		North margin of Yangtze platform					
			Outer margin	Inner margin	West Segment			Middle Segment		East Segment
					Outer margin		Inner margin	Outer margin	Inner margin	Outer margin
Upper Proterozoic	Sinian	Qingbaikouan System	Lieguliu Fm.	Lieguliu Fm.	Natuo Fm.	Dashigou Fm.	Guanyinya Fm.	?	Liantuo Fm.	Zhougang Fm.
			(shaded)	Wusidaqiao Gr. Bimodal volcanic terrigenous clastic rock association	(shaded)	Sunjahe Fm. Island-arc calc-alkaline volcano-clastic rock association	Tiechuanshan Fm. Bimodal volcanic rocks	?		
Lower Proterozoic		Jixian System	Yanbio Gr. Ebian Gr. Huangshuihe Gr. Ophiolite-flysch association in eugeosyncline belt	Huili Gr. Kunyang Gr. Association of sand-shale, black shale, stromatolite-bearing carbonates and bimodal volcanic rocks in miogeosynclinal belt	Bikou Gr. Ophiolite-flysch association in eugeo-synclinal belt	Xixiang Gr. Calc-alkaline volcano-clastic rock association in island-arc belt	Huodiya Gr. Shangliang Fm. Mawozi Fm. Association of sandy-pelitic rocks and stromatolite bearing carbonate rock	Wudang Gr. Suixian Gr. Hong'an Gr. Ophiolite-flysch association in eugeosynclinal belt	Shennongjia Gr. Dagushi Gr. Association of sandy-pelitic rocks and stromatolite bearing carbonate rocks in miogeosynclinal belt	Zhangbaling Gr. Association of spillite keratophyre, sandy-pelitic clastics and carbonate rocks in eugeosynclinal belt
Lower Proterozoic		Changcheng System								

Table 4.7. Classification and correlation of Middle-Upper Proterozoic (pre-Sinian).

terrigenous clasts; the detritus accounts for 35-50% of the total clasts, the rest are crystal fragments of which quartz amounts to 25-35%, feldspar 15-29%; and clayey matrix is 15-30% of the rock. Analysis of grain-size distribution, suggests an origin by the suspended transport of material. In general, during sedimentation of the upper part of the Yanbian Group, the basin was a widened bathyal-abyssal trough in which volcanic tuffaceous and argillic-sandy rocks were deposited. The detritus is dominated by acidic-volcanic components with less contourite (Zhang Yunxiang et al. 1988). It is characterized by typical flysch rhythms, and is therefore considered to result from turbidite flow. The Yanbian Group (and its equivalent sequences) have generally undergone a regional low-temperature dynamic-metamorphism to greenschist facies. This Group on the whole occurs as a folded sequence dipping steeply to NW. It strikes NE and is exposed for nearly 50 km.

The Ebian Group is equvialent to the Yanbian Group in age and has a similar sedimentary sequence. Its lower part is a metamorphic basalt, characterized by massive, porphyritic, and vesicular amygdaloidal structures; it is intercalated with black slate, metabasaltic and andesitic tuffs. Its middle-upper part is mainly argillic-sandy rocks, showing a graded rhythmic layering and containing intermediate-basin volcanic and carbonate rocks. Furthermore, in the Shimian area west of the Ebian Group are asbestos deposits, hosted in metamorphic peridotites. Analysis of the major elements, transitional metallic elements and REE indicates that the peridotites are residual-mantle rocks, strongly depleted in incompatible elements. The gabbro is in fault contact with the peridotites and is quite similar to the cumulus gabbro of the Yanbian Group ophiolite in petrological and chemical composition. In addition, the peridotites are unconformably overlain by the Sinian Kaijianqiao Formation, suggesting that they are contemporaneous to the Yangbiao Group ophiolite. They appear to be the products of an eugeosyncline on the west margin of the Yangtze Protoplatform. Therefore the area extending from Shimian to Xichang and to Yanbian is most likely to represent a Proterozoic ophiolite belt (Li Jiliang 1984).

Another sequence equivalent to the Yanbian Group is the Huangshuihe Group. Its lower horizons are composed of metamorphic basic volcanic rock (spilite) and dacite, whereas the middle to upper part of the Group consists of schist with acidic volcanic rock and graded rhythmic structure. Finally the uppermost part of the Group consists of carbonate rock. Therefore this succession is similar to the Yanbian Group as its characteristic sedimentary formation is also the product of an eugeocline on the west margin of the Yangtze Protoplatform.

The miogeocline belt is distributed east of the eugeoclinal belt and its sedimentary sequence consists of the Huili Group in Sichuan and the Kunyang Group in Yunnan. These two groups are essentially identical in lithology, sedimentary formation and ore-bearing potential as well as in stratigraphic subdivisions (Yang Xianhe and Zhang Honggang, 1984). The Huili (Kunyang) Group is overlain unconformably by strata. of Sinian age. Whole rock Rb-Sr dating of the dacite at the top of the Huili Group is dated as 907.6 ± 18.5 Ma (Liu Hongyun et al. 1981) and the upper part of the Kunyang Group is dated at 922, 933, 948 and 1002 Ma (Wu Maode, 1988; Duan Jinsun, 1988) by the same method. The age of the lower part of the Kunyang Group is 1736-1794 Ma (U-Pb age determined by Dong Shenbao et al. 1986) and the sequence

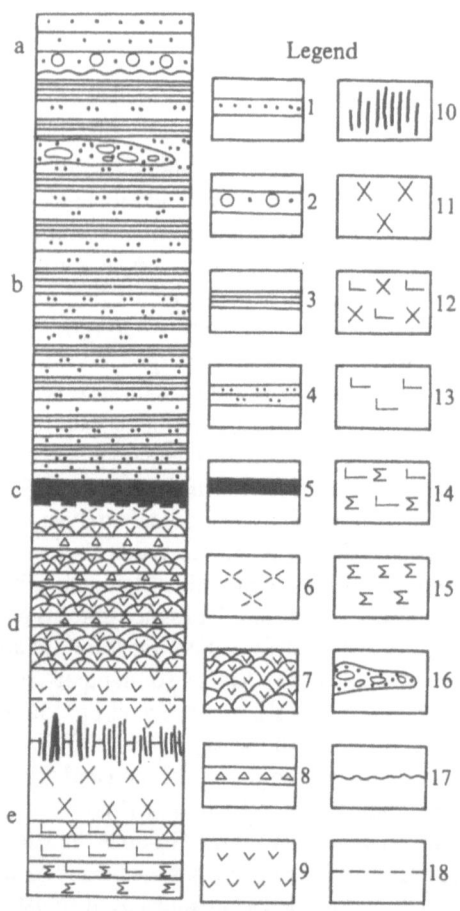

Fig. 4.13. Rocks sequence of the Yanbian Group, west Sichuan (after Li Jiliang 1984). a. Molasse; b. flysch; c. pre-flysch; d. lava of ophiolite; e. cumulate complex of ophiolite: 1. Sandstone; 2. gravel sandstone; 3. shale or slate; 4. siltstone; 5. pre-flysch carbonaceous slate; 6. siliceous rock; 7. pillow lava; 8. tuff; 9. massive lava; 10. diabase dyke; 11. gabbro; 12. norite; 13. anorthosite; 14. plagioclase-bearing peridotite; 15. harzburgite; 16. olistostrome; 17. unconfomity; 18. disconformity

underlying the Group yields an Rb-Sr whole rock age of 1720 Ma and a U-Pb age of 1900 Ma (Dong Shenbao et al. 1986). Recently, U-Pb ages of values of 2478 Ma and 2506 Ma have also been obtained (Wu Maode, 1988). Therefore, the Huili and Kunyang Groups must be of Middle-Upper Proterozoic age.

The evolution of the miogeocline exhibited by the Huili Group can be divided into three stages and corresponds to that of the Yanbian Group. The early-stage sequence i.e. the base of the Huili Group, includes purplish grey argillite, ferruginous sandy slate and quartz sandstone which are overlain by neritic dolomite, indicating a shallow-water environment in an early subsiding basin. The middle stage sediments are black carbonaceous slate, phyllite and volcanic rock with stromatolite-rich carbonate rocks. The presence of black carbonaceous argillite represents a stagnant-water, strongly reducing environment and reflects a sub-compensation succession, marked by the intensification of basin rifting and a greater subsidence rate than sedimentation rate. Therefore, during the middle stage of its development, the basin experienced strong down-faulting, followed by a quiet phase which was in turn followed by another strong episode of down-faulting. Trough depth alternated between deep and shallow, with rock facies changing vertically and horizontally (Fig. 4.13). The thickness of the sedimentary sequence is great, varying from 2600-8386 m. During this stage, intense subsidence was associated with volcanism represented mainly by basalt, K-Feldspar rhyolite, quartz-albite trachyte, agglomerate, breccia and tuff totalling almost 1000 m in thickness and forming two basic-acidic megacycles. The diagram of magmatic differentiation index and SiO_2 frequency (Fig. 4.14) indicates that these volcanic rocks form a typical bimodal volcanic association. The features of the volcanics denote an extensional tectonic environment.

Late sedimentation in the Huili Group is represented by a flysch association. The lower part of the sequence is a deep-water turbidit current sequence consisting of quartz, siltstone, sandy and carbonaceous phyllites, whereas the upper part is a flysch-wideflysch sequence composed chiefly of dark grey and black sandy (or sand-bearing) calcarenite, argillaceous calcarenite and argillaceous micrite. In detail, the lowermost part of the Group exhibits a Bouma sequence made up of terrigenous clastic sediments. Above this layer, the Huili Group can be divided into three similar cycles of carbonate gravity-flow sediments (Fig. 4.16) overlain by thin isobathic sedimentary beds (Weng Qiongying et al. 1988). These lithological associations show that during this stage, the sea basin underwent extensive and persistent isostatic subsidence, but that the differences within the basin decreased. The sedimentation of the Huili Group consists mainly of a metavolcanic complex, tuffite and silty-argillaceous rocks, constituting a littoral volcanic-sedimentary formation. The volcanics are chiefly thick rhyolitic-dacitic lavas of a non-alkaline or weak alkali series (Fig. 4.17). These show that the basin into which the Huili Group had been deposited was now in an inversion stage. During the folding and inversion caused by the Jinning tectonic movement, an overturned NE syncline was formed in the upper part of the Huili Group, with the axial plane striking NE and dipping NW. The lower part of the Group was more strongly deformed, resulting in the formation of a clear linear structure; an S_0 plane trends mainly in NW and is replaced by S_1 plane in many places (Cong Bailin. 1988). The metamorphism of the Huili and Kunyang Groups is closely related to fold deformation and is of sericite-chlorite grade, belonging to the lower greenschist (phyllite) facies type of regional, low-temperature, dynamic metamorphism (Dong Shenbao et al. 1986).

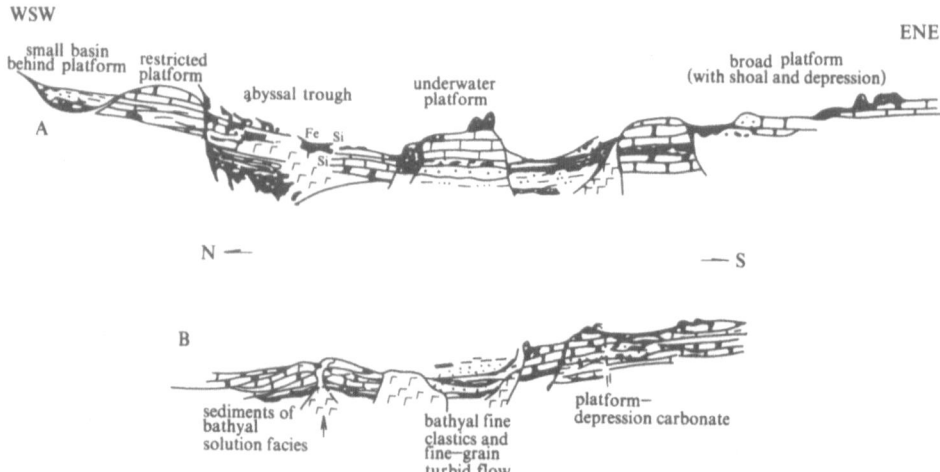

Fig. 4.14. Schematic cross-section along the long (A) and short (B) axes of the rift basin of the Huili Group, showing the relationship between the sedimentary environment and facies (Alter Wen Qiongying et al. 1988)

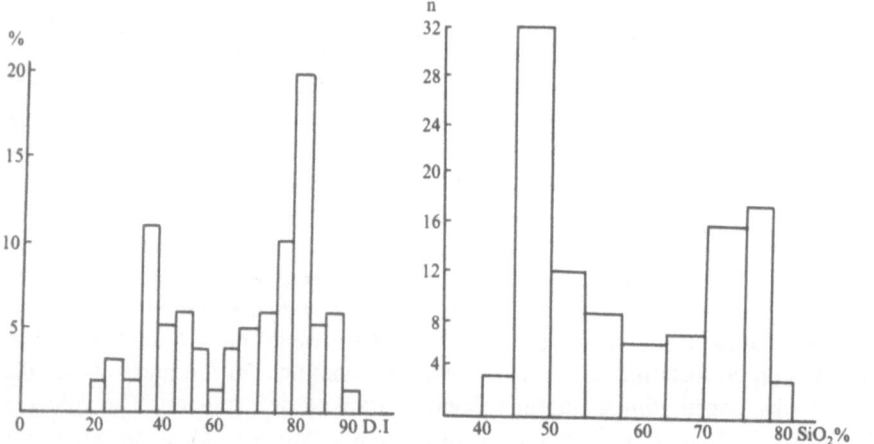

Fig. 4.15. Differentiation index (D.l) versus SiO_2 frequency diagram for basalt-rhyolite in the middle part of the Huili Group, Huidong, Sichuan. (after Wen Qiongying et al. 1988)

The eugeoclinal belt (represented by the Yanbian belt) and the miogeoclinal belt (represented by the Huili Group) are separated by the Kangding complex. The Kangding Group represents an Early Proterozoic-Archean crystalline basement which had mainly undergone middle-lower amphibolite facies metamorphism. The two belts developed on this rigid continental crust. In the Middle Proterozoic, approximately E-W extension perpendicular to the west margin of the Yangtze Protoplatform resulted in the Yanbian sea trough becoming an oceanic basin. Subsequently this basin became host to an ophiolite suite. In comparison the Huili sea trough was extended to form a rift basin within which bimodal volcanics were deposited (Fig 4.17). At the late stage, Yanbian oceanic crust on the west side was subducted eastward and, with the pre-flysch sediments, accreted onto the margin of the Yangtze Protoplatform. Subsequently, the formation of a deep trench environment resulted in the influx of flysch sediments and the formation of an olistostrome. In comparison, the Huili trough on the east side of the Yangtze Protoplatform was initially poorly developed, dominated by down-faulting and represented by numerous scattered second-order fault basins of varying depth, and associated with bimodal volcanic eruptions (Fig. 4.18a). This was followed by the deposition of abyssal terrigenous clastic turbidite and carbonate sediments on the inner side of the Huili back-arc basin (Fig. 4.18b). The subduction of the front margin of the oceanic crust beneath the Yangtze Protoplatform led to the eruption of non-alkaline volcanics in the arc and miogeosyncline area. The Jinning tectonic movement caused the folding and inversion of the miogeosyncline and its accretion to the western margin of the Yangtze Protoplatform (Fig. 4.18c).

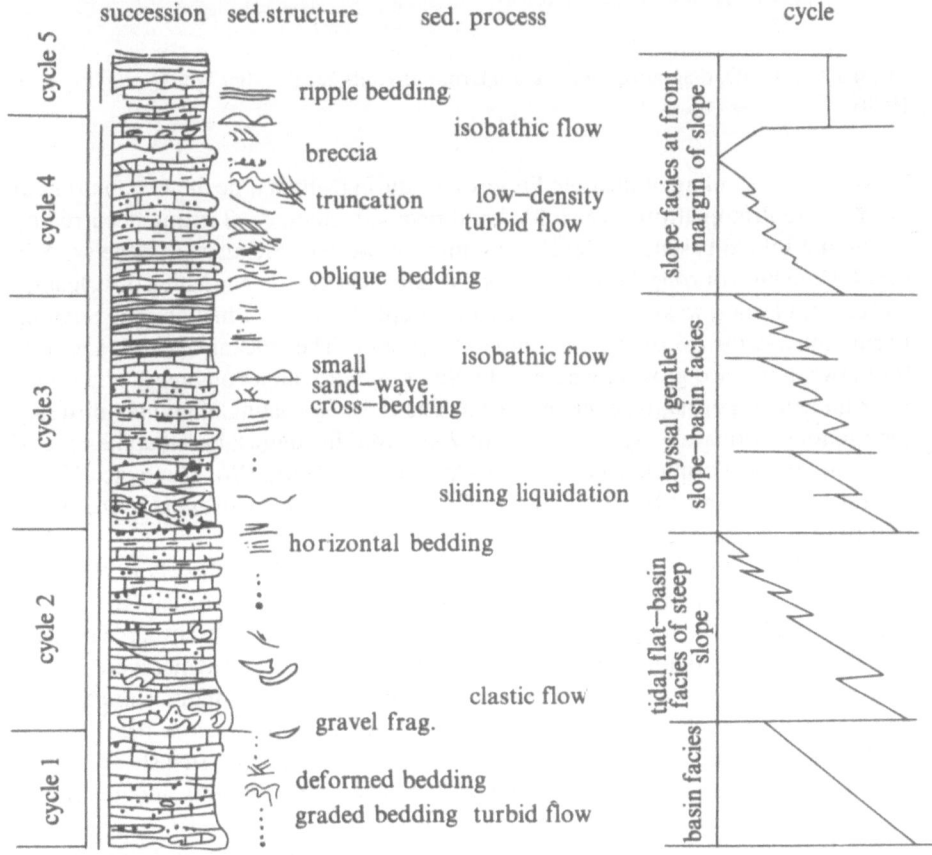

Fig. 4.16. Depositional environment and sequence of upper Huili Group (after Zhang Yunxiang)

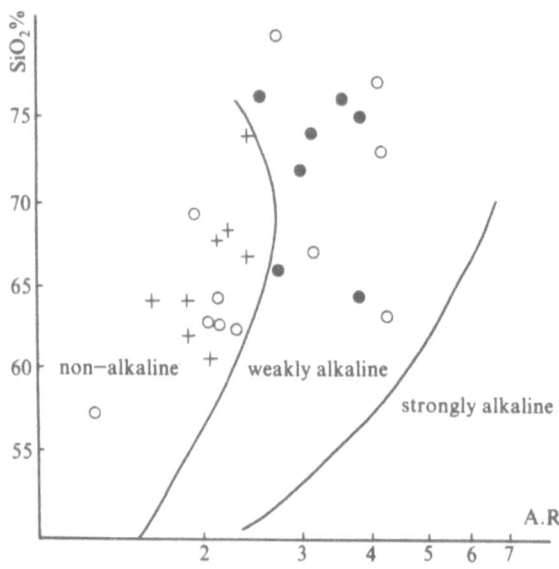

Fig. 4.17. A.R-SiO$_2$ diagram of upper Huili Group rhyolite-dacite (after Wen Qiongying et al. 1988)

After the beginning of the Late Proterozoic during which time the west margin of the Yangtze Protoplatform was folded and inverted, continental rifting occurred on the young folded basement, developing into the western Sichuan rift zone (4.-5 in Fig. 4.1). In this rift zone the 4200 m thick Wusidaqiao Group, consisting of volcanics and coarse clastics, was deposited. The Group consists of two formations: the Suxiong Formation and the overlying Kaijianqiao Formation. The volcanic assemblages differ between the two groups: whereas the Suxiong Formation contains acidic rocks, the Kaijianqiao Formation contains basalt. The K-Ar age dating of granite intruding the Suxiong Formation yielded an age of 795.7 Ma. In contrast basalt at the base of the Kaijianqiao Formation yields an K-Ar age of 759 Ma (Wu Gengyao, 1988). Therefore the main body of the Wusidaqiao Group must be equivalent to the upper part of the Qingbaikou System.

The REE partitioning pattern, abundances and petrogeochemistry of the volcanics suggest that the volcanism within the two formations is genetically related. The weakly alkaline basalt of the Kaijianqiao Formation is a product of early rifting. The fault controlling the Suxiongian volcanism are shallow in depth, and magmas were mainly derived from the melting of sialic crust. Subsequently, crustal faults were formed and became channels for the eruption of the Kaijianqiao Formation basalts. This evolutionary character shows two stages of crustal development: first shearing (uplifting) and then stretching (Wu Genyao. 1988), which represents another rifting stage within the rigid continental crust.

It follows that the western boundary of the Yangtze Protoplatform shifted further westward during the Late Proterozoic, but due to insufficient study the geology of the west part of this area is not clearly known. In recent years, the Regional Geological Survey Party under the Sichuan Bureau of Geology and Mineral Resources has made a 1:2 000 000 regional geological investigation in the Litang-Docheng area, western Sichuan. There the survey party distinguished a medium- to low-grade metamorphic sequence from the previously defined Devonian and Triassic and renamed it the Hansi Group. They found that this Group has a high angle unconformity with the overlying Sinian Dengying Formation, so bracketing the Hansi Group as pre-Sinian in age. It consists of metamorphic volcanics and metasedimentary rocks intercalated with minor carbonate and siliceous rocks, locally exhibiting migmatization. Analysis shows that the protoliths of trap rocks such as greyish green to greenish grey albite-epidote-actinolite schist in the Group are mainly metabasic volcanic rocks, albite leptite and schist, whereas those of the leucocratic rocks are chiefly intermediate to acidic magmatics. Chemically, the intermediate rocks are sodic whereas the acidic rocks are potassic (Liao Yun'an and Zheng Yumin 1986). These lithologies form the folded basement at the west margin of the protoplatform. It is therefore inferred that during the Late Proterozoic the western boundary of the protoplatform must have been to the west of the Litang-Daocheng area where the Hansi Group occurs.

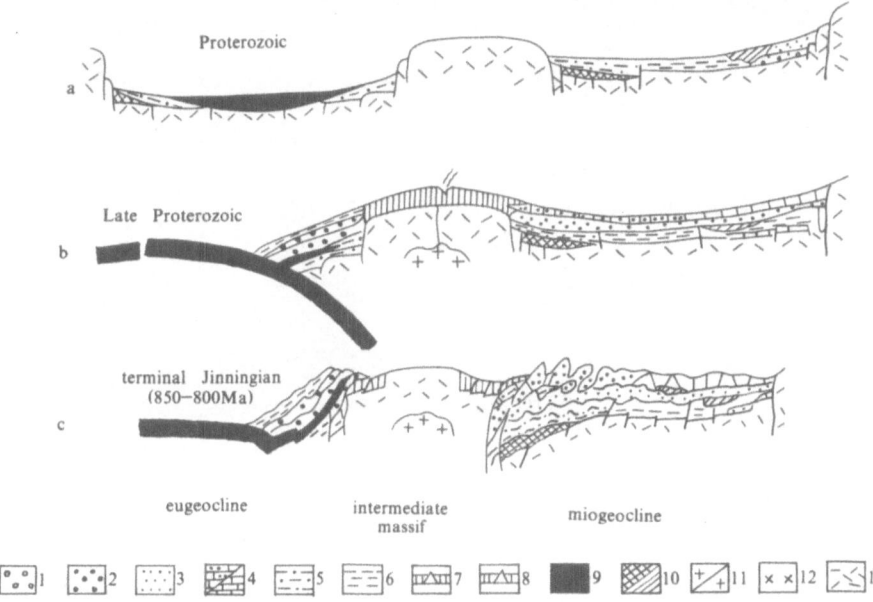

Fig. 4.18. Plate tectonic model showing the Middle Proterozoic tectonic evolution of the western margin of the Yangtze Protoplatform (modified after Zhang Yunxiang et al., 1988): 1. Molasse; 2. flysch; 3. flyschoid; 4. carbonate turbidite neritic carbonate rock; 5. abyssal-bathyal terrigenous clastic and carbonate sediments; 6. neritic-bathyal terrigenous clastic and carbonate sediments; 7. littoral volcanic-sedimentary formation (Jinningian); 8. rhyolite formation and fault-basin sediments; 9. tholeiite formation (marine); 10. spilite-keratophyre formation/bimodal volcanic suite; 11. Jinningian granite, Chengjiangian granite; 12. basic intrusive rocks; 13. Early Proterozoic-Archean crystalline basement.

North Margin of the Yangtze Protoplatform. In the Middle-Late Proterozoic, trench-arc-basin systems with differing characteristics also developed at the north margin of the Yangtze Protoplatform (4.-2 in Fig. 4.1). However, it is difficult to study the Middle-Late Proterozoic tectonic evolution of the ancient continental margin, due to the complexity caused by the long history of the Qinling tectonic belt. Nevertheless, through years of study by many researchers, the outline characteristics of the Middle-Late Proterozoic mobile belt have been reconstructed in terms of their relict paleo-lithologic record and tectonic features (Fig. 4.19).

1. Throughout the Lüliang orogeny, the ancient North China craton and Yangtze craton linked together to form a unified continent (see Fig. 3.20). In the Middle Proterozoic, this continent broke up, leading to the formation of an ocean basin, and the subduction of oceanic crust gave rise to the active continental margin and volcanic arc. Along the new Yangtze craton margin, the Wudang Group, its equivalents, and eugeoclinal assemblages, accumulated in the arc-trench. The miogeoclinal

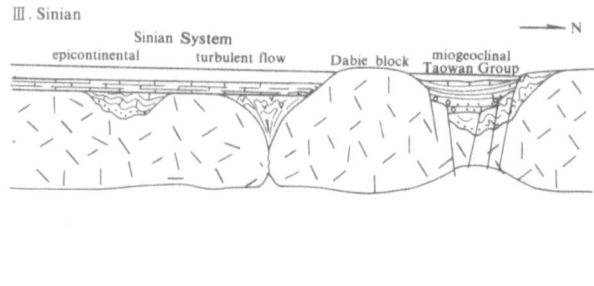

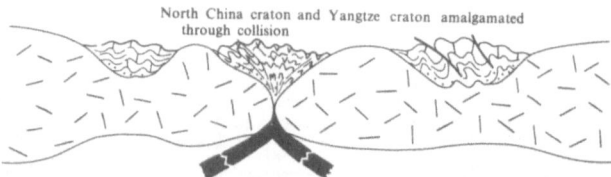

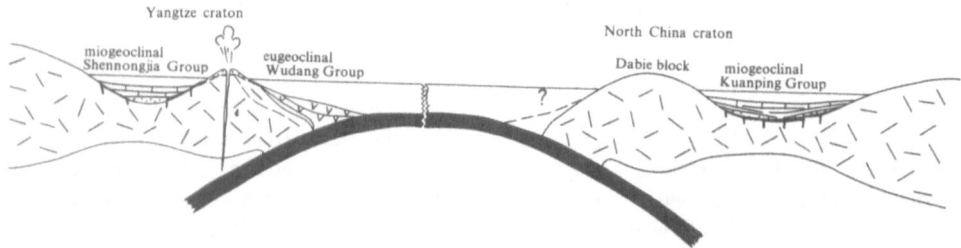

Fig. 4.19. Schematic model showing the evoultion of the collision zone between North China and Yangtze Protoplatforms in the Middle-Late Proterozoic.

Shennonongjia Group deposited in the back arc basin, while the marginal rift was generated along the reworked North China craton, in which the miogeoclinal Kuanping Group was deposited.

2. During the Jinning orogeny, the newer North China craton and Yangtze craton collided, and the rocks within the mobile belt in between them underwent intense deformation.

3. After the Jinning orogeny, the collision process was more or less continuous. Behind the Dabie block, a marginal rift was produced where the miogeoclinal Taowan Group accumulated. On the southern side of the Dabie block, the epicontinental sea transgressed and deposited the Sinian Series as platform cover.

In the western segment of the north margin of the protoplatform, the Bikou Group covers the junction of the Sichuan, Gansu and Shaanxi provinces. It consists of oceanic tholeiite (with pillow structures) overlain by a siliceous-argillaceous rock association and subsequently by flysch sediments. The sequence represents relict oceanic crust and associated sediments. The tholeiites, together with a series of well-developed diabase dykes, veins, sills and ultrabasic basic bodies distributed in the palaeo-trench and subduction zone, make up an ophiolite suite. The contact between the ultrabasic-basic bodies and their country rocks is structural, being often associated with melange, belonging to the ultramafic-mafic complex in the lower part of the ophiolite suite.

The metaperidotites are mainly composed of harzburgite, with sparse lherzolite, usually containing podiform chromite bands and lenses. Their petrochemical characteristics are as follows: Cr_2O_3=0.27-0.06%; NiO=0.24-0.36%; MgO/MgO+<FeO>=0.81-0.89; indicating that they are magnesian ultrabasic rocks. Pyroxenite, olivine pyroxenite, pyroxene amphibolite, occur approximately as layers on metaperidotite, locally with poorly-developed layered gabbro. The REE pattern shows LREE enrichment, with a positive Eu anoumaly (Eu/Eu*=1.81). Due to low-temperature high-pressure metamorphism, the rocks have been transformed into various greenschist and glaucophane schists. Chemical composition analysis shows that the rocks are rich in Na and are mostly low in K, CaO and MgO. They therefore belong to the oceanic tholeiite series (with the exception of one or two samples of alkaline basalt compositions). The REE pattern shows LREE enrichment, with ÝLa-Eu/ÝGa-Lu(Y)=1.08-4.5, La/Y=4.37-27.73 and $(La/Sm)_N$=1.31-2.42; so the rocks belong to type P or type T of oceanic tholeiite. The initial $^{87}Sr/^{86}Sr$ ratio is generally less than 0.706, also suggesting a mantle source for the magmas of the Bikou Group volcanics (Li Yaomin et al. 1988). The silicic-pelitic rock association in the middle of the Group consists mainly of black slate, phyllite, siliceous slate, silicalite and jasper rock, with minor intermediate-basic volcanics. Such an assemblage is characteristic of oceanic sediments. The flysch formation in the upper part of the Group is composed of volcanic flysch overlain by terrigenous-clast flysch. The volcanic flysch is a rhythmic sequence consisting of gravel-bearing coarse-clast tuffaceous sandstone, sandy conglomerate, lithoclast-crystalloclastic tuffaceous wacke, lithoclast-feldspar wacke, tuffaceous sandstone, tuffaceous sand-slate and black slate. The terrigenous-clast flysch is a clear Bouma sequence composed of conglomerate, gravel-bearing sandstone, lithoclastic feldspar wacke, feldspar lithoclast wacke, fine sandstone, siltstone, sandy slate and black slate. The sequence also has well-developed graded-bedding, scouring planes and sliding structure. Hence the Group is interpreted to represent a trench turbidity current deposit. The Bikou Group is unconformably overlain by strata of Sinian age and intruded by gabbro-diorite and subduction-derived plagiolase granites with a K-Ar age of 738.5 Ma and a U-Pb age of 1420 Ma respectively (Tao Hongxiang et al. 1986; Mao Yuian 1988). Therefore the Group must be Middle-Late Proterozoic in age, and equivalent to the Yanbian, Ebian and Huangshuihe Groups on the western margin of the Yangtze Protoplatform. The Bikou Group thus belongs to the initial oceanic crust-trench sedimentary association.

The main body of the palaeo-island arc on the southeast side of the trench is composed of the Xixiang Group and the overlying Sunjiahe Formation. Analysis of island arc structure and lithologic association suggest that the Xixiang Group includes a series of rocks ranging from oceanic island tholeiite to island-arc-tholeiite, with associated silicic argilite and island-arc flysch. In the Late Proterozoic Qingbakiouian Period, during the deposition of the Sunjiahe Formation, a change from marine to terrestrial volcanic eruptions took place, which was associated with intrusion of abundant intermediate-acidic rocks. The intrusions are dated at 860 Ma by the U-Pb isochron method and at 800 Ma by the Rb-Sr isochron method. This shows that the island arc became increasingly mature. During the Jinning movement, the crust beneath the arc was turned into a continental crust and at the same time the strata were folded and block-faulted (Tao Hongxiang et al. 1986).

Situated on the southeast side of the paleo-island arc is a relict back-arc basin. The basement of the basin is marked by the Lower Proterozoic Houhe Formation consisting of gneisses, schists and migmatites, on which the Mawozi Formation and Shangliang Formation of the Huodiya Group rest unconformably. The lower part of the Mawozi Formation is dominated by stromatolite-rich metamorphic carbonate rocks, with local intermediate-basic volcanics and quartz sandstone and metamorphic conglomerate at its base. The Shangliang Formation is an epicontinental, miogeosynclinal, sedimentary association consisting of potassium volcanics, pyroclastic rock, tuffaceous marble, tuffaceous-sericite calcareous schist and biotite-quartz slate. Judging from the K-Ar age of 1065 Ma (restricted publication: 'The pre-Sinian in Sichuan' by Zhang Honggang et al., 1983) for the olivine-amphibole pyroxenite intruded in the Mawozi Formation, the above formations are Middle Proterozoic in age, roughly equivalent to the Huili (Kunyang) Group. The Huodiya Group is unconformably overlain by the Tiechuanshan Formation, the middle and lower members of which comprise alkaline basalt, alkaline-feldspar rhyolite and ignimbrite. Indeed, the Group is characterised by the absence of andesite. Basic volcanic rocks are Na-rich, with ~4.08% total alkalis, whereas acidic volcanic rocks are K-rich, with ~8.26% total alkalis. Therefore these rocks belong to a continental alkaline basalt-rhyolite association. They exhibit a bimodal rift character and are quite similar to the Wusidaqiao Group on the west margin of the Yangtze Protoplatform (Tao Hongxiang et al. 1986) in that they also represent a further stage of rifting of the rigid continental crust after folding and basin inversion.

It follows that there was also a Middle-Late Proterozoic trench-arc-basin system on the northern margin of the Yangtze Protoplatform. The Bikou Group represents a residual slice of oceanic material which was a product of this environment. The intensity of tectonic deformation is reflected in a high-pressure low-temperature glaucophane schist belt. This belt is situated amongst the basic, intermediate-basic volcanics and pyroclastics of the Bikou Group and is accompanied by tectonic melange. A low-pressure high-temperature metamorphic belt distributed on the landward side of the paleo- island arc terrain and the side of the back-arc basin is characterized by a metamorphic mineral assemblage of andalusite-cordierite-sillimanite. These paired metamorphic belts are also associated with a double magmatic zone. It is thus inferred that the ancient oceanic-crust plate was subducted southeastward beneath the Yangtze Protoplatform. The Baiyigou Group (and subsequently strata of Sinian age) was deposited on the Middle Proterozoic folded basement. Volcaniclastic rocks within the Group yield Rb-Sr and U-Pb dates of 717-802 Ma. Analyses of major elements, REE and incompatible elements of the rocks all indicate that the Group was formed in a continental rift environment and is roughly equivalent to the Wusidaqiao Group on the western margin of the Yangtze Protoplatform. The evolution of the system and characteristics of subduction also suggest that the northwestern and western margin of the Yangtze Protoplatform are quite similar in tectonic style: at that time both were probably active continental margins on the same side of the protoplatform (Fig.4.1).

Other mid-Proterozoic sucessions occur in the central north region of the protoplatform: the Wudang Group is situated in the northwest Hubei-south Shaanxi area and composed principally of metamorphic basic lava, acidic lava, tuff, schist and quartzite. Samples from the Wudang Group were dated at 1304 Ma and 842 Ma by the U-Pb zircon method (Zhang Shuye and Kang Weiguo 1989). The Group probably represents the product of trench, subduction zone and island arc environments. In the northern Hubei area to the east a thick volcano-sedimentary sequence termed the Suixian Group occurs. Both the Wudang and Suixian Groups are mainly Middle Proterozoic, and partly of Late Proterozoic (Qingbaikouian) age. The Suixian Group has been dated at between 730-1228 Ma (Zhang Shuye and Kang Weiguo 1989) and varies greatly in lithology and thickness, reaching a maximum thickness of 9200 m. It is dominated by a sedimentary rock association in the south of its outcrop region and by a volcanic sequence with layers of spilitic basic lavas near Tongbai-Dabie Mountains in the north. Along the WNW regional structural trend in the north are widely-distributed basic-ultrabasic bodies of metagabbro-diabase and olivine-bearing gabbro, extending northwestward to connect with the volcanic series of the Wudang Group. The overall protolith assemblage is characteristic of an ophiolite suite: abyssal silicalite (metamorphosed into quartzite), pillow basic lava (metamorphosed to greenlandite amphibolite) and ultrabasic rock (metamorphosed to serpentinite) coexist in a partially ordered fashion (Yang Sennan et al. 1983). Petrochemical analysis of volcanics of this Group shows that they belong mostly to either the tholeiitic or calc-alkaline series. Some trace element concentrations suggest the character of basalts of mid-oceanic ridge, while others have the character of island-arc tholeiite. The REE patterns are slightly LREE enriched, similar to those of island-arc tholeiites (Zhu Xinren 1988). Therefore the Suixian Group also typifies a trench-subduction zone complex.

On the south side of this relict trench-arc system are miogeosynclinal sediments originally deposited in a back-arc basin environment. Successions include the Shennongjia and Dagushi Groups. They consist mainly of carbonate rocks, with sandstone, shale and minor basic to intermediate-basic volcanic lavas. Dolomite is the dominant carbonate rock, containing abundant plant microfossils, stromatolite and related reef structures. The dolomite contains silicic bands, often with intraclast texture, and show shallow-water signs such as ripple marks, cross beds, dry cracks and salt-bearing pseudocrystals in some horizons. These indicate that they are not abyssal sediments. Furthermore, these sedimentary sequences in this region reach great thicknesses (e.g. the Shennongjia Group is up to 12 596 m thick), suggesting that they were deposited during strong subsidence in a back-arc extensional enviroment. There are also slump breccias and disturbed bedding in the sequence, which indicate an active crustal environment. The back-arc basin was developed on an old basement, and the Shennongjia Group has been recently discovered to be unconformably underlain by the Lower Proterozoic Suiyuesi Group (Li Fuxi et al. 1989). The shale from the middle part of the Shennongjia Group has been dated at 1332 ± 67 Ma by whole-rock U-Pb method and a diabase within the upper part of the Group at 963 Ma by whole-rock K-Ar method (Li Quan and Leng Jian 1989). This Group is overlain unconformably by the Late Proterozoic Macaoyuan Group and deposits of Sinian age. Therefore it is considered to be Middle-Late Proterozoic in age, roughly equivalent to parts of the Changcheng System, Jixian System and Qingbaikou System. The Dagushi Group corresponds to the middle-upper part of the Shennongjia Group, and both are probably the products of the same basin (Zhao Yinsheng et al. 1987).

Following the formation of the orogeny on the north margin of the Yangtze Protoplatform, differing degrees of extensional rifting were initiated. For example, the lower part of the Macaoyuan Group unconformably overlies the Shennongjia Group and consists principally of a dolomite breccia sequence with basic-intermediate volcanic eruptives, lithoclast and volcanic ash. Such an assemblage is characteristic of a molasse-like formation. The upper part of the Group is composed of argillaceous and massively stratified dolomites containing stromatolite. The Huashan Group, which unconformably overlies the Dagushi Group, is also a conglomerate-volcanics-slate association and hence similar to the Macaoyuan Group. The volcanics (spilites) of the Huashan Group yield whole-rock K-Ar dates of 846 Ma whereas the pillow spilites yield whole-rock Rb-Sr dates of 831 Ma (Lu Xuemiao et al. 1988). Thus the Group is Late Proterozoic in age, approximately equivalent to the upper part of the Qingbaikou System. Both the Huashan and Macaoyuan Groups are overlain by parallel or slightly unconformable Sinian strata.

The trench-arc-basin system in the centre-north margin of the Yangtze Protoplatform underwent subduction and eventually collided with the North China Protoplatform during Late Proterozoic Jinningian-Chengjaingian time. Important evidence of subduction and collision is the occurrence of a blue schist-eclogite in the Yangtze Protoplatform. This high pressure low-temperature metamorphic zone is not only consistent in orientation with the eugeosyncline belt, linear ophiolite belt and regional structural line but is also present in the sequences below the Middle-Late Proterozoic Dengying Formation. Successions affected by this metamorphism include the Wudang Group, Suixian Group and Hong'an Group as well as the overlying Yaolinghe Group and Doushantuo Formation. As the Dengying Formation and its overlying Paleozoic sequence are controlled by both the suture and their stratigraphic position no blueschist has been found within them. The blueschists can be assigned to the metasedimentary, metabasic volcanic and meta-acidic volcanic types (Zhou Gaozhi et al. 1989; Kang Weiguo et al. 1989). Accompanying blueschist metamorphism in this zone was intense and polyphase structural deformation. The main phase was roughly simultaneous with the metamorphism of glaucophane greenschist facies, expressed by the existence of a progressive metamorphic belt of glaucophane greenschist, lower greenschist and higher greenschist facies. Therefore, it is considered that the high-pressure metamophic zone on the north margin of the Yangtze Protoplatform was formed in Late Proterozoic times. Geophysical data show that the steeply graded linear gravity and magnetic anomalies along the Dabie Mts. are all caused by the tectonic activity associated with an ancient subduction zone (Xie Doke et al. 1984).

The eastern segment of the Yangtze Protoplatform northern margin is essentially identical in development character to the central segment. Towards the east, the NW-trending structural boundary of the middle segment changes its direction abruptly to NNE, and overlaps the southern segment of the Tancheng-Lujiang fault, thus forming the eastern segment of the north margin. On the eastern side of the fault along the boundary are the NNE-trending Middle-Proterozoic Zhangbaling Group and the Upper Proterozoic Sinian sequence (Table 4.8). The lower part of the Zhangbaling Group consists of mafite-rich sand-pelitic clastic rock, acidic volcanic tuff and dolomite. The upper part of the Group is dominated by quartz keratophyre, with pilite and tuffs derived from volcanic lavas of the same material, intercalated with thin siltstone. The Group totals more than 3600 m in thickness and has yielded a variety of isotopic ages: a zircon U-Th-Pb age of 1026 Ma from quartz-keratophyre, a whole-rock Pb-Pb age on spilite of 1031 Ma (Bureau of Geology and Mineral Resources of Anhui Province 1987), a whole-rock K-Ar age on blueschist of 1175 Ma (Zhang Liangtian et al. 1989), and whole-rock Rb-Sr isochron ages of 736 Ma and 848 Ma (Liang Wantong et al. 1989). Therefore the Group is Middle-Late Proterozoic in age, and largely equivalent to the Wudang, Suixian and Hong'an Groups.

The Sinian sequence on the northern margin of the Yangtze Protoplatform is sporadically distributed, with a maximum thickness of 3700 m. The Lower Sinian consists of metamorphic wacke and tilly gompholite with meta-andesite, whereas the Upper Sinian is composed of argillo-arenaceous rock at the bottom overlain by argillaceous lithologies and subsequently by carbonates, which form the major part of the sequence. The succession is unconformable with the underlying Zhangbaling Group (Jing Yanren et al. 1989). During Late Proterozoic time, on the east side of the Tancheng-Lujiang fault, there also developed a high-pressure metamorphic zone striking NNE, more than 400 km long and 30-80 km wide, consisting of two metamorphic facies; greenschist and the glaucophane greenschist. The greenschist facies which does not contain glaucophane or epidote, is located primarily in the central part of the zone and consists of metamorphosed Sinian strata. The glaucophane greenschist facies constitutes the main body of the zone, made up of the Zhangbaling Formation. It has extensive outcrop (Liang Wantong et al. 1989) and is characterized by blueschist metamorphism and deformation resulting in the formation of planar, linear and superimposed folds. The occurrence of the Sinian greenschist metamorphic belt, as part of the high-pressure metamorphic zone, marks the end of the high-pressure low-temperature metamorphic event (Jing Yanren et al. 1989).

In summary, a 600-800 Ma Late Proterozoic high-pressure metamorphic belt lies along the north margin of the Yangtze Protoplatform and extends for nearly 2000 km in an E-W direction. It served as an important boundary between the Yangtze and North China Protoplatforms during the Late Proterozoic and belongs to the Jinning cycle of events. Its existence indicates that the Qinling-Dabie massif, derived from the North China Protoplatform, completed its accretion to the north margin of the Yangtze Protoplatform during Late Proterozoic time.

Southeast Margin of the Yangtze Protoplatform. On the southeast margin of the protoplatform, the Middle and Upper Proterozoic are separated by an extensive unconformity. The Middle Proterozoic succession comprises the Sibao, Fanjingshan, Lengjiaxi, Jiuling, Shuangxiwu and other Groups and forms the lower tectonic layer. This sequence underwent low-grade metamorphism and intense structural deformation resulting in the formation of a series of tight linear folds. Comparisons of isotope ages, micropaleobotany, sedimentary facies and formation features indicate that the Groups are roughly equivalent in age to each other, and comparable with the Changchengian and Jixianian Systems. Above the unconformity is a Late Proterozoic succession which includes the Danzhou, Xiajiang, Banxi, Xiushui, Qigong, Shangxi Groups and others. These Groups and others can be correlated with the Qingbaikouian System, and are characterized by essentially unmetamorphosed strata deformed into broad arcuate folds. This in turn is overlain disconformably by Sinian strata (Table 4.7). After being folded and inverted, the Middle Proterozoic sequence was accreted to the southeast margin of the protoplatform, becoming part of the basement. After this process was complete the belt of island-arc activity shifted southeastwards (4.-3 and 4.-6 in Fig 4.1).

Recent studies of the Sibao Group (northern Guangxi region) and Fanjingshan Group (northeastern Guizhou region) in the southwest sector of this margin (Wang Yangeng 1988; Dong Baolin 1988) suggest that these two Groups are roughly similar in rock association and order of development. The lower part of each Group is dominated by terrigenous clastics including conglomerate, lithoclastic quartz sandstone, sandstone, siltstone, sericite phyllite and tuff, with minor basic-ultrabasic rocks. The basic rocks are mainly relatively Mg-rich layered diabase and gabbro-diabase, with ultramafic rocks occurring only at the center of some rock bodies. The middle part is composed of thick volcanic lavas with dark-coloured sandstone and clays; the lavas are primarily pillow basic lava spilite of oceanic tholeiite, occuring as concordant layers in dark fine clastics and associated with komatiite. The komatiite occurs as three layers, each being subdivisible into three zones (in descending order, the chilled margin, the spinifex and the cumulus zones). The boundaries between zones are transitional to each other. The spinifex zone consists of basaltic komatiite (Photo 4.3), with a brush-like hollow, skeletal structure up to 1 cm long. Its petrochemical characterists are as follows: $SiO_2$46-53%, MgO>8-10%, TiO_2<0.9%, K_2O<0.9%, high NiO and high Cr_2O_3, FeO*/FeO+MgO<0.65 (Yang Lizhen 1990; Mao Jingwen et al. 1988); with the Al_2O_3content correlating positively with FeO*/FeO+MgO. All projected points fall into the komatiite field (Fig. 4.20).

The upper part of the Group is a flysch sequence deposited in a bathyal turbidite facies environment. The Group consists of sandstone, siltstone, tuff, and sericite phyllite, intercalated with marine volcanics such as spilite-keratophyre. A whole-rock analysis yielded an Rb-Sr age of 1667±247 Ma on basaltic komatiite from the Sibao Group in the northern Guangxi region (Dong Baolin 1988). Therefore the Sibao Group or Fanjingshan Group is inferred as being formed during the Changchengian Period. It should be pointed out that the bottom and basement of both the Sibao and Fanjingshan Groups do not outcrop. However, the conglomerate in the lower part of the Fanjingshan Group has a complex gravel composition: it consists mainly of assorted metamorphic lithologies, but also contains a small amount of granitic gravel. Furthermore, the determination of four Nd-Sm isotope systematics from the granodiorite of the Sibao Group, yield the range in values: [143]Nd/

Photo 4.3. The spinifex texture, olivine tabulate serpentinized (cross polarizer. X 200)

[144]Nd=0.511982-0.512055, [147]Nd/[144]Nd=0.1286-0.1634 and εNd=-10.0718, which suggest the granodiorite is a product of anatexis of old crust. The model age calculated from the four samples is 2513 Ma (Mao Jingwen and Chen Yuchuan 1988). The age of the granite which forms the Motianling body in northern Guangxi has

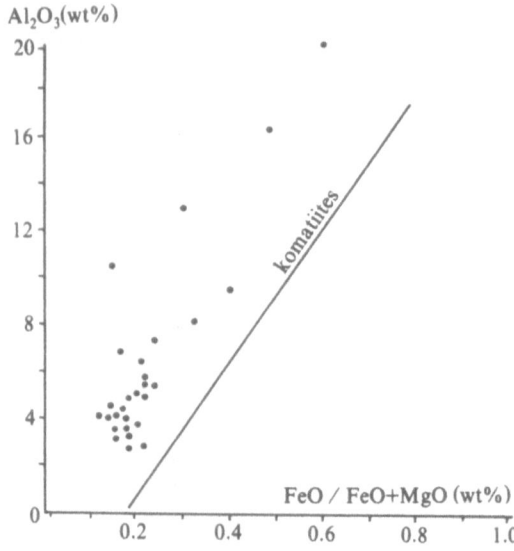

Fig. 4.20. Petrochemical diagram of layered basic-ultrabasic complex of the Sibao Group (after Xu Jun 1987).

also been determined by U-Pb zircon dating; a rounded type of zircon grain has an age of 2860 Ma. This age is thought to indicate the age of granitisation of an older protolith (Shi Shi 1976). Hence even older granitic crust may exist in this area.

The Lengjiaxi Group in northwestern Hunan is a flysch sequence composed of greenish grey to greyish green slate or tabular shale, sandy slate and metamorphic sandstone. In the lower part of the Group, there is also a metavolcanic sequence greater than 2000 m thick, dominated by basic lavas and lesser amounts of tuff, with tuffite and basaltic komatiites. Both sequences are gradational and transitional to each other. The komatiite has a microscopic, hollow, blastoskeletal spinifex texture. Olivine spinifex is mainly found in komatiite while the clinoaugite spinifex occurs chiefly in komatiitic basalt. The lavas show the typical pillow structure of underwater eruption, with the pillow pointing to a southwestward flow (Xiao Xidi 1988). Thus the Lengjiaxi Group is essentially similar in characteristics to the Sibao and Fanjingshan Groups.

There are different views on the tectonic setting in which this sequence originated: Guo Lingzhi et al. (1986) pointed out that the Sibao Group and its equivalents are flysch formations composed of greywackes, siltstones and pelites intercalated with a mafic-ultramafic ophiolite and calc-alkaline volcanic rocks. This assemblage indicates that they are the typical products of a Middle Proterozoic active continental margin on the southeast of the Yangtze Protoplatform. Wang Yangeng (1988), however, argued that mafic volcanism (komatiite) represents the rupture of the lithosphere rather than the formation of an ophiolite suite; thus the volcano-sedimentary sequence was deposited in a small newly born oceanic basin which resulted from intercontinental rifting. Wang Yangeng (1988) also considered that the Sibao Group was formed in an epicontinental aulacogen related to a divergent structure. Indeed all authors emphasised such an environment of deposition. If the tectonics and rock associations on the whole southeastern margin of the Yangtze Protoplatform are considered, then the view of Guo Lingzhi et al.(1986) probably corresponds most closely to the actual situation.

In the northeast segment of the southeast margin of the Yangtze Protoplatform, including northern Jiangxi, southern Anhui and western Zhejiang areas, occur a number of formations of similar age and characteristics. These include the Jiuling, Zhanggongshan and Shuangxiwu Groups. Regional fault zones are also host to suites of basic and ultrabasic rock types, including relatively complete ophiolite successions. A good example of these ultrabasic-basic rock suites is given by the ophiolite lying along the northeast Jiangxi deep fault zone. The ophiolite was structurally emplaced in the Middle Proterozoic low-grade metamorphic rocks. According to Bai Wenji et al. (1986), the ophiolite suite consists in ascending order of ultrabasic rocks (e.g. harzburgite, dunite), cumulus gabbro, diabase, and metamorphic basic lavas (spilite-keratophyre and basalt). The footwall country rock of the ophiolite suite is gneissic granodiorite in fault contact with metaperidotite whereas the hanging wall is Middle Proterozoic phyllite, faulted against pillow metabasic lavas of the upper part of the ophiolite suite. Petrochemical data indicate that when projected to the $MgO-Al_2O_3-CaO$ diagram, the metamorphic ultrabasic rocks from the lower part of the ophiolite all fall in the field of harzburgite and dunite, while those for the massive or layered, fine to medium-sized gabbro above them all occur in the mafic cumulite area defined by Coleman (1976). The volcanics in the upper part of the ophiolite are a spilite-keratophyre sequence, and display perfect pillow structure. They are evidently LREE enriched, with $(Ce/Yb)_N$=4.25-5.55, $(La/Sm)_N$=2.29-2.78, with a weak negative Eu anomaly. Sm-Nd isotopic work shows that the cumulate of the ophiolite in northeastern Jiangxi has the εNd of +5.1 to +5.6; an indication of island arc affinity. Moreover, 22 basalt samples from the upper part of the ophiolite mostly fall in the field related to island arc magmatic evolution using the discrimination diagram given in Pearce (1976). Whole-rock Sm-Nd age dating yielded dates of 929.73±33.6 Ma on the ultramafic rock of the northeastern Jiangxi fault zone (Xu Bei et al. 1989) and 1024±30 Ma on the ophiolite suite in Fuchuan, southern Anhui (Zhou Xinmin et al. 1989) clearly indicate that the ophiolite was formed in Middle Proterozoic times.

The Middle Proterozoic Jiuling, Zhanggongshan and Shuangxiwu Groups in the Jiangxi-Zhejiang-Anhui Provinces form a low grade metamorphic sequence consisting of two rock suites: an argillic-arenaceous assemblage with volcano-clastic flysch and an spilite-quartz keratophyre volcanic formation. The metasedimentary sequence is composed of widespread turbidity current deposits and the latter is distributed along fault zones. For example the Jiuling Group in northwestern Jiangxi is dominated by sandy argillaceous flysch, and belongs to the turbidite facies of a bathyal-abyssal environment. Judging from the preferred orientation (180°-190°) of paleocurrent data, the terrigenous clasts from the north formed a turbidite fan system that developed southward, and the general trend of development was characterized by the gradual decrease of sand and increase of argillaceous matter (Xu Bei 1986).

On the south side of turbidite basin, the spilite-keratophyre sequence along the Xifeng-Fantuoshan lies in a belt 5 km wide and more than 100 km long. Chemically this sequence belongs to the sodic spilite-keratophyre type and petrologically consists primarily of spilite, diabase spilite, diabase, diabase-porphyrite, basalt, basalt-andesite, keratophyre, dacite and quartz keratophyre. These rocks are layered and exhibit relict phenocrysts, residual vesicles, amygdaloidal structure and occasional pillow structure. Together with their country rocks, the sequence suffered strong folding and mylonitization. The petrochemistry suggests that this volcanic sequence belongs to the calc-alkaline and island arc tholeiite series and represents the magmatic differentiation trend of an island arc zone on an active continental margin. Hence the volcanic belt belongs to the island arc zone, whereas the flysch of sandstone, conglomerate, tuffaceous sandstone and slate, together with the terrigenous clastic turbidite, represent sediments in a back-arc basin deposited on the northern side of the arc. The Shuangxiwu Group distributed in the Xiqiu, Shaoxing, Zhejiang regions is also a spilite-keratophyre sequence. It can be divided into three eruption cycles, which are commonly dominated by keratophyric lava and pyroclastic rock, with a very small proportion of spilite at the bottom. Petrochemically it is rich in Na and poor in K and Ca, constituting a submarine sodic volcanic sequence on the outer side of the late Middle Proterozoic island arc (Qi Qu et al. 1986).

In summary, the northern segment of the southeast margin of the Yangtze Protoplatform was also evidently characteristic of an active continental margin in middle Proterozoic times and, although disrupted and disintegrated by subsequent tectonism, it still displays an incomplete trench-arc-basin system.

After the ending of this period of Middle Proterozoic deposition, the southeast margin of the protoplatform experienced widespread tectonism: these sequences experienced intense folding and regional low-temperature dynamic-metamorphism of lower greenschist facies. This tectonism is called by different names in different regions: namely the Sibao movement in northern Guangxi, the Fanjingshan movement in northeastern Guizhou, the Wuling movement in western Hunan, the Xiushui movement in northwertern Jiangxi, the Jiuling movement in northern Jiangxi and the Shengong movement in northwestern Zhejiang. All these movements are approximately coeval to the unconformity between the Middle and Upper Proterozoic on the west and north margins of the Yangtze Protoplatform, and are thus collectively called the Sibao movement. A number of Rb-Sr dating studies constrain the age of tectonic activity: the Bedong granodiorite emplaced into the Sibao Group in northern Guangxi yields an age of 1063±95 Ma (Zhang Taigui et al. 1988), whereas the the Pengshan spilite in De'an, Jiangxi is dated at 1515±241 Ma and the tuffaceous slate in the lower part of the Xikou Group in northeastern Jiangxi has an age of 1401 Ma. Based on these and other isotope age data, it is inferred that the times of the tectonism is equivalent to the Changchengian-Jixianian Period. The tectonic movement at the end of this period resulted in the folding and uplifting of the eugeosynclinal belt around the Yangtze Protoplatform, producing its basic configuration seen today.

At the beginning of Late Proterozoic time, the rock associations representative of an active continental margin were well-developed on the southeast margin of the

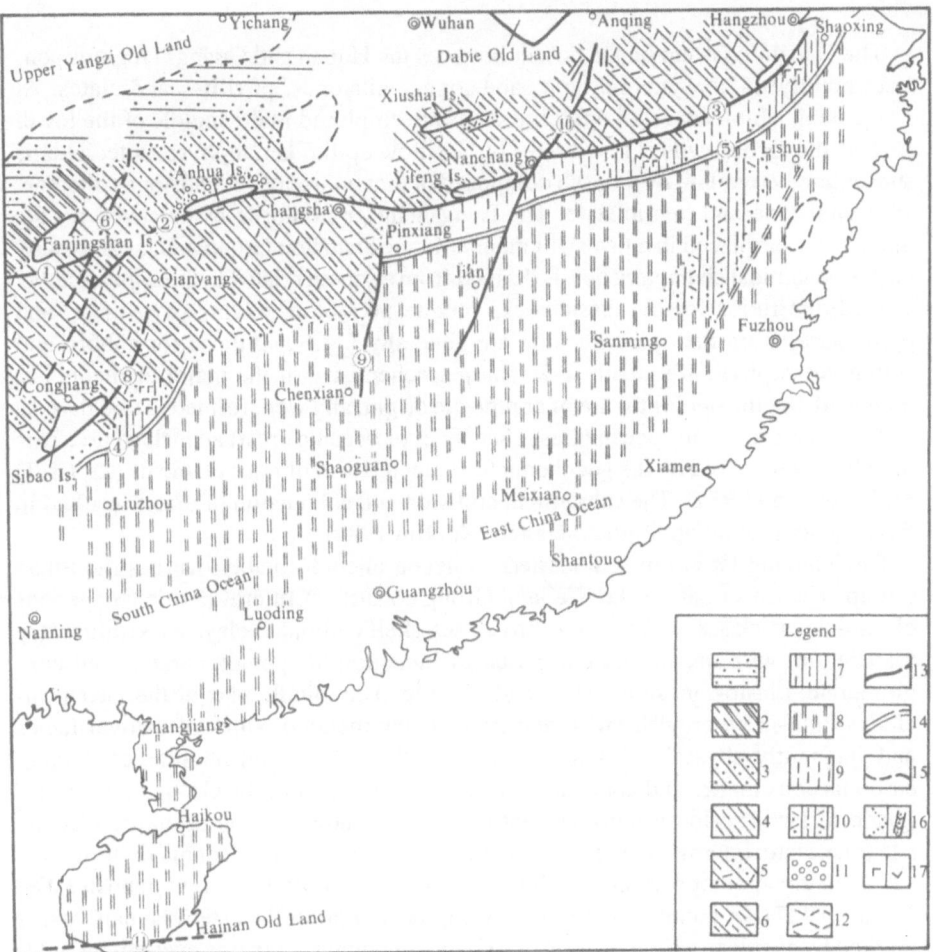

Fig. 4.21. Late Proterozoic (Pre-Sinian) tectono-paleogeographic map of South China: 1. Littoral and shallow sea clastic and argillaceous sediments; 2. Turbidite with predominant argillaceous sediment in back-arc basin; 3. Early stage littoral-shallow sea clastic argillaceous sediments, late stage turbidite; 4. Clastic argillaceous sediments in back-arc and intra-arc basins; 7. Clastic argillaeous turibidite in front of the arc; 8. Basin with oceanic crust (the nature of the sediment is not clear); 9. Marginal sea and oceanic deposits; 10. Volcanic clastic and tuffaceous sediments in marginal sea; 11. Clastic flow; 12. Volcanic tuffaceous rocks; 13. Deep faults (important tectonosedimentary boundaries): 1. Guiyang-Xupu fault; 2 Xupu-Anhua fault; 3 Yifung-Dexing fault; 4 Longsheng fault; 5. Yichun-Shaoxing fault; 6. Guzhang-Jishou-Lingshui fault; 14. Subduction zone; 15. Boundary of denudated old land mass; 16a. Facies boundary; 16b. Fault basin; 17a Basic volcanics; 17b. Intermediate, intermediate-acid volcanics.

protoplatform but their character of distribution and sedimentation showed remarkable changes when compared with those formed in Middle Proterozoic time. The Danzhou, Xiajiang and Banxi Groups in the Guangxi-Hunan-Guizhou Provinces represent rock associations each formed in three different tectonic environments, and their distribution is shown in Fig. 4.22.

The Danzhou Group on the border between the Hunan and Guangxi regions consists mainly of geosyncline-type sandstones, siltstones, phyllites and slates. At Longsheng, abundant relict ophiolite slices are emplaced in the middle of the lower part of the Group which is overlain by a pillow spilite-keratophyre suite. This is subsequently overlain by a flysch composed of metamorphosed sandstone, siltstone, silty mudstone and pelitic slate. The Longsheng granodiorite associated with the ophiolite is emplaced in a lit-par-lit manner in the Danzhou Group and, as shown by its petrochemical characteristics, is derived from a source related to the upper mantle. The Mg/Fe ratios for ultrabasic rocks (augite peridotite, olivine pyroxenite, and pyroxenite) from the ophiolite suite are generally 2-4.6 and those for gabbro and spilite-keratophyre are mostly 1-2. The petrochemistry of the major elements has indicated that the ophiolites are of composite origin, and were formed in two diverse tectonic environments: oceanic (as relict slices of oceanic crust) and island arc. This has also been proved by the geochemical characteristics of trace elements (Zr, Cr, Ti Sr, Ni, etc.) and REE. The subsequent collision and subduction of plates resulted in the formation of an ophiolitic melange (Xia Bin 1984).

The Xiajiang Group in southeastern Guizhou unconformably overlies the Sibao Group. The lower part of the Xiajiang Group consists of littoral-neritic terrigenous clastic-clay rock association and a broad, sea-shelf carbonate-clayrock sedimentary association, with the former composed of metamorphic conglomerate, sand-conglomerate, blasto-sandstone, slate and phyllite. The middle part of the Group includes light-grey, greyish-green and grey-clayey metamorphites of bathyal facies and stagnantly abyssal sediments of dark to black slates and phyllite, containing carbonaceous matter and abundant pyrite. The upper part of the Group comprises a flysch of large-scale turbidity current deposits including metamorphic sandstone, siltstone, slate, intermediate-acidic tuff and tuffite, displaying a Bouma rhythm. This has a chemical composition and characteristics of a turbidity current deposit (Wen Xiande 1987). In contrast, the Banxi Group occurs in northwestern Hunan, and is marked by purplish-red sandy slate and Ca-bearing lumps or lenses overlain by light-grey blasto-sandstone and tuffaceous sandstone, with a well-developed Bouma rhythm, repesentative of sediments of an inner turbidite fan.

Taken together, the three sedimentary assemblages of the three regions described above show the development of a turbidite fan from north to south. The sediment characteristics indicate a transition from a miogeoclinal to a eugeoclinal active epicontinental environment.

The northwestern Jiangxi area was folded, uplifted and eroded in the late Middle Proterozoic, and began to subside again in the Late Proterozoic, experiencing a littoral-neritic stage of development which rapidly changed into a bathyal environment. The Xiushui Group in northwestern Jiangxi and the Shangxi Group in southern Anhui and northeastern Jiangxi are coeval with the Danzhou, Xiajiang and Banxi Groups. Both consist of relatively thick turbidite sequences composed of argillic-arenaceous or volcanic-clastic-argillic-arenaceous materials deposited in a back-arc basin. The Xiushui Group is composed of littoral conglomerate, overlain by relatively pelite-rich, thin fine-grained terrigenous sediment and tuffaceous rock. This forms an argillic-arenaceous flysch with a clear Bouma sequence deposited in a transitional epicontinental-sea environment. Conditions are similar in the northeast segment of the southeast margin of the Yangtze Protoplatform.

The Qigong Group is approximately synchronous to the Xiushui and Shangxi Groups and is situated on the south side of the Shangxi Group. It consists of a sequence of volcanic-clastics and turbidites containing a spilite-keratophyre formation. Analysis of rare elements and the chemical compositions of the basic volcanics show that this Group is of oceanic and island arc basalt series. Tectonically, it is located near the southern oceanic crust area and seems to be composed of deposits of front-arc basins on the oceanward side of island arc (Zhou Hongrui 1986). The NE-ENE trending Dengshan Group, lying on the southeast of the the Xiushui and Shangxi Groups, consists of a set of volcanic-sedimentary series. The volcanics are characterized with varied types of basic-intermediate-acid rocks with andesite being the predominant rock type. Tectonic discrimination diagrams based on REE data demonstrate an island-arc environment represented by low-potassium tholeiite, calc-akaline basalt and peridotite; hence these lithologies belong to a mature island-arc volcanic formation. The sedimentary rocks are also a littoral clastic and island-arc turbidite series (Xu Bei, 1990). In general, the stable Yangtze Protoplatform was still

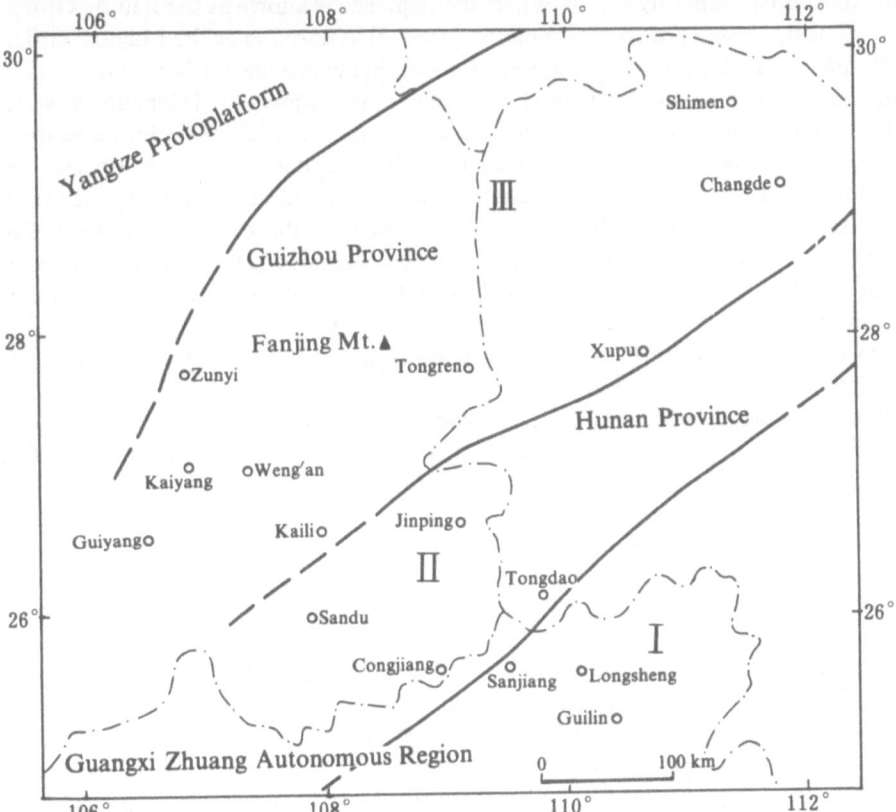

Fig. 4.22. Rock assemblages of the early Late Proterozoic on southeast margin of the Yangze Protoplatform

expanding during the beginning of the Late Proterozoic: - the ocean boundary gradually shifting southward due to the growth of a back-arc basin and the corresponding
marginal sea.

4.4.2 Cathaysia

In the Zhejiang-Fujian Provinces there is a discontinuously distributed metamorphic
sequence more than 10,000 m thick. The protoliths to the sequence are a
eugeosynclineal type volcanic-wacke flysch composed of carbonaceous and alumi-
nous-rich wacke, mudstone and carbonates. These protoliths suffered medium- to
high-grade regional progressive metamorphism, forming a metamorphic sequence
dominated by Al-rich gneiss, schist, plagioclase amphibolite and marble (He Tongxing
et al. 1988). This is mostly characterized by higher greenschist and amphibolite facies, with granulite facies and migmatization in some higher-grade metamorphic
zones. This sequence is called the Chencai Group in southern Zhejiang, and was
divided by Shui Tao (1987) into lower and upper subgroups on the basis of an
unconformity. In northwestern Fujian, the sequence is known as the Jian'ou Group,
whereas the disintegrated pre-Sinian sequence is referred to as the Mayuan Group
(Bureau of Geology and Mineral Resources of Fujian Province, 1985). It is found to
be at an angular unconformity to its overlying Sinian sequence (Li Genkun 1989). In
the Mayuan Group, two types of metamorphic volcanic rocks are well-developed:
one is acidic, and represented by muscovite-albite leptynite; the other is basic, and
includes highly mafic rocks represented by albite actinolite-epidote-biotite schist. In
the basic volcanics, no typical pillow flows have been discovered in the field. The
petrochemistry indicates that the metamorphic volcanic sequence mostly consists of
calc-alkaline volcanic rocks, of which the REE partitioning characteristics are as
follows:

(1) Eu/Sm is high for basic rocks, ranging from 0.308-0.339, but low for acidic
rocks;

(2) acidic volcanics all have evident negative Eu anomalies;

(3) mafic rocks are characterized by enrichment of LREE and depletion of HREE,
with LREE/HREE=5.69-14.08;

(4) $\acute{Y}$REE=116.03-541.53, and tends to increase with decrease in the MgO content.

These results show that the volcanic sequence of the Mayuan Group seems to represent an island arc or continental margin environment (You Zhendong et al. 1986). In
recent years, a series of Precambrian isotopic age data have been obtained from
southern Zhejiang and western Fujian (Table 4. 8 and 4. 9). They indicate that an
ancient Precambrian basement does exist in the Zhejiang-Fujian region. The age
data can be divided into three groups: 600-1000 Ma (Late Proterozoic); 1400-1700
Ma (Middle Proterozoic); and older than 1800 Ma (Early Proterozoic); which contains a maximum age of 2713+112/-106 Ma. Therefore it is inferred that the metamorphic volcanic sequence includes not only the Early Proterozoic but also indicates the existence of Archean rocks, which are not yet properly distinguished due to
inadequate study.

Stratigraphic unit & rock	Dating method	Age (Ma)
Chencai Group: leptynite, gneiss, etc.	Whole-rock Rb-Sr	674±24
Chencai Group: pegmatite in metamorphic sequence	Biotite K-Ar	720
Chencai Group: pegmatite in metamorphic sequence	Biotite K-Ar	641
Chencai Group: gneissic granite in metamorphic sequence	Monazite U-Pb	644
Chencai Group: augite peridotite	Augite K-Ar	892
Chencai Group: metamorphic rock	Whole-rock Rb-Sr	901
Chencai Group: metamorphic rock	Zircon U-Pb	1438
Chencai Group: metamorphic rock	Whole-rock Rb-Sr	1569
Chencai Group: metamorphic rock	Whole-rock Rb-Sr	1813
Chencai Group: metamorphic rock	Zircon U-Pb	2005

Table 4.8. Isotopic ages of metamorphic rocks from ancient basement in southeastern Zhejiang (Xu Butai 1987)

The term, "Cathaysia", has been controversial since Grabau (1924) proposed the concept to define the crystalline basement located along the south-east coast of China. That the Chencai Group and Mayuan Group belong to the pre-Sinian basement has been proved not only by the isotopic data quoted above but by the Mesozoic volcanics widespread in the Fujian-Zhejiang area and the source areas characteristics of various types and ages of granitoids. This indicates that the deep basement in areas covered by volcanics is of pre-Sinian age. Isotope study of granitoids in southeastern China by Jahn et al. (1984) yielded the following results: $\acute{Y}Nd(T)$= -2 to -12, initial Sr (Isr)= 0.7055-0.7010, and Nd model age (T^{DM})=1000-2000 Ma, which suggest derivation from the Proterozoic metamorphic basement (Shui Tao 1987). In addition, more than 300 m of an old metamorphic sequence beneath the Tertiary has recently been found by drilling on the East China sea continental shelf, southeast of Shanghai, and yields an Rb-Sr date of 1680 Ma (Liu Guangding 1986). Similarly metamorphic migmatite from a borehole on the Yongxing Island of the Xisha Islands in the South China Sea yielded an Rb-Sr date of 1465 (Shui Tao 1987). Furthermore, a residual zircon xenocryst age of 2516±6 Ma (Li Xianhua et al., 1989) has been analyzed from the Tanghu granite of the northern part of the Zhuguangshan composite rock body which intrudes the Cambrian across the Hunan-Jiangxi border. In summary, more and more data demonstrate that a Cathaysian oldland composed of pre-Sinian metamorphic basement did exist in the Zhejiang-Fujian-Guangdong area, South China: it formed part of an ancient rock sequence extensively outcropping along the west margin of the Pacific Ocean.

Stratigraphic unit & rock (dating sources in brackets)	Dating method	Age (Ma)
Mayuan Group: metamorphic volcanoclastic rock (1)	Zircon U-Pb	1960.5 ±1
Dikou Formation: biotite-plagioclase leptynite (1)	Zircon U-Pb	1805
Dikou Formation: gneiss (1)	Zircon U-Pb	1822
Dikou Formation: gneiss (1)	Zircon U-Pb	1851
Granodiorite (1)	Zircon U-Pb	2713^{+112}_{-106}
Porphyritic granite (1)	Zircon U-Pb	2280
Meta-tuffite (1)	Zircon U-Pb	2412
Porphyritic monzonitic granite (1)	Zircon U-Pb	2063^{+78}_{-76}
Porphyroid granite (2)	Zircon U-Pb	1686 ±253
Sinian meta-tuffite (3)	Zircon U-Pb (upper intercept)	2412^{+95}_{-87}
Longbeixi Formation (1)	Zircon U-Pb	924
Sinian metamorphic silicalite, sericite phyllite, etc (3)	Whole-rock Rb-Sr	605.7±57.6

Table 4.9. Isotopic ages of Precambrian metamorphic sequences in Fujian Sources: 1) Li Genkun et al. 1988; 2) Li Genkun et al. 1989; 3) Zhu Yulin et al. 1986.

It is clear that the Cathaysian oldland and the Yangtze Protoplatform collided and coalesced with each other as early as Proterozoic time. Shui Tao (1987) studied the Shaoxing-Jiangshan convergence belt between the two terrains: in a zone between more than ten to tens of metres wide, there is a highly compressed, unique rock series. Amphibolized ultramafic rocks in the zone occur as scattered rootless fault blocks of different sizes. These blocks contain residual layering, being composed of lenticular pyroxene peridotite overlain by amphibole pyroxenite and pyroxene amphibolite. The sequence displays a typical cumulus texture rocks and is cut by numerous diabase and gabbro dykes and sills. The ΥREE is low, with flat REE patterns. The lanthanide elements are commonly low in content, with (La/Sm)=0.485-0.810, similar to the enrichment factor (0.47-0.72) for mantle source magma in the Mid-Altantic Ridge. In Valashov's La/Yb versus ÝREE diagram, all the samples fall in the oceanic floor and rift field. The greenschist-plagioclase amphibolites from the zone mostly fall in the subalkaline basalt field in the Zr/TiO_2^{-Ga} and Ce diagram of microelements. The muscovite schist, muscovite-quartz schist and leptite are derived from a volcanic flysch. The rocks mentioned constitute an ultramafic-mafic-flysch formation similar in petro-stratigraphic succession to an ophiolite suite and representing an oceanic crustal remnant. In the coalescent zone there is also an intermediate-acid magmatic belt that can be traced for several hundred kilometres, of which the diorites have a slightly smooth REE pattern with no Eu anomaly, and yields an I_{Sr} value of 0.705-0.708. These geochemical characteristics indicate that the diorites were produced by anatexis. Such a magmatic series is common in plate marginal active zones and indicates a coalescent zone. In the coalescent zone, phengite is quite well developed and serves as an indicator of increase in metamorphic facies and deformation intensity towards the frontal margin of Cathaysia. It is therefore inferred that Cathaysia overthrust on the southeast side of the Yangtze plate and overlapped to the north. Based on the 868 Ma age of the acidic intrusion accompanying the folding and orogeny of the Shangxi Group, and the 844 Ma metamorphic age of the ultramafic rock (Shui Tao 1987), it is presumed that the collision between the two oldlands began in Jinningian times.

There are still different views on the evolution of Middle-Late Proterozoic crust in the South China domain. Some scientists such as Guo Lingzhi et al. (1984) and Shu Liangshu et al. (1987) hold that the Proterozoic ancient Jiangnan island arc was composed of allochthonous terranes mainly of oceanic affinity and of varying age. Five such terranes have been recognized from their remarkable differences in stratigraphy, lithology, structural pattern, volcanism, mineralization and geophysical characteristics. By the end of the Jinning movement, they had coalesced into the unified Jiangnan island arc at the margin of the Yangtze Protoplatform. The arc was integrated with the platform and the whole area was covered by the typically relatively stable and unique Sinian sequence. However, Hsu K J et al. (1988) have produced a totally different explanation of crustal evolution in the South China area. They argue that the tectonic configuration in the area was composed of three major units: the South China terrane, the Yangtze terrane and the Hunan-Jiangxi-Zhejiang ocean. Due to the influence of the Indosinian movement, the two terranes amalgamated, forming the Hunan-Jiangxi-Zhejiang suture zone and the South China orogenic belt. The Banxi Group is widespread in the northern Guangxi, eastern Guizhou, northwestern Hunan, northern Jiangxi, southern Anhui and western Zhejiang regions. It is regarded as a possible flysch nappe of trench sediments having been thrust onto the top of a carbonate platform during the Indosinian orogeny. In many places, Paleozoic limestone is found in windows, whereas in some mountainous areas (e.g. Fanjingshan) the Banxi flysch is overlain by an ophiolitic melange that in turn is covered by Precambrian granite, all of which are erosional remnants of a huge, overthrust complex. The melange zone in the Shaoxing-Jiangshan convergent belt, exhibits tight folding of the Paleozoic and Lower Triassic as well as well-developed thrusts on its west flank: this must be the product of the Early Mesozoic collision and suturing, in which the Precambrian rock blocks are allochthonous, rather than representative of the timing of orogenesis (Li Jiliang et al. 1989).

Further south on Hainan Island the 1700-2600 m thick Precambrian Shilu Group is a medium to light metamorphic rock series largely composed of schists, phyllite, quartzite, dolomite and carbonaceous slate, bearing several to ten layers of strataform hematite. The age of the Group is a long disputed issue. In recent years mega algae Chuaria-Tawuia fauna has been found in the Group, and accordingly its main sequence is defined as Precambrian and correlated with the Qingbaikou System (Zhang Renjie, et al. 1989; 1990). Alternatively, the Baoban Group which outcrops in Dongfang County, Hainan is another metamorphic series, made up of quartz-two mica-schist, hornblende schist, amphibolite and striped, augen and homogeneous migmatite. A recent zircon U-Pb study shows that the age of the migmatite in this Group is 1145+40/-25 Ma. Hence the age of the rock series is no younger than the early Middle Proterozoic and corresponds to the Jixianian Period (Ye Bodan et al. 1990). Therefore it is evident that there are Mid-Late Proterozoic rocks in the Hainan Province.

4.4.3 Large Fault Systems

In South China (including the Yangtze Protoplatform and Cathaysia) three large fault systems have developed, trending N-S to NNE-SSW, E-W to ENE-WSW and WNW-ESE (Fig. 4.22; Ma Xingyuan et al. 1987). Some of the faults were initiated in Early Proterozoic time, with a long history of activity. Most of the faults are moderate to deep ductile features, closely related to the entire Middle-Late Proterozoic crustal evolution, forming tectonic or tectonic-lithofacies boundaries of different dimensions, and representing important structural patterns of South China.

The N-S or NNE-SSW fault system can be subdivided from west to east into several fault zones. The Xikang-Yunnan fault zone, striking nearly N-S, is composed of such large faults as the Xiangyun, Luzhijiang, Anninghe-Yimen, Xiaojiang and Hezhang-Panxian faults. Some of them are paleoplate suture lines or subduction zones, others are boundaries of front-arc basins, paleo-island arcs and back-arc basins (Luo Yaonan 1983), controlling the development of ophiolite belts, magmatic intrusions and even the general tectonic evolution of the area. The easternmost Hezhang-Panxian fault has a gravity anomaly gradient zone about 40 km wide with an amplitude of 25-50 mgal. In a sector about 40 km wide near Yiliang, the Conrad and Moho discontinuities are lowered by about 2.7 km. Hence some of the N-S faults in the western Sichuan-Yunnan area are mantle-tapping fractures. However the structural deformation pattern and basement shear faults of the Kunyang Group show that some large faults become increasingly gentle downward, which is probably an indication of a detachment structure.

The Songtao-Rongjiang fault is located west of a large fault zone of the Xupu-Sanjiang fault on the east side of the Xuefeng Mountain, representing the south end of the Greater Khingan-Taihang Mts. - Wuling Mountain gravity gradient belt extending from north to south in China. Within Guizhou, it generally trends NNE-SSW, with the gravity gradient belt being about 40 km in width and the amplitude fluctuation reaching 30-50 mgal. Accordingly, the crustal thickness varies considerably: westward from Tongren to Jiangkou, the Moho depth changes from 40 to 43.7 km. Ophiolite material accumulated along this fault during Middle to Late Proterozoic times. Sedimentation is clearly differentiated between the east and west sides of the structure, and the deformation and metamorphism are remarkably clear.

The Hukou-Ji'an fault is most likely to mark the southern extension of the old Tancheng-Lujiang shear zone, exerting a marked influence mainly in the Jiuling area to the west, whereas Late Proterozoic Sinian volcanism was primarily restricted to the northeastern Jiangxi and Anhui-Zhejiang areas in the east. There are also clear structural indicators. In the Lushan-Yongxiu-Nanchang area, the Middle-Late Proterozoic strata have been strongly mylonitized or show marked schistosity, forming various protomylonites, mylonites, and conjugate strain slip cleavages displaying or cutting mylonitic schistosity. At Zhoutian in Yongxiu County, the mylonitic schistosity strikes SE 90°-105° and dips 40°-46°, whereas the conjugate-strain slip cleavage trends NW 325° and dips 70°. A considerable portion of the amphibolite gneiss on the south side of the Lushan Mts., west of Nanchang, has been homogenised by intense deformation. Near the Hukou-Ji'an fault, the Middle-Late Proterozoic structural line in northwestern Jiangxi changes in strike from nearly E-W to NE-SW. This change in trend results in the formation of a southeastward protruding arc in the Gao'an-Jing'an-De'an area, clearly indicating the sinistral shear nature of the fault. This direction of motion is also consistent with that of the Tancheng-Lujiang shear

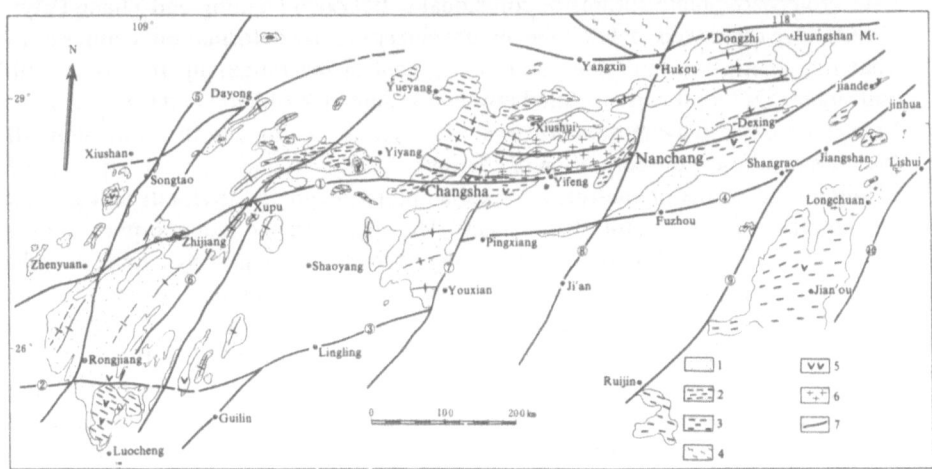

Fig. 4.23. Geological sketch map of Pre-Sinian outcrops and tectonic features in southern China (after Ma Xingyuan et al. 1987).
1. Upper Proterozoic; 2. Middle Proterozoic; 3. Mid-Upper Proterozoic; 4. Archean-Lower Proterozoic; 5. basic-ultrabasic rocks; 6. intermediate-acid intrusives; 7. deep fault.
① Guiyang-Xupu-Yifeng-Dexing fault; ≠ Duliujiang fault; ③ Guilin-Lingling fault; √ Pingziang-Jiangshan fault; ∞ Songtao-Rongijiang fault; ≈ Xupu-Sanjiang fault △ Yonxian fault ⑧ Hukou-Ji'an fault; ⑨ Shaowu-Heyuan fault; μ Lishui-Haifeng fault.

zone. It is self-evident that since the Phanerozoic, especially the Meso-Cenozoic, faulting has been more active and has controlled basin development, deformation, magmatism and mineralization. It should be pointed out that the so-called Hukou-Ji'an fault is actually not a simple fault but a fault zone 50-60 km wide, consisting of many strong strain belts and weak strain domains. Therefore, it is often difficult to determine the major representative faults by conventional geological methods. A trace of the Hukou-Ji'an fault can also be recognized on the tectonic map of the deep crust beneath Fujian and Jiangxi.

The Lishui-Haifeng fault is more than 1000 km, extending from the Hangzhou Bay in the north, through Lishui in Zhejiang, Songzheng and Longxi in Fujian, to Haifeng in Guangdong Province. Between this and the Shaowu-Heyuan fault was a tectonically active zone or a deep sea trough of Precambrian age, which now contains an ophiolite suite. These fault structures had an evident control on the Phanerozoic sedimentation, magmatism and metamorphism. In particular, they made up the marginal fault of the west Fujian aulacogen in Late Paleozoic times. However, they were strongly modified and obscured by Meso-Cenozoic tectonism, and hence many problems related to them are still under discussion.

Faults trending nearly E-W or ENE-WSW include the Chengjiang, Zhanggongshan, Guiyang-Xupu-Yifeng-Dexing, Duliujiang and Pingxiang-Jiangshan faults. The Duliujiang and Pingxiang-Jiangshan faults exerted a major control on the tectonic evolution of the south and southeast margins of the Yangtze Protoplatform. They represent important structural boundaries in South China. Sedimentary formation, geochemical and structural indicators all suggest that they possibly represent Middle-Late Proterozoic plate sutures or collisional belts. Zhao Deming and Zhang Peiyao (1983) have discussed in detail the texture and structural deformation features of the Shaoxing-Jiangshan fault (i.e. the east segment of the Pingxiang-Jiangshan fault) and suggested that it is the suture between the Jiangnan oldland and Cathaysia, or a Late Proterozoic plate subduction zone. The strong folding and thrusting along the fault has resulted in the formation of complex, strongly dynamic-metamorphic zones of sillimanite-glaucophane-felsic gneiss, amphibolite gneiss, mylonite and structurally emplaced basic and ultrabasic sheets. The deformed and metamorphosed zones are symetrically distributed on the two sides of the fault, suggesting a long, complex structural evolution. To the west, in Jiangxi, the fault constitutes the boundary between the Yangtze Platform and the South China geosynclinal system, which is clearly reflected in the magnetic and gravity fields. There is considerable difference in the deep crustal structure between the north and south sides of the fault: on the north side, the Moho undulates considerably, is slightly uplifted and a little inclined to the south; on the south side, it is relatively flat with occasional depressions. Further to the west, the fault is displaced southward by the Youxian fault and probably connects to the Duliujiang or Guilin-Lingling faults, for which there are abundant available geophysical data and sedimentation and deformation indicators.

The Guiyang-Xupu-Yifeng-Dexing fault, besides exerting a remarkable control over Middle-Late Proterozoic sedimentation and crustal evolution, plays an evident role in deformation. In the eastern Hunan and northwestern Jiangxi regions, this is shown by the presence of a strong structural deformation zone more than 50 km wide, in which the mylonitic schistosity caused by ductile shear deformation has obscured or transposed the original stratification and early schistosity. In the most intense strain zone, almost no original stratification can be found and the principal foliation is marked by bands or compositional layers resulting from mylonitization. Such bands or layers serve as slide planes and form small folds or microfolds with the axial planes being conjugate strain-slip cleavages. In Yifeng, the conjugate strain-slip cleavages dip gently north or steeply south. Differentiation stratification developed in the north-dipping cleavage producing new compositional layering. Subsequently, extensive conjugate kink bands were formed, redeforming the mylonitic schistosity and conjugate strain-slip cleavages. Thus a complete deformation sequence is formed which includes mylonitic schistosity, strain-slip cleavage and kink folds. Such sequences of deformed rock exist as zones which also contain indicators of shearing and strain softening such as sheath folds, recrystallization features and restored structure. Such features are distinguishable from those of the regional fold deformation sequences. Generally, the mylonitization is not identical everywhere in the deformation zone. Instead, the alternate zones of strongly mylonitized, protomylonitized and weakly strained rock or lenses occur. For example, a zone about 5 km wide along the Yifeng-Liuyang area is totally composed of mylonite or tectonic schist, characterized by intense strain and is roughly equivalent in position to the main fault. The area south of the zone is unknown because it is covered by Meso-Cenozoic red basin accumulations. However, the features of the Middle-Late Proterozoic phyllitic slate at Qibaoshan, Shanggao County show that such a strong strain zone could not extend to the south. From the north of the zone to the Xiushui Mt. and the Lianyun Mt. in Liuyang, strong strain zones alternate with weak strain zones. In contrast, in northeastern Hunan and the Jiangxi-Hubei border area, regional deformation and metamorphism of rocks predominate, and strongly

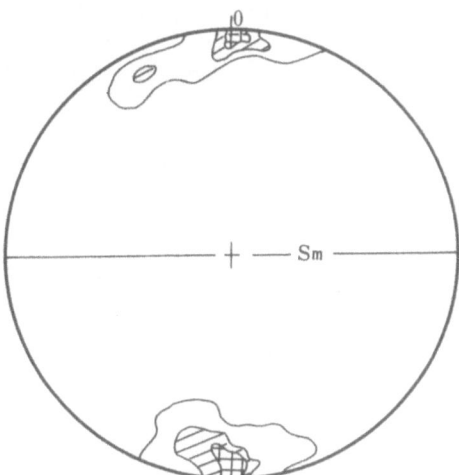

Fig. 4.24. ¶– Sm equal-area projection of Zhanggongshan ductile fault.The ¶– Sm (21 poles) contours are at 4.7, 14.2 and 23.3%

mylonitized ductile faults, such as the Shishou-Linxiang and Hanshou-Pingjiang faults, are only shown by narrow strips several centimetres to tens of metres wide. In complex cases, multiphase mylonitic schistosities crosscut or join with each other. Alternatively they are superimposed by transitional and brittle faults, adding to the complexity of the fault structure and rock deformation. Some of the ductile faults are possibly the expression, at depth, of brittle faults in the Phanerozoic cover. However, regional comparison and analysis of deformation history show that the faults and their related products mainly formed in Middle-Late Proterozoic time: for example, the southward-thrust nappes in Xishan or Nanchang and south of the Yifeng City are the result of Meso-Cenozoic structural deformation. Although, having somewhat reworked the Middle-Late Proterozoic ductile deformation structures, they represent a tectonic event quite different from the one which created the structures. The analysis of the planar geometric relation of the nearly E-W trending Zhanggongshan ductile fault and mylonitic schistosity (Fig. 4.24) in northeastern Jiangxi, together with the structures of formations on either side, suggests that the faults are also characterized by right-lateral motion. Therefore they may form, together with the Hukou-Ji'an fault, a paired conjugate shear system.

The WNW-ESE fault system principally includes faults on the north margin of the Yangtze Protoplatform, such as the Guangji-Xiangfan fault and the Guangji-Yingshan-Danfeng-Fengxian fault. The second fault, one of the main boundary faults between the North China and South China tectonic domains, is clearly reflected by gravity and magnetic geophysical data as well as by stratigraphic and tectonic indicators. The region situated between the two faults is termed the North Yangtze epicontinen-

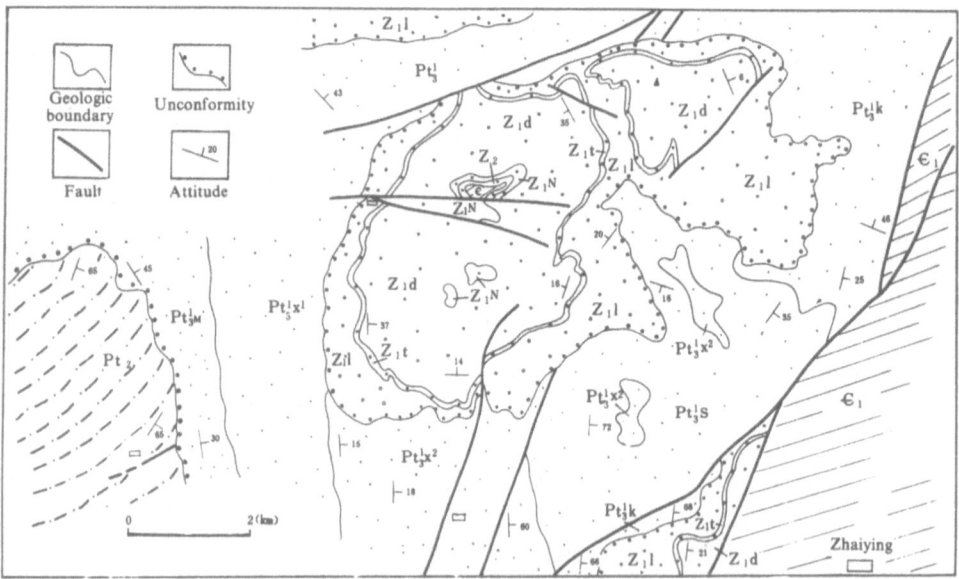

Fig. 4.25. Simplified geological map of the Zhaiying district, northeastern Guizhou (After Ma Xingyuan et al. 1987). Pt_2, Middle Proterozoic Fanjingshan Group; Pt^1_3, lower Upper Proterozoic Xiajiang Group; Z, Sinian; C, Cambrian

tal zone, which may have been formed in Late Propterozoic time and evolved into an Early Paleozoic mobile belt (Yang Sennan et al. 1983). This indicates that the faults have a long, complex history of development.

4.4.4 Tectonic Deformation Style

A "tectonic deformation style" refers to the three-dimensional characteristics of structures formed by large-scale tectonic movement. It includes the study of the structural relationship between folds and faults, the relationship between folds of different sizes, the development and association of various small planar and linear structures, the three-dimensional form and orientation of folds, the assemblage and microscopic characteristics of related metamorphic minerals and the overall spatial distribution and variation of structures. Different tectonic deformation style groups reflect tectonic environments and distinct stress and strain states. In certain circumstances, such groups can be used for regional stratigraphic comparison and for identifying superimposed deformation and establishing structural succession.

Based on the work by a number of geological workers, we may divide the Middle-Late Proterozoic in South China into three structural environments, which are separated by two regional unconformities (Figs. 4.25, 4.26). They possess different tectonic deformation style groups: the Middle Proterozoic group was formed by the Sibao movement and is marked by a thrust zone with associated tight, linear folds; the early Late Proterozoic group is a broad, arcuate fold zone formed by the Jinning movement (Fig. 4.23); and the late Late Proterozoic (i.e. Sinian) group is characterized by large-scale slip-overlap structures or multi-level decollement structures. This tectonic style is similar to the tectonic style of Palaeozoic movements.

The Sibao (or Fanjingshan) movement had a widespread influence on the surroundings of the Yangtze Protoplatform. During this movement the oceanic crust was subducted along a trench-arc zone, causing strong folding and faulting of the Middle Proterozoic sequence, forming a thrust belt of tight, linear folds. The event was accompanied by regional low-grade metamorphism and magmatic intrusion.

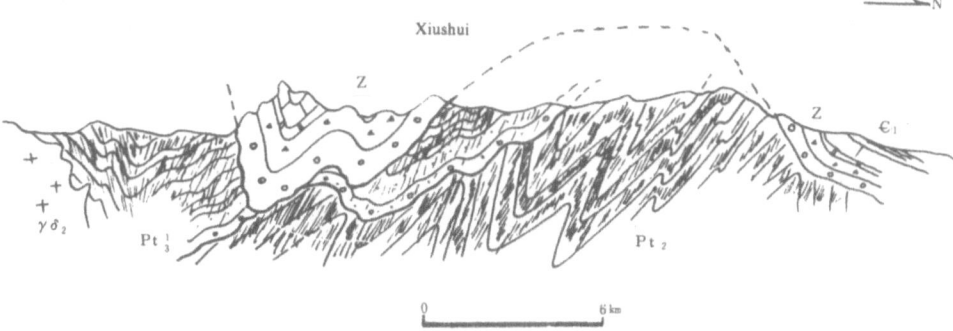

Fig. 4.26. Geological cross-section in Xiushui, Jiangxi Province, showing the unconformities between the Middle and Upper Proterozoic, and the Upper Proterozoic with the Sinian (after Ma Xingyuan et al. 1987).

On the west margin of the Yangtze Protoplatform is the N-S-trending Xikang-Yunnan thrust belt of tight, linear folds. The most characteristic of these structures is a comparatively small fold thrust zone within the Kunyang Group, which occurs between the Xiaojiang and Luzhijiang faults in the Yunnan region. This N-S trending tectonic belt, more than 100 km wide, can be divided into five second-order compound fold zones separated by four large thrusts. The direction of thrusting is generally eastwards, with the thrust planes mostly dipping to the west. Associated with the folding and faulting are basement shears and the development of variously shaped breccia diapirs. On the southeast margin of the Yangtze Protoplatform, (i.e. the vast area comprising the Guangxi, Guizhou, Hunan, Jiangxi, Anhui and Zhejiang regions) nearly E-W or ENE-WSW thrust zones containing tight, linear folds are developed in the Middle Proterozoic sequence, indicating strong N-S shortening and thickening. The linear folds of the Sibao Group strike nearly E-W, with the axial plane dipping to the south and overturning to the north. Among them, at least three large anticlines and three synclines are identified, which, together with a series of south-dipping thrusts, form northward-thrusting imbricate structures. The Fanjingshan Group is featured by ENE-WSW trending and east-overturning tight folds. In contrast, the Lengjiaxi Group at Yangjiaping, Shimen County at the Hunan-Hubei border is characterized by tight E-W folds (Fig. 4.27) with a north-dipping and south-overturning axial plane. In Zhijiang (western Hunan), Yuanling (Hanshou region) and Yueyang (northern Hunan region), the Lengjiaxi Group's main structures all strike E-W. The Jiuling Group in Xiushui and Wuning, north-western Jiangxi also constitutes an E-W striking, tight, linear fold thrust zone (Fig. 4.28), which differs in plan distribution, cross-section shape and orientation from that of the Xiushui Group (Fig. 4.29). Single folds are mostly isoclinal, overturned or chevron folds with pen-

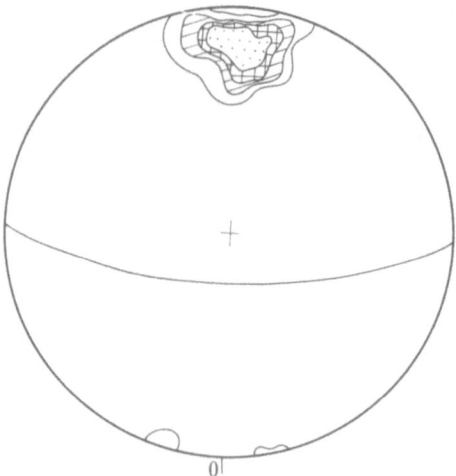

Fig. 4.27. The equal-area stereographic projection of ¶–So in the Lengjiaxi Group at Yangjiaping, Shimen, Hunan Province (after Ma Xingyuan et al. 1987)The ¶–So (38 poles) contours are at 2.6-7.9, 13.2-18.4 for one percent area

Fig. 4.28. Profile of the Jiuling Group from Huangjingling to Shenshinao in Xiushui County, Jiangxi Province (after Ma Xingyuan et al. 1987), showing the style of the tight, folds and thrust belts

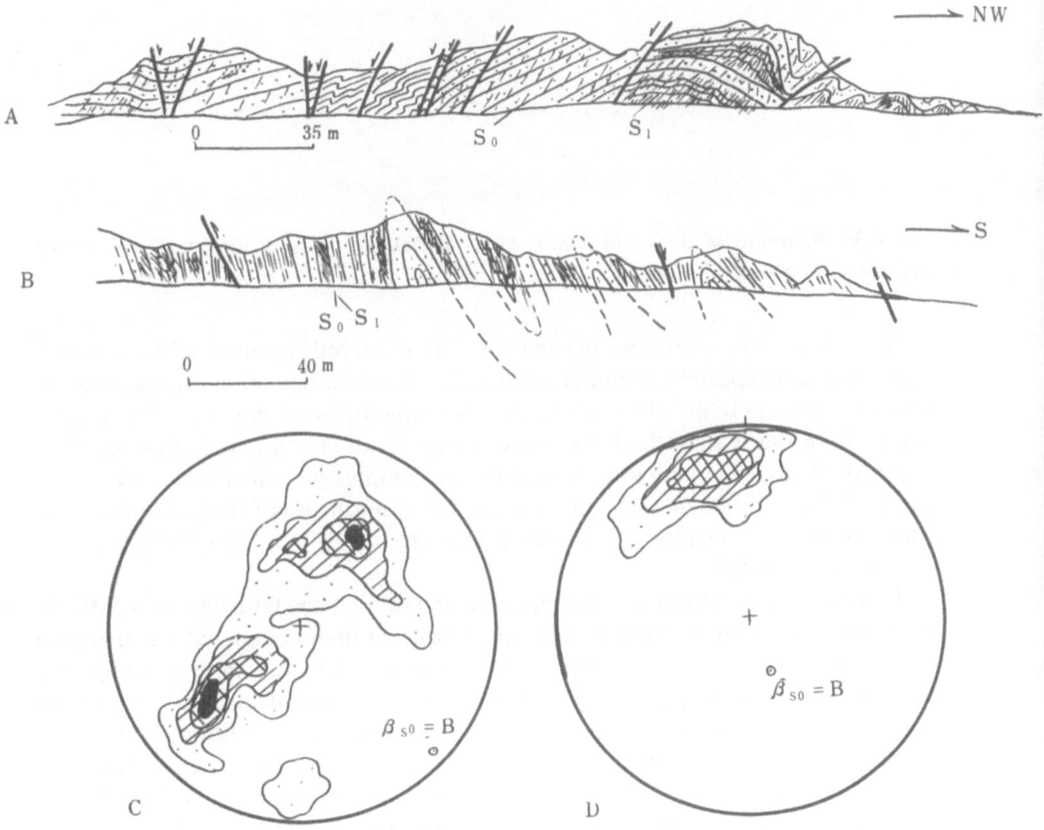

Fig. 4.29. Deformation-style groups of the Xiushui Group and Jiuling Group in northwestern Jiangxi Province (after Ma Xingyuan et al. 1987): A. Cross section of Xiushui Group; B. Cross section of Jiuling Group; C. equal-area stereographic projection of ¶–So (154 poles) in section A, with contours at 0.06, 0.32, 0.65 and 0.97% for one percent area, ßSo= SE 134° – 10°; D. equal-area stereographic projection of ¶–So (106 poles) in section B, with the contours being at 0.09-0.45-0.94-1.41% for one percent area, ¶¬So = SE 161° – 66°

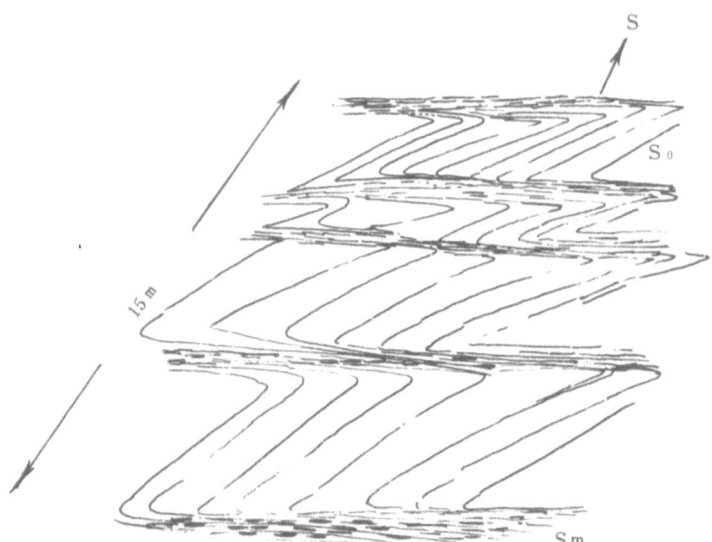

Fig. 4.30. Planview of the small ductile faults at Dakeng, Xiushui, Jiangxi Province Sm: Mylonitic schistosity

etrative axial-plane slaty cleavage and monoclinic-triclinic symmetry fabric. Due to complex superimposed deformation caused by strong compression and shear strain, plunging vertical folds, oblique folds and isotropic folds are common. Progressive strain along the thinned limb has led to mylonitization and ductile faulting (Fig. 4.30) which, together with folding, constitute a complex pattern of strain (Fig. 4.31). In the Middle Proterozoic sequences in Dexing of northeastern Jiangxi and western Zhejiang, the principal tectonic feature is also a nearly E-W or ENE-WSW complex of tight, linear folds.

It should be pointed out that in most areas, the folding and thrusting of the Middle Proterozoic is north-oriented, not exactly consistent in direction with the marginal structural style typical of the Yangtze Protoplatform plate margin. For example, in the Hunan-Jiangxi area, five E-W large composite fold structures may be recognized. Listed from north to south these are: the Mufushan anticlinorium, the Shankou-Xijiang synclinorium, the Qiping anticlinorium, the Songmen-Zhangfang synclinorium and the Yifeng anticlinorium. Their axial planes and associated thrusts generally dip to the south and overturn and thrust to the north. Guo Lingzhi et al. (1980) also discovered such "anomalous phenomenon" in the Sibao Group of northern Guangxi, where isoclinal and overturned folds with a south-dipping in axial planes occur. These structures may be explained by northward-thrusting of the Sibaoian island arc along the Duliujiang fault in the Jiangnan region (i.e. south of the Yangtze). The authors hold, however, that although such a possibility cannot be ruled out, more reliable data should be collected. Furthermore, the complexity of crustal evolution at the various temporal stages and structural boundaries and within different areas should be considered in the discussion of complex tectonic phenom-

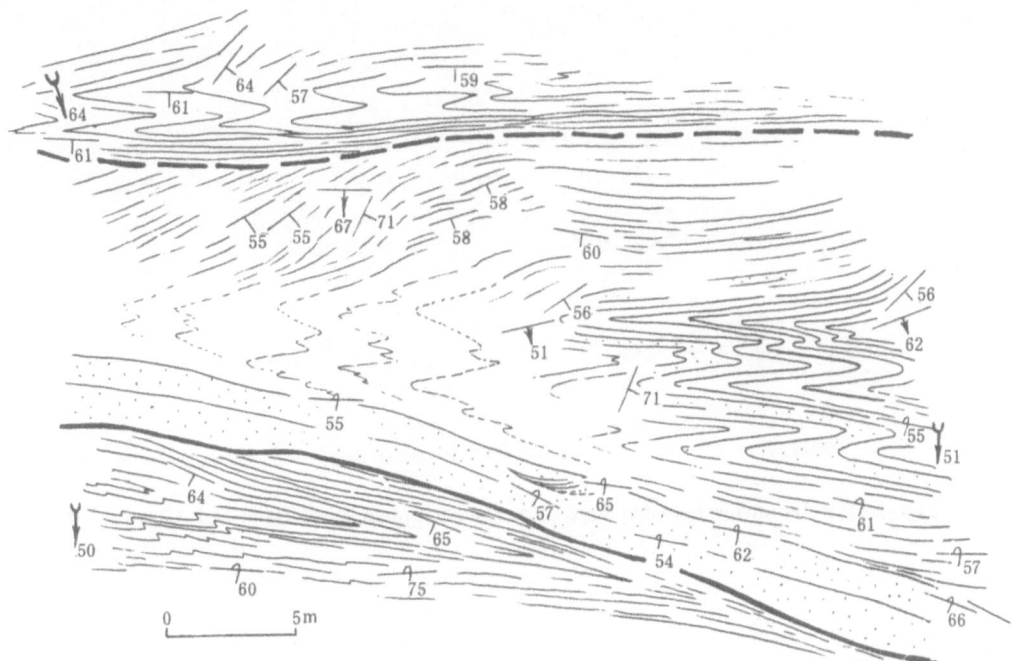

Fig. 4.31. Plan view of the deformation style of the Jiuling Group, northwestern Jiangxi Province. The folds are consistent in strike and dip with axial-plane schistosity and therefore are oblique folds

ena within such a large region. Potential models which may fit the types of structural evolution outlined above can be constructed by studying the relationship of Precambrian shear zones with plate collision in South Africa. Coward (1980) established two models (Fig. 4.31). In model A, the rock mass is detached from the subducting plate, and the shear zone gradually propagates in front of the obducting plate, forming an imbricate thrust zone directed toward the subducting plate. In model B, which is based on the flake tectonic model developed by Oxburgh (1972), the shear zone extends in front of a flake plate obducted onto the subduction zone, producing an imbricate thrust zone parallel to the direction of subduction. Whether the tectonic deformation style group of the Middle Proterozoic on the south or southeast margin of the Yangtze oldland coincides with model B can only be proved in further study.

Owing to the Jinning movement, the mobile belt and the Late Proterozoic sequence beneath the cover sequence of Sinian age were folded and deformed, and at the same time, low-grade regional metamorphism and large-scale magmatic intrusion took place. The large E-W trending Jiuling granodiorite batholith intrudes the Jiuling Group and Xiushui Group in the Hunan-Jiangxi border area. It is unconformably overlain by quartz sandstone of the Sinian Dongmen Formation. Age dating studies of other intrusions indicate that they are products of the Jinning movement and also serve as a bracket on its timing: For example, the K-Ar age of

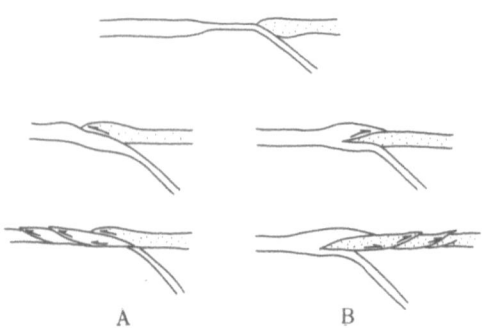

Fig. 4.32. Relationship between large shear zone and plate suture (after M.P. Coward 1980)

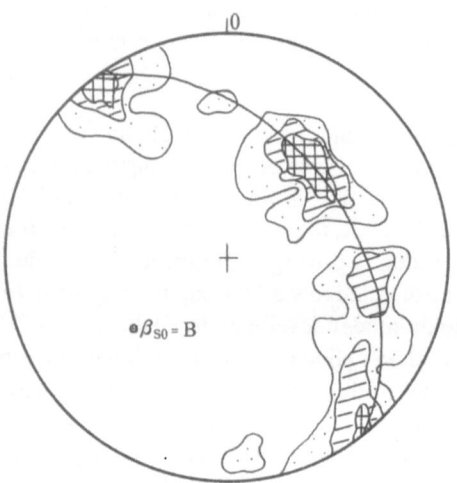

Fig. 4.33. Equal-area stereographic projection of ¶–So in the Xiushui Group at Chengshen, Xiushui County, Jiangxi. The ¶–So (62 poles) contours are at 1.6-4.8-8.1 bSo=234° – 46°

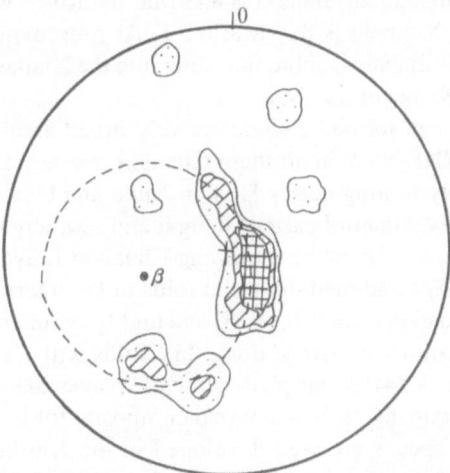

Fig. 4.34. Equal-area stereographic projection of ¶–So of the Xiushui Group at Luoxi, Wuning County, Jiangxi (after Ma Xingyuan et al. 1987).The ¶–So (24 poles) contours are at 4.1-8.3-12.5%, baxis= SW 255°°– 59°

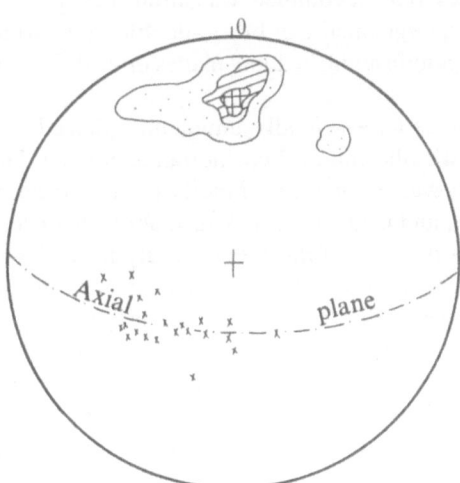

Fig. 4.35. Equal-area stereographic projection of structural elements of the Xiushui Group at Yubei, Xiushui, Jiangxi (after Ma Xingyuan et al. 1987). x - Small fold axes (22 in mumber)

the Xiaguan Granite in Gao'an, Jiangxi is 843 Ma; the Rb-Sr whole rock age of the Huangling granite in Xiadong is 819 Ma; the K-Ar muscovite age of a pegmatite dyke is 805 Ma and the diabase-gabbro intruding into the Xiajiang Group in Sanmen, Guangxi yields a U-Pb age of 837 Ma.

The Jinning movement formed a comparatively broad arcuate fold belt, with its main body trending NNE-SSW in northern Guangxi, western Hunan and northeastern Guizhou, gradually turning nearly E-W in Xupu and Guzhang, Hunan, and reverting to a NE-SW trend in northeastern Jiangxi and southern Anhui. Such an arcuate tectonic style is also evident in the Hengyang-Zhuzhou-Liuyang area at the Hunan-Jiangxi border. The large and medium-sized folds in the interior of the arcuate fold belt are simple and broad in form (Fig. 4.33) and highly symmetric (mostly of orthorhombic-monocline symmetry). Broad dome-like folds with a steep conical axis occur occasionally (Fig. 4.34). Axial-plane slaty cleavages are developed and show evident refraction, occurring as fans in broad composite folds. Small tight folds are found locally. Faded specks are well-developed in the Xiushui Group in Xiushui County, Jiangxi, which act as a good indicator of an approximate planar strain, with shortening perpendicular to the axial-plane slaty cleavage of 47%. Slaty cleavages are parallel to each other on outcrops but small fold axes and lineations tend to be dispersed in a common plane, showing an anisotropic relationship (Fig. 4.35).

In Jiangkou (northeastern Guizhou region) and in Tonggu and Jing'an (Jiangxi region), second-generation small folds were formed after the main folding stage with conjugate crenulation cleavages as their axial planes and later conjugate shear joints. These structures show a complete transition from plastic to brittle deformation. At the same time, large ductile or brittle-ductile faults were produced, and detachment faults are common along bedding planes or the basal unconformity in some places.

The Sinian constitutes an essentially unmetamorphosed cover on the Yangtze Protoplatform and marks the end of Precambrian tectonism. The tectonic deformation style of the cover was influenced primarily by the Phanerozoic, especially the Indosinian movement, and is identical in deformation character to Paleozoic strata: Jura-type ejective or trough-like folds are generally formed and box folds also developed. Most striking are the huge gliding structures in the Hunan, Hubei and Jiangxi Provinces (Fig. 4.35). These display a unique tectonic pattern produced by the gravitational instability of a superficial layer of the crust and by the ductility difference between the cover and basement. On the south flank of the Wugong Mountain, the Sinian, together with the lower Paleozoic, was pushed and slid southward during the Caledonian orogeny, forming an interference pattern of complex multiphase folds.

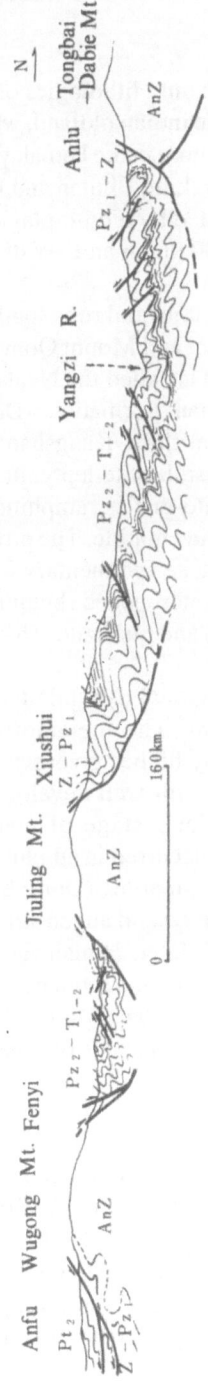

Fig. 4.36. Cross section of the huge slip structures in Hunan, Hubei and Jiangxi.

4.5 Basic Characteristics of the Northern Margin of Gondwana

In the south of the Qinghai-Tibet Plateau only lithologies of Middle-Upper Protero-zoic age are reported. This contrasts to the indian oldland, where Archean and Lower Proterozoic rocks are developed. Furthermore in the Himalayan area along the bound-aries between China and India, Nepal, Sikkim, Bhutan and Burma, there is a 20-100 km wide metamorphic sequence dominated by biotite-plagioclase gneiss and schist, extending discontinuously for about 1700 km in an E-W direction, and divided into the east and west segments within China.

The western segment is about 850 km long and runs southeastward from the west and south of Gar through Burang, Nyalam and Mount Qomolangma to Yadong. The sequence in Nyalam is well exposed and is called the Nyalam Group, which can be subdivided into the Kangshanqiao and Zham Formations (Dong Shenbao 1986; Xiao Xuchang et al. 1988). The 3500-3600 m thick Kangshanqiao Formation consists mainly of garnet two-mica (quartz) schist, biotite leptynite, sillimanite-biotite-pla-gioclase gneiss, garnet-biotite monzonite-gneiss, amphibolite, thin quartzite and amphibole-plagioclase gneiss with diopside marble. The protoliths are intermediate-acidic volcanic flysch and tuffaceous wacke sedimentary formation. The overlying Zham Formation is composed chiefly of interbedded rhythmic two-mica schist (con-taining kyanite, staurolite and silimanite) and quartzite. The protoliths are of argillo-arenaceous flysch type.

The Nyalam Group has undergone regional dynamic-thermal metamorphism of an intermediate-pressure facies system. The metamorphic facies range from greenschist to amphibolite, dominated by higher greenschist facies and lower am-phibolite facies. Deformation is strong, with well developed schistosity and com-mon isoclinal-overturned folds. At a later stage of metamorphism, extensive migmatization and granite emplacement occurred in amphibolite facies strata, form-ing banded and porphyritic or augen migmatites (Dong Shenbao et al. 1986). In view of the strong deformation, such banded and augen structures are considered as the products of intense shear strain. Still later, Himalayan deformation and retro-grade metamorphism were superimposed on this sequence.

A whole-rock Rb-Sr isochron age of 640-660 Ma has been obtained from this sequence (Dong Shenbao. 1988). At the current state of research, although the influ-ence of subsequent tectono-thermal events on the rehomogenization of the initial Rb-Sr system can not be precisely estimated, we may at least be certain that the metamorphism represented by this age took place in Late Proterozoic times. The zircon from the biotite-garnet gneiss in Yadong and the gneiss near Mount Qomolangma has been dated at 718±158 Ma and 1250-750 Ma respectively by the U-Pb isochron method (Xiao Xuchang et al. 1988; Xu Ronghua et al. 1985).

In the Gangdise massif, an equivalent sequence to the Nyalam Group developed in the Nyaiqentanglha Mountain Range termed the Nyaiqentanglha Group. A U-Pb isochron age of 1250 Ma has been acquired for zircon from the augen biotite gneiss in this Group (Xu Ronghua et al. 1985). This date is consistent with the one obtained from the Mount Qomolangma area gneiss and thus is considered to be representative of rock-forming age. The inherited lead from zircon in Himalayan gneiss has yielded a maximum age of 2.2 Ga, indicating that both the Himalayan and Gangdise massifs have old basement (Allegre et al. 1984). The Group is called the "Central Crystallines" in India (Thakur 1980), in which biotite-muscovite augen gneiss has yielded a whole-rock Rb-Sr isochron age of 1830 Ma (Bhanot et al. 1977), also suggesting the presence of an old basement.

In the east segment of the belt, Precambrian metamorphic rocks occur in an arc about 500 km long and 100 km wide, located south of Moindawang and Medog. The metamorphic sequence has largely identical features to the western segment of the belt, and consists predominantly of phyllite, sericite-quartz schist, two-mica (quartz) schist, garnet-biotite schist, biotite leptynite, biotite-plagioclase gneiss, actinolite-albite schist and amphibolite, with a little diopside marble. The protolith formation consisted of a tuffaceous greywacke layer containing intermediate-basic volcanics overlain by an argillo-arenaceous flysch containing carbonate rocks.

This metamorphic sequence distributed on the north margin of Gondwana strikes roughly E-W (consistent with the elongate terrane) and generally dips north. The distribution of metamorphic facies zones is in agreement with the trend of the strata. However, the metamorphic sequence is reversed, with weakly metamorphosed strata occurring in the south and underlies the more strongly metamorphosed strata exposed to the north. Dong Shenbao et al. (1986) considered that this phenomenon could be explained by a thrusting process which has disrupted the normal progression of metamorphic grades within the belt. A second explanation which takes into account the paleogeothermal distribution, is that the facies zones with comparatively high metamorphic temperatures represent zones of greater uplift.

A combination of protolith formation, deformation, metamorphism, paleogeothermal distribution and age data suggests that an epicontinental active belt existed in this region in Middle-Late Proterozoic times. Although the features of the tectonic environment cannot be completely explained at the present time, this active belt is evidently related to the generation and early evolution of the India plate because it is parallel to the north margin of the plate. Even if this sequence belongs to the crystalline basement and is exposed as a thrust slice due to subsequent nappe formation, and no matter how great a distance it has migrated, the spatial distribution and orientation of the metamorphic facies zone are still the same as the spatial distribution and orientation of the active belt.

4. 6 Paleomagnetic Evidence

The apparent paleomagnetic polar wandering path has been used as an important means for reconstructing paleocontinent positions and plate motion. For highly deformed and metamorphosed Precambrian terranes, the multiple deformations and metamorphisms may unfortunately partly or totally change the isotope age and orientation of magnetization, making it difficult to determine apparent polar wandering

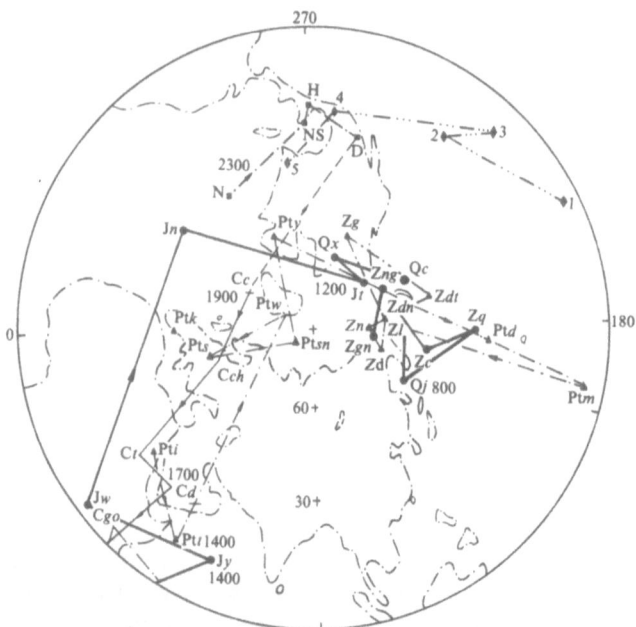

Fig. 4.37. Proterozoic apparent paleomagnetic polar wandering paths for the North China, Yangtzi and Tarim Protoplatforms (stereographic projection) (after Zhang Huimin et al. 1989): N - Nanzhangsi Fm; NS - Nansi Fm; H - Haoting Fm; D - Dongjiao Fm; Cc - Changzhougou Fm; Cch - Chuanlinggou Fm; Ct - Tuanshanzi Fm; Cd - Dahongyu Fm; Cg - Gaoyuzhuang Fm; Jy - Yangzhuang Fm; Jw - Wumishan Fm.; Jh - Hongshuizhuang Fm.; Jt - Tieling Fm.; Qx - Xiamaling Fm; Qc - Changlongshan Fm; Qj - Jing'eryu Fm; Zq - Qiaotou Fm; Zc - Changlingzi Fm; Zng - Nanguanling Fm; Zgn - Ganjingzi Fm; Pti - Luanshigou Fm; Ptt - Taizi Fm; Ptw - Dawokeng Fm; Pts - Shicaohe Fm; Ptsn - Songziyuan Fm; pty - Yanwodong Fm; Ptm - Macaoyuan Fm; Z1 - Liantuo Fm; zdt - Datangpo Fm; Zg - Gucheng Fm; Zn - Nantuo Fm; Dd - Doushantuo Fm; Zdn - Dengying Fm. Solid line - North China Protoplatform (1900-600 Ma); dot dash line - North China Protoplatform (2300-2000 Ma); dash line — Yangtzi Protoplatform (1700-600 Ma); three-dot dash line — Tarim Protoplatform (800-600 Ma).

paths prior to deformation. However, for the Middle-Upper Proterozoic sequences that are essentially unmetamorphosed, simple in deformation history and easy to evaluate in structural orientation, it is possible to define reliable paths. Paleomagnetists performed systematic paleomagnetic studies on continuous, clearly outcropped, structurally simple and representative sections on lithologies formed in stable environments. Examples of these are stable-type and transitional-type sedimentary regions, such as the Middle-Upper Proterozoic section in Jixian (Chen Jinbiao et al. 1980; Lin Jinlu 1988), the Middle-Upper Proterozoic section in Shennongjia, western Hubei (Li Quan and Leng Jian 1987) and the Upper Proterozoic section in Fuzhou, Liaoning (Gao Rongfan et al. 1983). On such a basis, these authors explored the relationship between the pre-Sinian in North China and Yangtze Protoplatforms (Zhang Huimin and Zhang Wenzhi 1984, 1985, 1989), providing an insight into the Precambrian tectonic evolution of the continent in China. Based on presently available data, Zhang Huimin and Zhang Wenzhi (1989) have established a preliminary Middle-Late Proterozoic polar wandering paths for the North China and Yangtze Protoplatforms and produce a Sinian polar wandering path for the Tarim Protoplatform (Fig. 4.37).

In summary, research shows that the Middle and Late Proterozoic apparent polar wandering path of the North China Protoplatform is similar to that of the Yangtze Protoplatform in overall direction and characteristics, whereas the Sinian path of the Tarim Protoplatform differs considerably from those of the other two protoplatforms: the Middle-Late Proterozoic (1900-600 Ma) apparent polar wandering path for the North China Protoplatform runs from the south end of Greenland through western Europe and northern Africa into Sudan south of the equator, then executes a polar loop winding northwestward and finally returning westward to western Africa. During the deposition of the Lower Maling Formation in Late Proterozoic times, the path migrates northwestward again into Alaska, forming another loop, then continuing westward into the south of the Kamchatka of the former USSR reaching the Bering Strait at about 600 Ma. Similarly, the Middle-Late Proterozoic apparent polar wandering path for the Yangtze Protoplatform began to form approximately coeval to the deposition stage of the Luanshigou Formation of the Shennongjia Group (or the Dahongyu Formation of the Middle Proterozoic Changcheng System). Its initial position is in Sudan, northern Africa, near to the pole position of the Tuanshanzi and Dahongyu Formations of the Changcheng System in the North China Protoplatform. During the deposition of the Taizi Formation (i.e. the middle of the Shennongjia Group), it migrated into Arabia, its position being adjacent to the pole position of the Yangzhuang Formation (Jixian System) in the North China Protoplatform. Subsequently it drifted into the north of the Atlantic, forming a loop trajectory; then moved westward to the central part of the north Pacific via North America. It formed another loop during the period between deposition of the Dayanping Formation and the Macaoyuan Formation at the bottom of the Shennongjia Group. Finally it shifted towards the Arctic region and about 600 Ma ago reached the Bering Strait. In contrast, the apparent polar wandering path of the Tarim Protoplatform during Sinian time (800-600 Ma), is simple and distinctly different from those of the North China and Yangtze Protoplatforms (Fig. 4.37). It is entirely located in Africa south of the Equator and at about 20°S in the Indian Ocean, having shifted eastward, and formed a z-shaped trajectory at the time equivalent to the deposition of the Nantuo and Doushantuo Formations. (Unfortunately the trajectory of the apparent polar wandering path is not yet constrained for Pre-Sinian times.) The presently available data therefore suggest that the protoplatform lay in low-latitude (10°-25°N) region for a long time in the Late Precambrian (Li Yong'an et al. 1984). Hence it most likely drifted and evolved as an independent plate.

About 1.7-1.6 Ga ago, the North China Protoplatform and Yangtze Protoplatform were in close proximity to each other, but not physically joined. In the Yanshan taphrogenic trough of the north part of the North China Protoplatform and the Shennongjia back-arc basin on the north side of the Yangtze Protoplatform, the deposition of abundant stromatolite-bearing carbonates and volcanic rocks took place. As a result the Xiong'er Group on the south margin of the North China Protoplatform and the Sibao and Fanjingshan Groups in the Yangtze Protoplatform all contain relatively more basic to intermediate-acidic volcanic rocks. This may indicate that in this period, the mechanisms for global crustal expansion, rifting and aulacogen generation were well-developed and active. As a result, magmas rose from the mantle and erupted and effused along deep faults, forming extensive volcanic-sedimentary sequences. Subsequently, between approximately 1.4-1.3 Ga ago, great thicknesses of high-Mg carbonate rocks rich in stromatolites were deposited in the Yanshan and Shennongia areas. After this time until 700 Ma ago, the two protoplatforms markedly drifted and rotated and their latitudinal position changed appreciably, indicating that crustal movement was frequent during this period. This is reflected in the diversification of sedimentary rocks and the frequent hiatuses or unconformities between stratigraphic units. For example, the Qinyu and Weixian upliftings and the Jixian movement have a widespread influence within the North China Protoplatform, and the Jinning movement within the Yangtze Protoplatform took place mainly in this period. About 800 Ma ago, at the late stage of the deposition of the Qingbaikou System, the Yangtze and Tarim Protoplatforms were so close to each other that they might have formed an integrated oldland: they were situated relatively far from the North China Protoplatform which by now had migrated to high latitudes. The obvious difference in paleolatitudes inevitably led to a discrepancy of paleoclimatic conditions and the formation of different sedimentary sequences. For instance, on the Yangtze Protoplatform volcanic-sedimentary sequences (e.g. the Wusidaqiao and Baiyigou Groups) and molasse formations formed (e.g. the Macaoyuan Group). In contrast on the North China Protoplatform comparatively thin, cold-type carbonate and clay sediments developed (e.g. the Jing'eryu Formation).

The conclusion that in the late Qingbaikouian the Yangtze Protoplatform was relatively far from the North China Protoplatform contradicts the hypothesis that the two coalesced during the Late Proterozoic, which has been deduced on the basis of the development of the Middle-Late Proterozoic trench-arc-basin system and the Late Proterozoic glaucophane schist zone on the north margin of the Yangtze Protoplatform and discussed in Section 4.4.1. If the paleomagnetic determinations are correct, the amalgamation would be between the Yangtze Protoplatform and the Qinling-Dabie massif. However the Qinling-Dabie massif was separated at that time from the North China Protoplatform. Further study is required in order to solve this problem.

4.7 Characteristics of Supracrustal Rocks

Sedimentary and volcanic sequences and some hypabyssal intrusive rock associations in the Middle-Upper Proterozoic of China can be roughly grouped into the following types on the basis of their marked differences and Condie's (1982) classification of rock associations of Proterozoic supracrustal rocks. These sequences are described in the following sections.

Photo 4.4. The interference ripple mark on the bedding of fine grained sandstone, from the Gaoyuzhuang Formation, Qinglongshan, Hebei Province

4.7.1 Quartzite (Quartz Sandstone)-Shale-Carbonate Association

This is the most common association in the Middle-Late Proterozoic (especially in the Late Proterozoic), accounting for about 70% of the Middle-Upper Proterozoic sedimentary rocks succession. The main characteristics are that the association was formed on a crystalline basement and that its constituent quartzite (quartz sandstone), shale and carbonates often make up one or several megacycles.

The quartzite (quartz sandstone) usually occurs in the lower part of a megacycle and on a regional unconformity or hiatus. The lowermost part of the quartzite succession consists of conglomerates or feldspar-quartz sandstone and wacke containing more feldspar and some lithoclasts, which are different in degree of development. Such a composition suggests that the terriginous clasts are derived from the underlying sequence of old basement. Upwards a pure quartzite (quartz sandstone) occurs, which is a product of a stable littoral zone and is high in both compositional and textural maturity. This rock occurs chiefly as medium-thick layers, in which trough, tabular and other crossbeds are developed and in some places tidal stratifications (e.g. l-shaped crossbed) with ripple marks on bedding planes predominate (Photo 4.-4). Further upward siltstone grades into intercalated shale and displays wavy and lenticular stratifications with occasional dry cracks on bedding planes.

The shales are grey, black, greyish-green, green and purplish-red in color, different in sedimentary environment and composition and dominated by horizontal stratification. The lower part of the sequence shale contains fine sand and silt bands whereas the upper part contains carbonate lenses.

Carbonates often occur in the upper part of a megacycle, comprising dolomite (70% or more), with lesser amounts of lime dolomite and dolomitic limestone and minor amount of pure limestone. The carbonates contain abundant stromatolites, oncolites and algal mats, with frequent silicic (chert) bands, lumps or nodules. Generally the stromatolites constitute reefs of different sizes and the oncolites form dams. There are also special stromatolites such as the Cu-bearing one found in the Dongchuan copper deposit, Yunnan and the phosphatic one found in the phosphorite beds of the Sinian Doushantuo Formation in the Guizhou, Hunan and Hubei regions. Another remarkable characteristic of the carbonates is that they often have an intraclastic texture. These characteristics suggest deposition in a clear-water epicontinental sea.

There are often more than two sedimentary cycles in the same area, with the uppermost cycle being generally the thinnest. Between the cycles is a parallel unconformity, and within them small hiatuses are common. This rock association type is developed in depressed belts on the protoplatform, in an epicontinental sea or in large taphrogenic troughs, forming a stable-type sedimentary cover.

4.7.2 Bimodal Volcanic-Clastic Rock Association

This association is also common in the Middle-Upper Proterozoic of China. The volcanic rocks are typically bimodal in character. The clastics include pyroclastic and terrigenous clastic types, with the latter comprising conglomerate, sandstone, feldspar sandstone, and shale. Tectonically, this association occurs in taphrogenic troughs or rifts within the protoplatform. Volcanics in different tectonic units are dissimilar in development and character: some aulacogens or rifts were accompanied from the very outset by large-scale volcanic eruptions (e.g. the Xiong'er-Hangao aulacogen), the products of which dominate the stratigraphy. In contrast, in other aulacogens passive volcanic eruptions occurred at the late stage of taphrogeny (e.g. the Yanshan aulacogen). In this case the rock association is dominated by sedimentary rocks, with subordinate volcanics. In a dead aulacogen or rift, such a rock association is often followed by a quartzite (quartz sandstone)-shale-carbonate association.

4.7.3 Sandstone-Shale-(Silicalite)-Carbonate Flysch Association

This association is characterized by well-developed flysch rhythms and dominated by sedimentary rocks intercalated with minor volcanic rocks. It is stratigraphically continuous, reflecting ongoing crustal subsidence. On the margins of oldlands characterized by the influx of abundant terrigenous clastic material (e.g. in the early Late Proterozoic Xiajiang Group and Banxi Group on the southeast margin of the Yangtze Protoplatform), a thick flysch of interbedded sandstone and shale as well as gravity flow deposits were formed. When the sedimentation rate equalled the subsidence rate, a thick sequence of dolomites with sandstone and shale (e.g. the Middle Proterozoic Shennongjia Group on the north margin of the Yangtze Protoplatform) was formed in a mainly clear water environment. The dolomites have stromatolites, algal mats, granular and intraclastic textures, as well as occasional crossbeds, ripple marks, cracks and rock salt pseudocrystals. In non-compensated basins (e.g. at the triple junction of Hunan, Guangxi and Guizhou on the southeast margin of the Yangtze Protoplatform), a closed deep-water environment was produced by strong subsidence during Late Sinian time. In this environment a black shale-silicalite association is often formed. It occurs in long, narrow back-arc basins or miogeosynclinal basins on passive continental margins. Generally, sediments exhibit a sharp facies change and great variation of thickness in the direction perpendicular to the long axis of the basin. They usually suffered strong tectonic deformation and greenschist facies metamorphism after their formation.

4.7.4 Ophiolite (Komatiite)-Flysch Association and Calc-Alkaline Volcanic-Flysch Association

These two associations are distributed in the Middle-Late Proterozoic mobile belts around the Yangtze Protoplatform but have become scattered due to later tectonic disruption and denudation. On successive occasions in recent years, a discontinuous ophiolite suite representing relict slices of oceanic crust has been reported from the Middle-Late Proterozoic subduction zone along the protoplatform (Li Jiliang 1984; Xia Bin 1984; Bai Wenji 1986; Shui Tao 1987; Xu Bei 1989; Zhou Xinmin et al. 1989). The lower part of this suite is incompletely preserved, but the sequence can still be reconstructed: it is comprised upwards of ultrabasic rocks, accumulative gabbro, swarms of sheeted dykes to pillow basalts. The upper part of the suite is composed of an abyssal pre-flysch to flysch association. In the Middle Proterozoic active belt situated to the southwest on the southeast margin of the protoplatform, komatiite-type rocks have also been discovered. These occur in a metamorphic basic volcanic sequence in the lower part of the Sibao and Lengjiaxi Groups (see Sect. 4.4.1.), forming a komatiite-flysch association. In the Middle-Late Proterozoic mobile belts around the Yangtze Protoplatform, a calc-alkaline volcanic-flysch association has been discovered and is considered by many researchers to be of island arc affinity. These two associations are similar not only to Archean greenstone belt sequences but also to sequences found on the convergent margins of Phanerozoic plates. They both suffered strong deformation and metamorphism of greenschist or lower greenschist facies after their formation.

4.8 Anorogenic Magmatism

4.8.1 Anorthosite and Rapakivi Granite Events in North China Protoplatform

Rapakivi granite and granite were intruded, and K-rich intermediate-basic volcanic rocks extruded, in the north of the North China Protoplatform during Middle Proterozoic time. The Damiao-Heishan complex in Chengde, Yanshan Mt. area, is representative of the anorthositic intrusions. It was controlled by E-W trending faults and intruded into Archean gneisses. The rock body has a nearly east-west strike and a northerly dip with an E-W length of 18 km and N-S width of 9 km. The complex, a multiple intrusion of anorthosite and gabbro-norite, is the product of deep-seated differentiation of magma derived from the same source. Early anorthositic plutons, making up the majority of the complex, later experienced thermal alterations resulting in saussuritization, epidotization and uralitization. White-coloured saussuritized anorthosite is most widely distributed and makes up the main part of anorthositic plutons. It is a kind of greyish white medium- to coarse-grained (average grainsize of 3-5 mm, occasionally 15-20 mm) rock with mostly equigranular texture and some heterogranular or porphyroid textures. The rock contains 80-90% of plagioclase (An_{34-46}) with regular platy crystals. The plagioclase has no primary cataclastic texture, which indicates the stability of its conditions of origin. It has been commonly saussuritized and in places shows polysynthetic twinning. The stronger the saussuritization is, the smaller the anorthite content of plagioclase. Hence the primary composition of the plagioclase is more calcic than An_{46}. Anorthosite commonly contain 3-5% dark-coloured minerals, i.e. chlorite and titanomagnetite, and occasional apatite. Epidotized anorthosite, characterised by a grey-white colouration with green spots, is scattered widely throughout the region. Chlorite filled and replaced white-coloured plagioclase along joints and fissures, commonly forming 7-10% of the rock with a maximum content of 40%. A scattering of gabbronorites, commonly with irregular and ramified shapes, are closely associated with the anorthosites. They have different sizes and occur in a suite with a close spatial and genetic relationship to iron and apatite deposits. According to mineral associations gabbronorites can be divided into norite, gabbro, tilaite and olivine norite, of which norite is principal. Dark grey norite consists of plagioclase 40-45% An_{45-54} and 20-23% hypersthene with a small amount of augite, ilmenite and apatite. In addition there is monzonite and adamellite in the north and northeast parts of the complex. The Damiao anorthosite complex has been dated at 1686±193 Ma by the Rb/Sr whole-rock isochron method, 1735±239 Ma by Sm/Nd whole rock isochron method (Xie Guanghong et al. 1988) and 1656±15 Ma by the $^{40}Ar/^{39}Ar$ method (Hu Shiling et al. 1988). The age of emplacement calculated by the various dating methods are therefore within error of each other. It is therefore apparent from field and age data that the anorthosite complex is very similar to other Proterozoic anorthosite events in the world in terms of both rock associations and timing of emplacement.

When compared chemically with similar complexes elsewhere in the world, the anorthosite of the complex contains slightly high concentrations of SiO_2, K_2O and

Na$_2$O, slightly low Al$_2$O$_3$ and CaO, and variable MgO. Norite and gabbro are low in SiO$_2$ and CaO, and high in Al$_2$O$_3$, Fe$_2$O$_3$, FeO, Na$_2$O and K$_2$O; the monzonite and adamellite are low in SiO$_2$, and high in Fe$_2$O$_3$ and FeO (Zhai Yusheng, 1963; Jia Bingwun unpublished report, 1986). The anorthosite is similar to the gabbronorite in transitional element concentrations as well as in the pattern and contents of REE, being characterized by high values of Ti, Fe, V and Mn, low values of Cr, Co and Ni, normal Eu, a positive correlation between Eu and Ce and a negative correlation of LREE/HREE ratio and Ce concentration. The gabbronorite is higher in Mg, Fe, Ti and Ca and much higher in HREE than the anorthosites (Qi Changmou et al. 1988). The initial values of ^{87}Sr/^{86}Sr and ^{143}Nd/^{144}Nd for the Damiao complex are 0.70402 and 0.51013 respectively, in accordance with the corresponding rock bodies world-wide, which demonstrate that the magma of the complex derives from the upper mantle.

Rapakivi granite, found in a close relationship to the Damiao anorthosite rock body, is located around Shachang, Miyun County, Beijing. It forms an EW-trending, multiple-intruded complex with a length of 10 km and a length width ratio of 6:1, and was emplaced into the Archean high-grade metamorphic rocks of the Miyun Group. Lithologically it mainly includes rapakivi amphibole-biotite granite, prophyritic biotite granite and medium- to fine-grained biotite granite. Mineralogically it contains large amounts of potash feldspar, plagioclase (andesite and albite) and accessory minerals of fluorite and zircon. The rock is rich in K$_2$O (5%) and mostly belongs to an aluminium-oversaturated series. The rapakivi granite is characterized by orbicular and rapakivi textures (Photo 4.5) with predominant phenocrysts

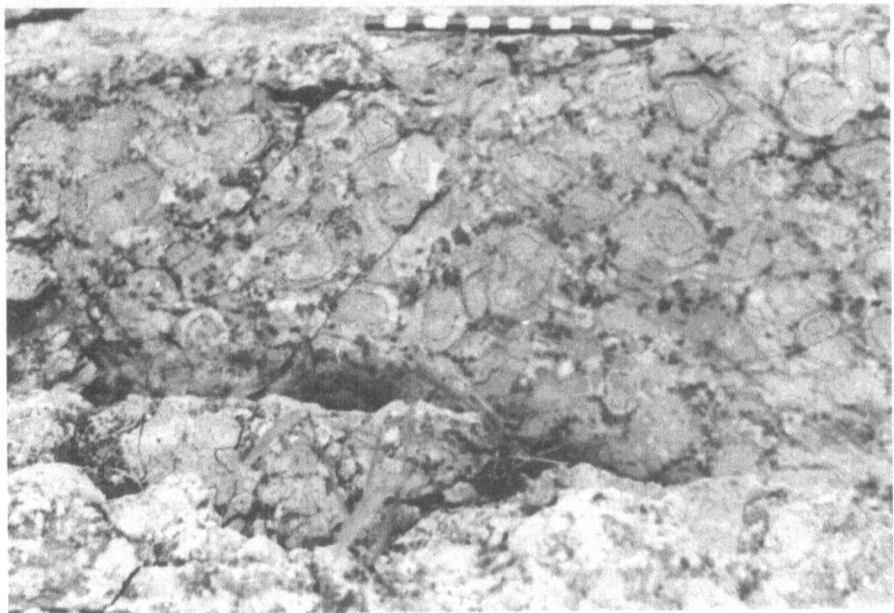

Photo 4.5. An outcrop view of rapakivi granite, Miyun County, Beijing (by Yu Jianhua).

of oval potash feldspar with diameters between 2-12 cm, most commonly measuring between 6-8 cm. An orbicular grain comprises individual or several potash feldspar crystals with some fine-grained dark-coloured minerals such as hornblende, with occasional magnetite and biotite. About one third of orbicular phenocrysts retain the external rings which constitute rapakivi texture (Zhao Chonghe, unpublished report 1963; Gao Jiarui, 1986). Geochemically the rock is rich in incompatible elements especially F and REE, possesses a LREE/HREE ratio of 17.4 and a small negative Eu anomaly (Eu/Eu*=0.09). The REE pattern and some element ratios suggest that the magma was derived from the lower crust. The rapakivi granite has been dated as 1638 Ma by K/Ar method on mica and 1767-1741 Ma (Song Biao 1988); and 1715±31 Ma (U/Pb zircon age; Yu Jianhua, 1990). It is the same age the Damiao anorthosite complex and constitutes a typical anorogenic magmatic rock association. This type of rapakivi granite is quite rare - the biotite granite which outcrops out in Kuandian, Liaoning region on the cratonic margin in the northern North China Protoplatform, constitutes one of the few known occurrences. The granite shows egg-porphery textures with phenocrysts 5-6 cm in diameter, made up of orthoclase-perthite and microcline perthite. The plagioclase content is generally 50-55% and locally 70-80%. The rapakivi granite can be chemically classified as kaligranite with $Na_2O/K_2O=0.39$ and a ΎREE = 287.7 ppm. It is a LREE integrated type (La/Yb(N)=18.4) and has a clear negative Eu anomaly (Eu/Eu*=0.42). The single zircons from the rapakivi granite yield Pb/U ages of 1163.8 Ma ($^{206}Pb/^{238}U$); 1405.2 ($^{207}Pb/^{235}U$) and 1793 Ma ($^{207}Pb/^{206}Pb$). The 1793 Ma date is the commonly accepted age for the granite (Zhang Qiusheng, 1988).

The Shicheng granite and the Bajiazhai rock body in the Songshan region were intruded into the Archean Dengfeng complex and Early Proterozoic Songshan Group, and are unconformably overlain by the Wufoshan Group of Mid-Late Proterozoic age. They have the K/Ar ages of 1524, 1547 and 1632 Ma (Ma Xingyuan et al., 1981b) which are somewhat older than the basic dyke swarms, together with the much later quartzporphyry dykes, which make up a granite-diabase-quartz porphyry series.

4.8.2 Basic Dyke Swarms

Vast swarms of basic dykes were intruded into the early Precambrian terranes after 1.7 Ga and particularly in the period between 1.6-1.2 Ga (Ma Xingyuan et al, 1987). The most abundant dyke swarms occur in the central north part of the North China craton (Ma Xingyuan et al 1981a, 1984), including the central south part of Inner Mongolia, northern Shanxi and western Henan Provinces, and also in parts of the Tarim craton. A study of these basic dyke swarms may help to reveal the tectonic environment and crustal condition of the craton, the secular changes in mantle composition and the spatial relationship between the Earth's continental cratons in Mid-Proterozoic times (Halls, 1982). Therefore some part of the basic dyke swarms have been studied in further detail in recent years (Chen Xiaode and Shi Lanbin, 1984; Qian Xianglin et al 1987).

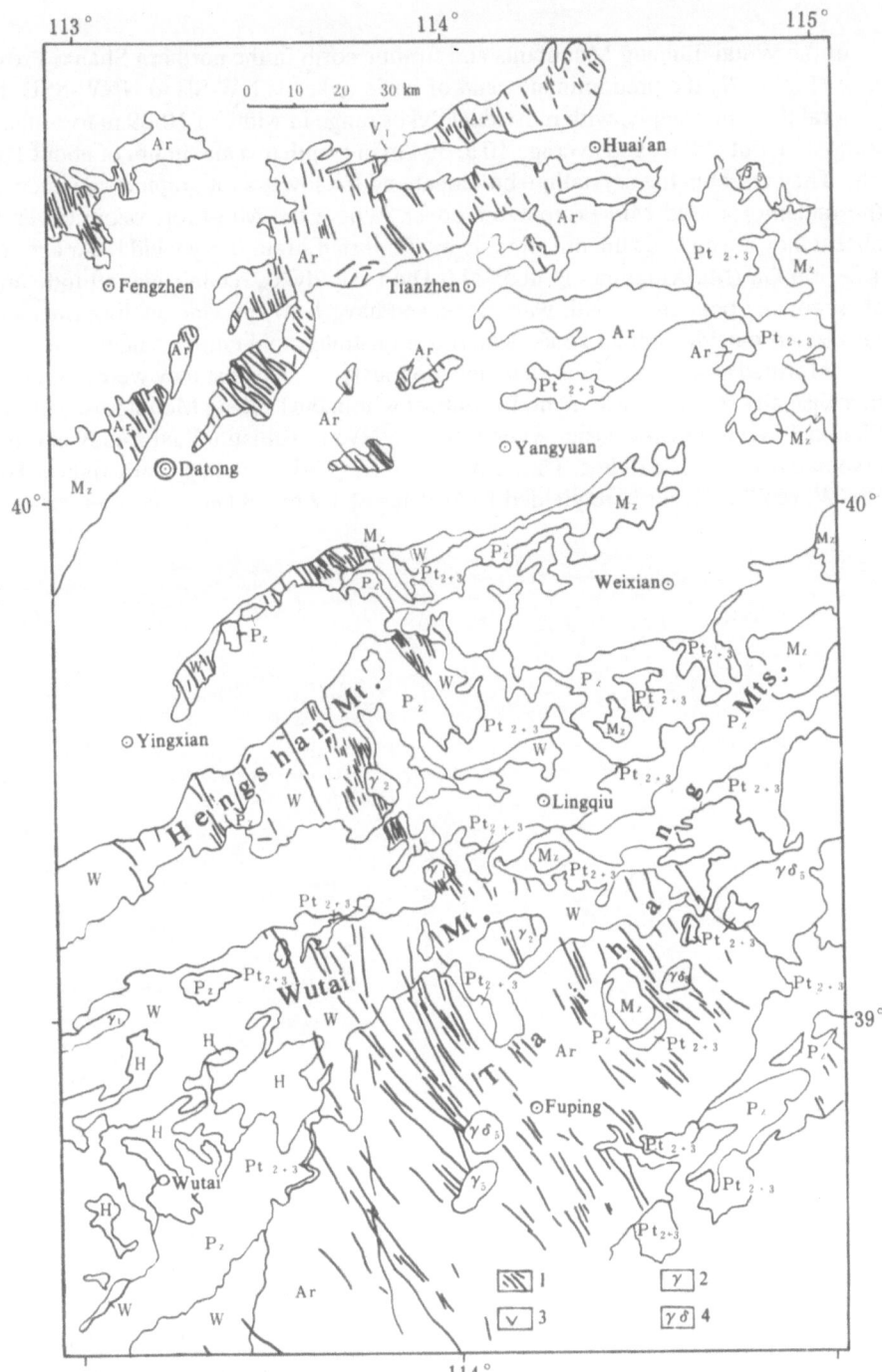

Fig. 4.38. Map showing diabase dyke swarms in Wutai Mt, Hengshan Mt., and Taihang Mt. areas Mz-Mesozoic; Pz-Paleozoic; H-Hutuo Group; W-Wutai Group; Ar-Archean 1. Diabase dyke; 2. granite; 3. gabbro; 4. granodiorite

In the Wutai-Taihang Mountains and furthur north in the northern Shanxi Province (Fig. 4.37) the predominant trend of basic dykes is NW-SE to NNW-SSE. In general they dip steeply, with individual dykes range in width of 10-39 m to a maximum of about 100 m, and average 10 to 30 km in length to a maximum of about 100 km. They cut both the crystalline basement and the lower stratigraphic sequence of the metamorphosed Mid-Proterozoic cover. Where the Mid-Proterozoic cover is absent they are covered unconformably by Cambrian strata. They yield K-Ar ages of 1.2-1.66 Ga (Ma Xingyuan et al 1987). The basic dykes retain steep attitude and clear contact boundaries with wall rocks, and have 1-10 cm wide cooling rims and inside and outside contact zones which demonstrate rapid emplacement. They are not deformed or metamorphosed, which obviously indicate that they were emplaced in a basically stable environment. In contrast within the Lüliang Mountains, western Shanxi Province major basic dykes follow WNW to sublatitudinal trends and are associated with felsic dykes. They cut a set of long NW-trending mafic dykes. The WNW-trending dykes here yielded K-Ar dates of 1.2 to 1.3 Ga.

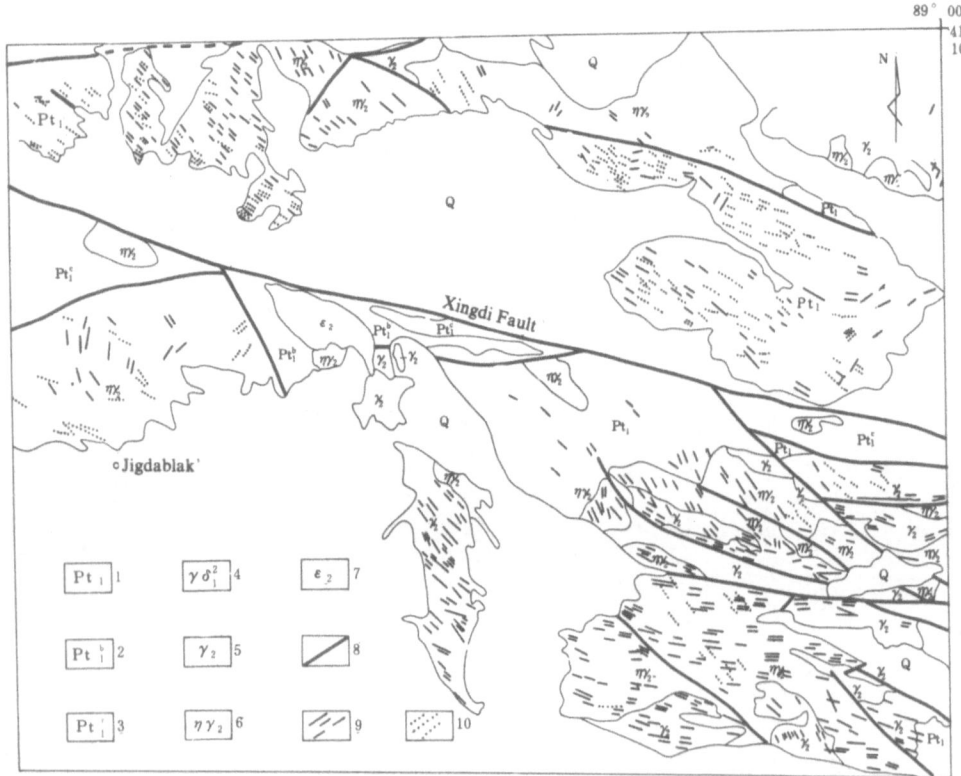

Fig. 4.39. Basic dyke swarms in the Koksu area, Xinjiang. 1 (Pt_a^1). Amphibolite, hornblende gneiss and migmatite; 2 (Pt_b^1). marble with schist intercalated in the lower horizons with lenses of magnetite; 3 (Pt_c^1). quartzite with muscovite quartz schist and lenses of marble; 4 (gd_2). gneissic granodiorite; 5(g_3). granite; 6 (hg_2). adamellite; 7 (e_2). syenite; 8. fault; 9. basic dykes; 10. intermediate and acid dykes.

Basic dykes are dominantly diabase in type with lesser amounts of gabbroic or porphyritic diabase intrusions. Trachybasalt, olivine-basalt dykes are also found occasionally within the Wutai-Taihang Mountains. Perfect diabasic texture is commonly preserved, with gabbroic texture being found in the centre of relatively wide dykes. Mineralogically they mainly consist of plagioclase, pyroxene and small amount of titaniferous iron oxide, quartz or olivine. Plagioclase amounts to 40-60% of the mineral content of the rock, and is mainly labradorite (An_{50-52}) with a degree of order at d=0.4-0.8. Pyroxene (mainly augite), forms 15-30% of the dyke mineralogy. The dykes have a chemical composition of 48-52% SiO_2, 2.6-3.8% K_2O+Na_2O, 0.4-1.0% P_2O_5, and 5-8% MgO,. They have low aluminium contents but are evidently iron-rich, which indicates that the dyke magma belongs to continental tholeiite type. This in turn may be subdivided into quartz tholeiite and olivine tholeiite magma types. They are of 3-4‰ dS^{34} and 4.6-7‰ dO^{18}. Quartz tholeiitic and olivine tholeiitic dykes have an ÝREE of 160 ppm and 130 ppm respectively, and $(La/Sm)_N$ of 2.1-2.6, $(La/Yb)_N$ = 6-6.1, and high LREE without an obvious Eu anomaly. In summary the chemical characteristics of the dykes imply that the magma feeding the dyke swarms may be derived from partial melting of the upper mantle (Chen Xiaode and Shi Lanbin 1984).

Paleomagnetic study of the samples from both NW- and WNW-trending dykes in the Lüliang Mountains yield an average paleomagnetic pole position of lat. 50.7 N, long. 263, 2 E, thus indicating that the North China craton was located near the equator in Late Precambrian time (Qian Xianglin et al 1987). The pole position is similar to the that from the coeval sedimentary sequences associated with the final stage of the NE-trending Yanshan aulacogen (Zhang, W and Li, P. 1980). Thus it indicates that the dyke swarms have formed in connection with the Mid-Proterozoic rifting and were intruded in association with stress systems that formed during early abortive attempts to breakup the continental plates.

The dyke swarms in Songshan Mt. area are intruded into the Archean Dengfeng complex and the Early Proterozoic Songshan Group and overlain by the Late Proterozoic Wufoshan Group. They can be divided into diabase-porphyrite, diabase and vitrodiabase based on their compositions and textures. Through detailed mapping and statistics, dykes with six directions have been found, i.e. E-W, NE-SW (~025°), NE-SW (~070°), N-S, NW-SE (~340°) and WNW-ESE (~290°). The E-W trending dykes are dominant and approximately perpendicular to the main structural trend of the Early Proterozoic Songshan Group, which suggests that crustal shortening has happened along the direction of the principal dyke swarms and the strain imparted by regional extension occurred perpendicular to the dyke swarms (Ma Xingyuan et al. 1987).

In the Koksu and the Altun Mt. of the Tarim Protoplatform basic dyke swarms intrude in the Archean-Lower Proterozoic rocks have similar features to those in North China. They extend mainly in a northwest direction, perpendicular to the strike of the Early Proterozoic orogenic belt (Fig. 4.39). The swarms indicate a unique tectonic environment and significant tectonic-thermal events. The swarm is comprised mainly of diabase dykes with small amounts of granitic pegmatite and diorite veins and were injected during multiple periods of emplacement.

In summary the occurrence of large-scale dyke swarms indicates a major change in crustal conditions and stress and strain states. Swarm injection represents an important geological event in an anorogenic setting: - the fissure system formed by transverse spreading of a rigid crust which resulted in the rising of mantle materials and high heat flow values. In terms of the plate tectonic concept it seems like that the basic dyke swarms correspond to periods of continental dispersal and reassembly (Sawkins, 1976).

4.9 Fossils and Stromatolites

By Mid-Late Proterozoic times remarkable developments had taken place in biological evolution. Research has shown that the Middle Proterozoic biota was marked by the appearence and flourishing of eukaryotes; the start of the Late Proterozoic (Qingbaikou period) by numerous occurrences of macroscopic and advanced algae (e.g. Phaeophyta) and the end of the Late Proterozoic (the Sinian) by the gradual development of multicellular algae such as Phaeophyta and Rhodophyta as well as the first appearance and gradual expansion of metazoa such as Coelenterata, Vermes and Porifera.

At the start of the Middle Proterozoic Changcheng period at 1.8 Ga, the biota is dominated by prokaryotes such as Cyanophyta and Bacteriophyta. The microflora preserved in shales have the general features of a very thin test, simple costellae and a relatively primitive structure. Examples include *Leiopsophosphaera minor* Schep., *Trachysphaeridium simplex* Sin, and *T. hyalinum* Sin et Lin. They have the characteristic elements of *Leiominuscula pellucentis* Sin et Liu, *L. incrassata* Sin et Liu, L.L. *orientalis* Sin et Liu, *Dictyosphaera macroreticulata* Sin et Liu, D.D. *sinica* Sin et Lin etc. (Xing Yusheng et al., 1985). Advanced eukaryote algae also appeared at this time: - for example Sphaeromorphido, (Photo 4.-6, III-7), Scaphomorphida (Photo 4.-6, 1), Triangumorpha, Taeniatum (Photo 4.-6, 2) are found in the shale of the Changzhougou and Chuanlinggou Formations, at the bottom of the Changcheng System in Yanshan Mt. area of North China. The occurrence of algae fossils of eucaryotes with common diameters of more than 100 microns and tests is demonstrated by the presence of the core of *Nucellosphaeridium,* and *Favososphaeridium.* Also present are a multitude of individuals with large grains (hundreds or more than a thousand microns), complicated texture (double-text) and various shapes (Luo Qiling et al. 1985; Yan Yuzhong 1982, 1985). The chert layers (including cherty stromatolite) in the Changcheng System carbonate rocks contain algae fossils in different quantities, such as elements of *Gunflintia* and *Cyanonema* in the lower part of the Tuanshanzi Formation (Zhu Shixing 1982), together with elements of *Rhicnonema, Siphonophycus, Eomycetopsis, Palaeolynghya, Halythrix Eoentophysalis, Coniunctiophycus, Myxococcoides, Glenobotrydion, Nanococcus* and *Eosynechococcus* in the Gaoyuzhuang Formation (Zhang Yun, 1981; Zhang Zhongying et al., 1985; Zhu Shixing, 1982; Zhang Pengyuan, et al., 1984; Zhang

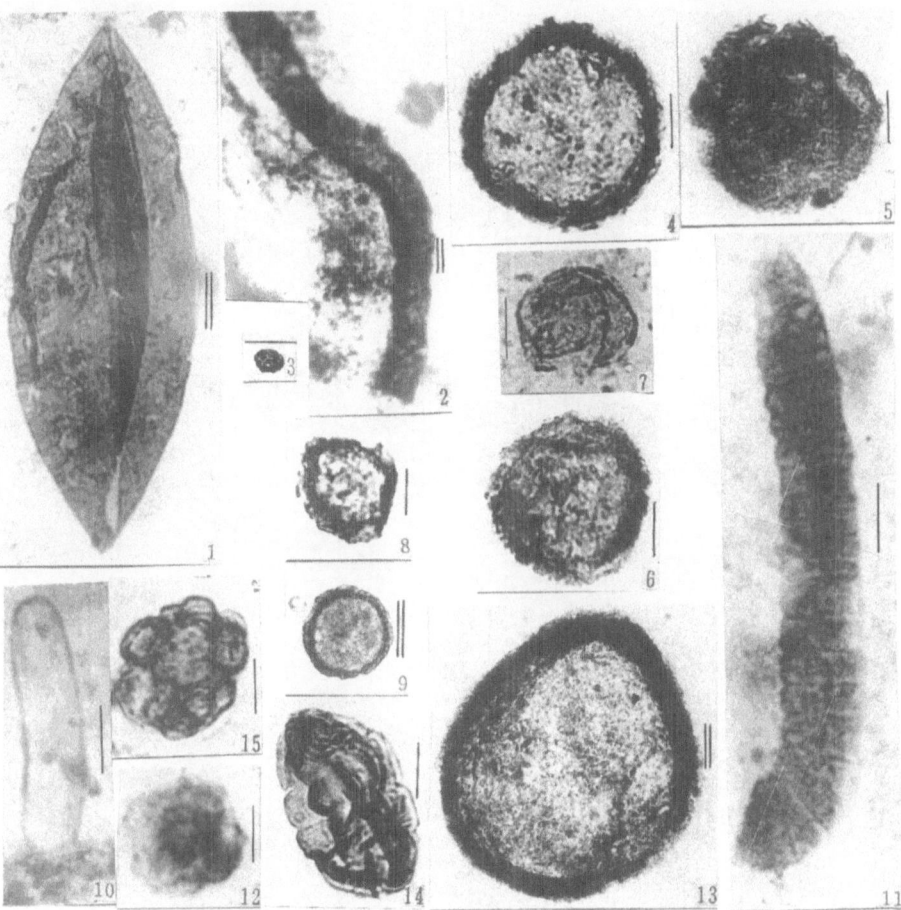

Photo 4.6. Mid-Late Proterozoic algae-fossils (provided by Luo Qiling). In addition to the sections (indicated), all of maceration. *Single and doubled scale bar in 10μ and 20μ respectively*: 1. *Leioarachnitum* sp., from the Changzhougou Formation, Kuancheng, Hebei Province; 2. *Siphonophycus capitaneum* Nyberg et Schopf, from the Chuanlinggou Fm., Xuanhua, Hebei Province (section); 3. *Leiominuscula incrassata* Sin et Liu, from the Chuanlinggou Fm., Jixian, Tianjin; 4. *Pseudozonsphaera verrucosa* Sin et Liu, from the Chuanlinggou Fm., Jixian, Tianjin; 5. *Asperatopsophospharea umishanensis* Sin et Liu, from the Tuanshanzi Fm., Jixian, Tianjin; 6. *Asperatopsophospharea partialis Schep.*, from the Dahongyu Fm., Jixian, Tianjin; 7. *Trachysphaeridium hyalinum* Sin et Liu, from the Gaoyuzhuang Fm., Jixian, Tianjin; 8.9. *Pseudozonosphare verrucosa* Sin et Liu, from the Yangzhuang Fm.(8) and Wumishan Fm.(9), Jixian, Tianjin; 10. *Archaeoellipsoides grandis* Horodyski et Donaldson, from the Wumishan Fm., Chicheng, Hebei Province (section); 11. *Oscillatoriopsis luazhuangensis* B.Zhang, from the Wumishan Fm., Chicheng, Hebei Province (section); 12. *Quadratimorpha ordinata* Sin et Liu, from the Honggshuizhuang Fm., Jixian, Tianjin; 13. *Pseudozonosphaera grossa* Luo et Li(MS), from the Tieling Fm., Jixian, Tianjin (section); 14. *Microconcentrica paratusiformisa* Lo et Sun, from the Xiamaling Fm., Jixian, Tianjin; 15. *Coneospharea inaequabalis* Luo(MS), from the Changlongshan Fm., Huailai, Hebei Province

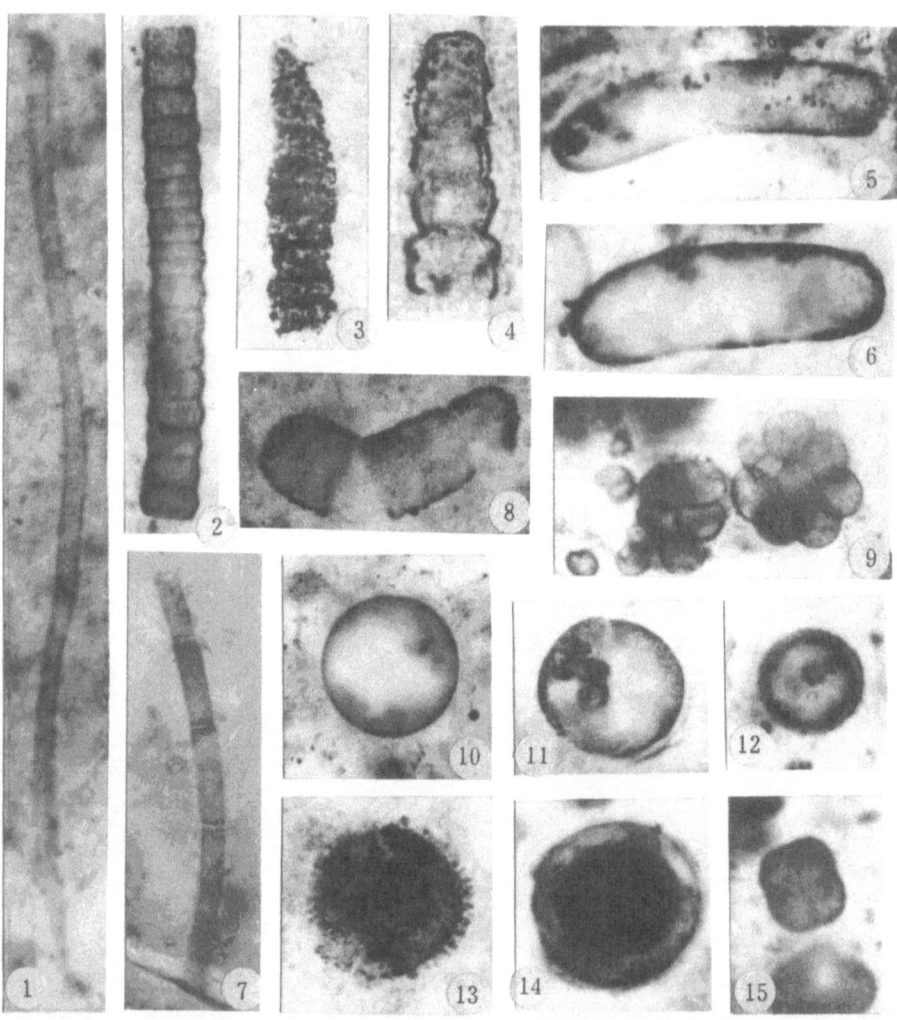

Photo 4.7. The Middle Proterozoic microfossils in Silicolites, Yanshan Mt. region (provided by Zhang Pengyuan): 1. *Eomycetopsis robusta* Schopf (x600); 2. *Hyvalothecopsis sinica* P.Zhang(x400); 3. *Oscillatoriopsis aculeata* P.Zhang(x360); 4. *Palaeolyngbya elliptica* P.Zhang(x500); 5. *Archaeoellipsoides grandis* Horodyski et Donaldson(x400); 6. Archaeoellipsoides granides Horodyski et Donaldson (x 400); 7. *Oscillatoriopsis subtilis* P.Zhang (x 300); 8. *Palaeoborizia cervicalis* P.Zhang (x450); 9. *Palaeoanacystis taihangshanensis* P.Zhang (x650); 10. *Myxococcoides grandis* Horodyski et Donaldson (x900); 11. *Melasmatosphaera magna* Hofmann (x560); 12. *Gloeodiniopsis lamellosa* Schopf (x1000); 13. *Palaeochroococcus jixianensis* P.Zhang (x600); 14. *Globophycus rugosum* Schopf (x1000)k; 15. *Tetraphycus hebeiensis* Liu (x500). 1,3,9,12 and 15 from the Gaoyuzhuang Formation, Laishui, Hebei Province; 2,4,8 and 13 from the Wumishan Formation, Jixian, Tianjin; 5,6,7,11 and 14 from the Wumishan Formation, West Hills, Beijing; 10 from the Wumishan Formation, Ming Tombs, Beijing

Pengyuan, 1987). There are also primitive macroscopic algae during the Changcheng period, including a multitude of *Taeniatum* and *Ulothrix* and secondary *Sphaeromorphide*. During the 1980s geologists also found fossils suspiciously similar to Chuariaceae in the lowest part of the Changcheng System, i.e. the Chuanlinggou-Tuanshanzi Formations (Hofmann and Chen 1981). They also found platy and banded carbonaceous fossils in the Chuanlinggou and Tuanshanzi Formations of Jixian County, and spiral, banded and other shapes of macroscopic algae fossils in the Gaoyuzhuang Formation, such as *Sangshuania sangshuanensis* Du and *Sangshuania linearis* Du (Du Rulin et al. 1985). This provides a reliable confirmation of the presence of macroscopic algae fossils in the Changcheng period.

During the late Middle Proterozoic Jixian period, the fossils that appeared in the early Middle Proterozoic Changcheng period were joined by thirty-one new species of microflora. These are preserved in argillaceous sediments, and include Quadratimorpha (Photo 4.-6, 12) *Leiopsophosphaera hyperboreica* (Tim). Xing et Liu, *L. bullata* (Andr.) Sin-et Liu, *L. solida* (Liu et Sin) Sin et Liu, *A speratopsophosphaera partialis* Schep., *A. bavlensis Schep.* etc., and fourteen new genera, such as *Favososphaeridium, Nucellosphaeridium, Orygmatosphaeridium*, most of which belong to eukaryote cell algae, i.e. Chlorophyto, Cyanophyta and Rhodophyta. Plant fragments of *Lignum striatum* with hard, dense textures and costa-like textures also occur. Their axes of protobranchlets have similar appearances with those of modern advanced algae (Phaeophyta, Chlorophyta and Rhodophyta). The *Nucellosphaeridium* in the Jixian System commonly has a differently-sized hard-dense round pit, this may be the cell nucleus of an algae cell, which is regarded as one of the characteristics of eukaryotes (Xing Yusheng et al. 1985). In the Wumishan Formation black chert in the Jixian System, geologists have found a large number of various well-preserved species and genera of microfossils, such as *Myxococcoides, Sphaerophycus, Cephalophytarion, Oscillatoriopsis, Siphonophycus, Rhicnonema, Archaeoellipsoides, Bactrophycus, Callosicoccus* and Clonophycus of Cyanophyta, Templuma sinica of Chlorophyta and *Eotetrahedrion* of Rhodophyta (see Photo 4.7; Photo 4.6, 10-11; Zhang Pengyuan, 1979, 1981, 1982, 1984, Zhang Pengyuan et al. 1986, 1989; Liu Zhili, 1982; Zhang Yun 1985). In addition, there are a plenty of carbonized filmy fragments of algae in the Hongshuizhuang Formation in Jixian County. After treatment by acid solution, under a microscope the filmy algae can be seen to contain copious cell structures, suggesting that they are a kind of multicellular algae, probably Phaeophyta.

During the Qingbaikou period at the beginning of Late Proterozoic time obvious changes in flora took place: the outline of the microflora changed and the number of the species and genuses of the microflora assemblage increased. The flora is characterized by some new types of Ulothrix such as Microconcentrica (Photo 4.6, 14) *Arctacellularia, Polyspharoides, Volyniella* and Sphaeromorphide. Sphaeromorphide contains large elements which have more complicated costellae and structures than those in the underlying beds. These include, for example, Pseudofavososphaera with a false-alveolus texture on its surface, *Brochopso-phosphaera* with very small caves, *Tasmanites* with a test cut through by dense holes, *Zonosphaeridium* with narrow-ring-like and thick-edged test, and *Reticulum* which has a braided texture. Secondly

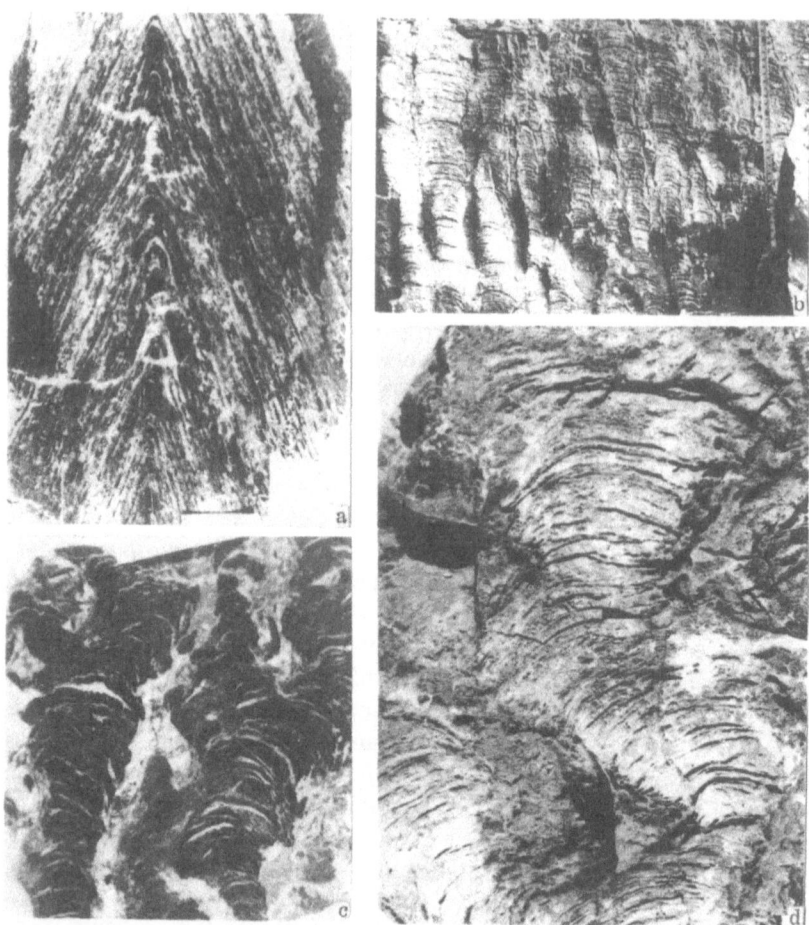

Photo 4.8. The Mid-Late Proterozoic stromatolites (in vertical section; provided by Zhu Shixing): a. *Conophyton cylindricum* (Grabau) Maslov, from the Gaoyuzhuang Formation of Changchengina System, Jixian, Tianjin; b. *Chihsienella chihsienensis* Liang et Tsao, from the Tieling Formation of Jixianian System, Jixian, Tianjin; c. Linella simica Krylov, from the Xiamaling Formaiton of Qingbaikou System, Western Hebei Province; d. *Baicalia cf. baicalica* Krylov, from the Tieling Formaiton of Jixian System, Jixian, Tianjin

the occurrence of *Laminarites antiquissimus* is another feature of the microflora. The presence of this species perhaps indicates that advanced algae entered into an exceptionally flourishing period (Xing Yusheng (1985); Duan Chenghua et al. 1985). The Qingbaikou period is also the time when macroscopic algae (e.g. Phaeophyta etc.) occurred on a large scale. The elements of Chuariaceae have been found in the Qingbaikou System of the Yanshan aulacogen in North China and the JiaoLiao-XuHuai downwarping belt. The Yanshan aulacogen also contains elements which can indicate the age of the strata, such as *Glossophyton*, *Phascolites* and

Longfengshania which are the earliest macroscopic fossils with evident separate organisms.

4.9.1 Stromatolites

Stromatolites are a biogenic sedimentary structure formed by Cyanophyta and/or Bacteriophyta absorbing precipitated minerals or capturing mineral grains during their growth or metabolism (Walter 1976; Monty 1976). The extensive occurrence of stromatolites is another remarkable feature of Mid-to Late Proterozoic times which is the time when stromatolites were most abundant during geological history (Photo 4.8). Through primary study and analysis, the stromatolites of the Middle Proterozoic Changcheng System and the Jixian System and the Upper Proterozoic Qingbaikou System can be divided into five assemblages (Zhu Shixing et al., 1987).

Stromatolite assemblage in the lower Changcheng System: This assemblage in the Jixian section is marked by abundant tabular and columnar stromatolites and the absence of Conophyton. The representative elements include *Gruneria sinensis* Zhu et al., *Xiayingella xiayingensis* Zhu et al., *X. primigenia* Zhu et al., *Djulmekella tuanshanzensis* (Liang et Tsao), *Yanshania simplex* Zhu et al, with predominant banded and occasional linear microstructures as well as a common spotted microtexture. Based on the regional data from the Yanshan and Taihang Mountains, the assemblage changes in various areas. The major features of the assemblage are described as follows: (1) although tabular columnar stromatolite is dominant in general, *Colonnella*, Komar (such as *Gaoyuzhuangia*, *Eucapsiphora*, and *Kussiella*), and partial regular or irregular *Conophyton* Maslov, (such as *Conophyton*, *Cornoconophyton* and *Tabuloconigera*) types also occur; (2) major ankeritic stromatolites form the major component of the stromatolite assemblage but special ferruginous stromatolites also occur, for example *Gruneria yantongshanensis* Zhu, *Eucapsiphora multiramis* Zhu and *Pangjiapuella palmata* Zhu; (3) Both stromatolites, such as *Gruneria*, *Katernia* and *Nordia* which occurred only within the Lower Proterozoic are present as well as others such as Parmites, Nucleella and Inzeria which are usually regarded as Late Proterozoic forms. All these are newly discovered characteristics of the stromatolite assemblage in the lower Changcheng System.

Stromatolite assemblage in the upper Changcheng System: The typical elements of the assemblage include *Tabuloconigera paraepiphyta* Zhu et al., *Confusoconophyton multiangulum* Zhu et al., *Conophyton cylindricum* (Grabau) Masl., *C. garganicum* Kor, *C. dahongyuensis* Liang, *Colonnella laminata* Komor, and the elements of *Kussiella* and *Gaoyuzhuangia* Groups. These are of generally big or medium scale in size, and form major linear as well as minor banded and massive microstructures. *Conophyton* began to develop on a large scale during this time. The distribution of this assemblage is regionally persistent, but there are a few small-sized stromatolites in some areas. *Conophyton* can extend into the stromatolite assemblage in the lower Jixian System.

Stromatolite assemblage in the lower Jixian System: This assemblage is regionally persistent and marked by the development of tiny *Colonnella* Komar which is represented by *Microstylus* and *Pseudogymnosolen* Groups. Its typical elements are *Pseudogymnosolen mopanyuensis* Liang et Tsao, P. *epiphytum* Zhao et al., *Microstylu zhaizhuangensis* Zhu et al., *Yangzhuangia columnaris* Zhu et al., *Conophyton concellosum* Liang et Tsao, C. lituum Masl., C. *shanpolingensis* Liang, *Jacutophyton furcatum* Zhao et al.. The microscopic stromatolites develop in the lower part of the assemblage and are the extension of the assemblage in the lower Changcheng System, and *Conophyton* Maslov occurs mainly in the upper part of this assemblage which can be divided into upper and lower subassemblages.

Stromatolite assemblage in the upper Jixian System: This assemblage is represented by the stromatolites of the Tieling Formation in the upper Jixian System of the Jixian section. They are generally undeveloped and non-persistent in the Yanshan Mt., Taihang Mt. and the east Liaoning regions. However, they are persistent in parts of southern, northwestern and southwestern China, and are characterized by a large amount of middle- to large-sized divergent *Colomella* Komar minus a wall but with a sheath. *Conophyton* Maslov occurs on a small scale. The microstructures include not only special banded types, such as a shell microstructure which retains light and dark laminations and markedly variable thicknesses, but also mixed microstructures comprising banded, linear and small massive ones.

Stromatolite assemblage in the Qingbaikou System: This assemblage is represented by the stromatolite in the Xiamaling Formation of the Qingbaikou System in the west section of the Yanshan Mountains. Its elements, mainly medium- to small-sized, have a variety of complex shapes, i.e. columnar, tumular, tuberose and tuberlike; intricate modes of devergence including normal ones and special tumulose, finger-like and bud-like ones; and wall, sheath, tumour and thorn on their sides. They possess mainly banded microstructures and vague spotted, lumpy and wiry microtextures. The typical elements of the assemblage are the genera of *Gymnosolen*, *Linella*, *Jurusania*, *Inzeria*, *Minjaria*, *Qingbaikounia* and *Katuikania* Groups. Conophyton occurs locally.

4.10 Tectonic Framework

The North China, Yangtze and Tarim Protoplatforms, and to a lesser extent Cathaysia, the south margin of the Siberia-Mongolia plate, and the north margin of Gondwana, are different from each other in terms of rock associations and deformational and metamorphic features, indicating that these tectonic provinces have experienced diverse tectonic evolutions. These are summarized below:

(1) The North China Protoplatform is the largest and earliest-consolidated rigid massif formed by the Lüliang-Zhongyue movement. During the Middle Proterozoic, taphrogenesis died out in its interior but the north and south margins became severely depressed. Although large-scale extension once occurred in the interior of the platform during an early stage of the Yanshan aulacogen and the Xiong'er-Hangao aulacogen, it did not proceed to the extent of causing splitting of the continental crust and the initiation of the oceanic crust. Instead, extension died out and gave way to sinking on a larger scale, leading to the formation of sediments of a stratigraphic cover. But on the north and south margins, severe depression-forming processes persisted. In particular, taphrogenesis and divergence of the northern Qinling-northern Huaiyang trough on the south margin resulted in the separation of the Qinling-Dabie massif from the North China Protoplatform. In addition, in the presence of gravity instability factors locally in the interior of the protoplatform, large-scale gravity tectonics represented by the Wufoshan gravity gliding structure took place along some incompetent planes. Anorogenic magmatism, marked by the anorthosite - rapakivi granite association and large swarms of basic dykes, also suggests the completion of cratonization of the North China Protoplatform.

(2) The Yangtze Protoplatform was the centre of the Middle to Late Proterozoic crustal movement in China. Around the Late Archean-Early Proterozoic stable craton centred on Sichuan a trench-arc basin system developed with active continental margins and rock assemblages associated with mobile tectonic environments. Complex orogenic deformation, strong orogenic-type magmatism and regional metamorphism, tight linear fold-belts, and a large network of faults all reflect a complicated tectonic evolution through which a unique tectonic style was formed. However the general tendency was that under the influence of the Sibao, Jinning and Chengjiang movements, the continental crust was progressively consolidated as tectonism gradually weakened. After the Jinning movement, the main body of the Yangtze Protoplatform entered a relatively stable stage of tectonic development.

(3) The Qaidam massif, also characterized by the development of mobile belt sediments around an old basement, was reworked to varying degrees by deformation and metamorphism which resulted from the Tarim movement (equivalent to the Jinning movement). This is shown by lateral accretion of the continental crust.

(4) The presence of "Cathaysia" is indicated by the abundant isotope age data acquired for Precambrian rocks and the discovery of the unconformity between Sinian and pre-Sinian age strata. "Cathaysia" is composed of a pre-Sinian metamorphic sequence along the southeast coast of China (including parts of the Fujian, Zhejiang and Guangdong regions as well as the continental shelf of the East China Sea and the South China Sea). Cathaysia was finally consolidated by the Jinning movement. During Sinian times, taphrogenesis occurred in the interior, forming bimodal volcanic sequences, whereas its northwest margin was a passive continental margin.

(5) The Middle to Late Proterozoic mobile belt of southern Tibet was located on the north margin of Gondwana, and was probably completely consolidated after the

"Qomolangma movement". Since the Phanerozoic it has drifted northward along with the India plate to its new location in southern Tibet.

(6) During Late Proterozoic time, the north margin of the Yangtze Protoplatform converged with the Qinling-Dabie massif which had separated from the North China Protoplatform. Proof of this convergence is the existence of a nearly 2000 km-long Precambrian glaucophane schist zone and other high-pressure metamorphic features along the north margin of the Yangtze Protoplatform. The southeast margin of the Yangtze Protoplatform and Cathaysia are also present in the western Zhejiang region.

In summary, the Middle to Late Proterozoic (1800-600 Ma) represents an important turning point in tectonic evolution. The whole process, from the Early Proterozoic primary craton to the creation in the Sinian (Late Proterozoic) of lithospheric plates similar in scale and composition with those of today and from the occurrence of modern-type convergent plate boundaries to the complete "opening" and "closing" (Wilson cycle) of plates, fully reflects the operation of a modern plate tectonic regime.

Major "leaps" also took place in biological evolution during the Middle-Late Proterozoic. The first leap is marked by appearance and mass multiplication of eukaryotes during Middle Proterozoic time, the second leap by the appearance of mass macroalgae and higher algae (e.g. brown algae) at the beginning of the Late Proterozoic, and the third leap by the occurrence of the metazoan fauna towards the end of Late Proterozoic (Sinian) time.

5 Sinian Crust

Huang Xueguang
Tianjin Institute of Geology and Mineral Resources
Tianjin, China, 300170.

5.1 Distribution and General Features

As a result of the 800-850 Ma Jinningian Orogeny, the Yangtze and Tarim cratons were fully consolidated. The pronounced unconformity which developed at that time subsequently formed the base of the Sinian System (850-570 Ma).

The Sinian System is invariably represented by a stratigraphic sequence deposited on stable platforms. In many places it is comformably overlain by Lower Cambrian, and for this reason many geologists in China tend to regard it as the first system of the Paleozoic. However the Sinian system is well developed only in the Yangtze Platform and its margin, whereas Sinian successions in the North China and Tarim Platforms are incomplete and of limited distribution (Fig. 5.1, Fig. 5.2).

Three types of Sinian succession may be distinguished, and are represented by the sections in the Upper Yangtze, the Kuruktag of Xinjiang and the eastern part of North China respectively (Xing Yusheng, 1976; Liu Houngyun et al. 1991). Semi-stable to mobile types of Sinian strata are developed along the Jiangnan Uplift and further to the southeast. In the southeast maritime provinces the Sinian occurs in a NE-SW-trending belt roughly parallel to the Jiangnan Uplift. The Sinian here is continuous with underlying and overlying strata, and is composed mainly of silicolites and volcanics, which were probably formed in an ocean floor environment (Wang Hongzhen, 1984). Conditions are similar in northeastern China, where on the western border of the Jiamusi massif, northeastern Heilongjiang Province, an Ediacara fauna including Glaessnerina and Arumberia was found near Jixi in metamorphic carbonates (Liu Xiaoliang, 1981).

5. 2 Sinian Crust in South China

Towards the end of Late Proterozoic time, Sinian sediments were extensively developed in southern China: both on the Yangtze Platform and Cathaysia and in the relict oceanic basin between them (Fig. 5.3). Their stratigraphic division and correlation as well as lithologic associations are shown in Table 5.1.

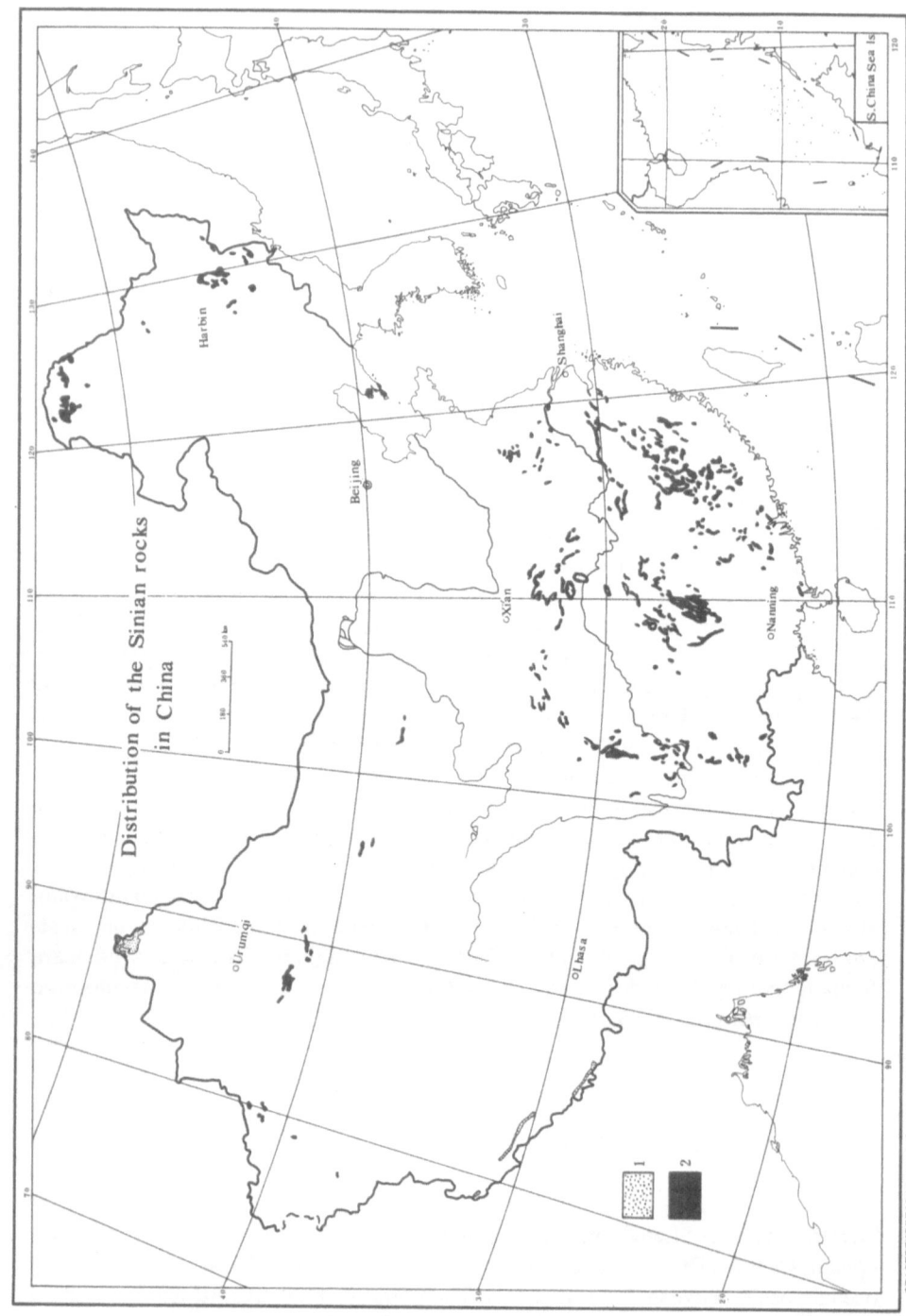

Fig. 5.1. Distribution of Sinian rocks in China

5.2.1 Yangtze Platform

After being strongly influenced by the Jinning movement, the Yangtze Platform became comparatively consolidated and stable, consisting of the Early Proterozoic-Archean crystalline basement and the Middle Proterozoic metamorphic, folded basement, upon which a stable platform Sinian sedimentary cover, represented by the stratigraphic section near Yichang in the eastern Yangtze Gorges, was formed. Since 1924 when J.S. Lee systematically studied the section and defined its upper and lower boundaries and its internal division (see Lee and Chao, 1924), the section has been regarded as a stratotype of the Sinian in southern China. Later a number of geologists (Zhao Ziqiang et al. 1980 1988; Xing Yusheng 1979, 1984; Liu Hongyun et al. 1991) studied it in detail using lithostratigraphic, biostratigraphic, chronostratigraphic, magnetostratigraphic and geochemical-stratigraphic methods. Large amounts of data now indicate that the sequence represented by the Sinian of the eastern Yangtze Gorges is characterized by persistent and widespread distribution. It contains horizons which enable clear stratigraphic division and correlation, shows evident features of paleobiota, good comparability with localities elsewhere, and exhibits a fair transitional relationship with overlying Cambrian strata. The Sinian is now defined as a chronostratigraphic unit of system rank in the uppermost part of the upper Precambrian in China (Commission on Stratigraphy of China, 1983). It is also of significance in global stratigraphic correlation.

The Sinian System is divided into two series, and further subdivided into a number of formations. In the eastern Yangtze Gorges area near Yichang, the Lower Sinian includes the Liantuo Formation and Nantuo Formation which are separated by a parallel unconformity. The Nantuo Formation is unconformably overlain by the Upper Sinian, which comprises the Doushantuo Formation and Dengying Formation. The Upper and Lower Sinian represent two different stages of the structural evolution of the Yangtze Platform.

The folding and uplifting of the main body of the Yangtze Platform during the Jinning movement resulted in much of the platform being covered by clastic sediments (Fig. 5.3) at the beginning of the Early Sinian (Liantuo epoch). In the eastern Yangtze Gorges area, the lower part of the Liantuo Formation, which unconformably overlies the Late Archean Kongling Group is dominated by fluvial feldspar-quartz sandstone with sandy shale. The thickness of the sequence is 50-260 m, and the terrigenous clasts with a low maturity in the rocks were derived from the nearby Huangling oldland to the north (Zhao Ziqiang et al. 1980). In the eastern Yunnan area on the west part of the Yangtze Platform, the N-S trending rift further evolved into a graben-type basin in the Late Proterozoic Qingbaikou epoch. The Chengjiang Formation was deposited in this basin at the beginning of Early Sinian times, and consists of the continental fluvial sediments unconformably overlying the Middle Proterozoic Kunyang Group. It is mainly comprised of a sequence of purplish-red, coarse clastics. The base of the Formation often consists of thick conglomerate and gravel-bearing sandstone. This is overlain by purplish red feldspar-quartz sandstone and lithoclastic sandstone intercalated with thin siltstone and tuff, with acidic or intermediate volcanics and pyroclastic rocks occurring locally. The sandstones vary

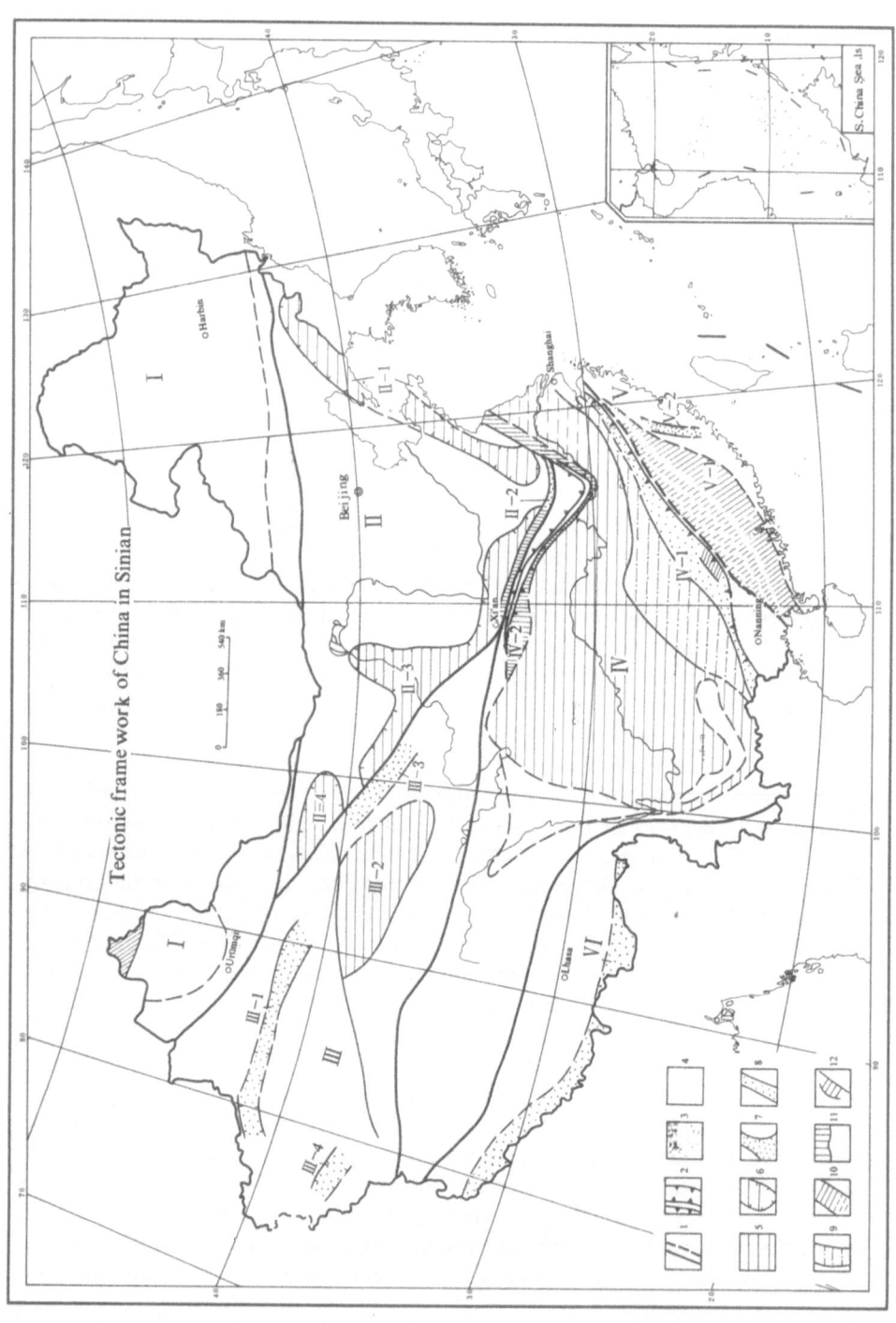

Fig. 5.2. Tectonic framework of China in Sinian times: 1. Prephanerozoic suture; 2.subduction belt; 3.oceanic crust; 4.oldland; 5.cover of platform; 6.cover of platform in downwarping belt; 7.aulacogen; 8.mobile belt; 9.back-arc basin; 10.passive continental margin; 11.sediment on suture; 12.island-arc volcanic; I.South margin of Siberia-Mongolia plate; II.North China Platform; II-1.Jiaolian-Xuhuai depressed belt, II-2.north Qinlin-northern Huaiyang ocean trough; II-3.depressed belt in the southwestern North China Platform, II-4. Beishan depressed belt; III.Tarim Platform; III-1.Tianshan rift, III-2.depressed belt in the northern Qaidam block; III-3.Qilian Mt. aulacogen, III-4.western Kunlun mobile belt; I5.Yangtze Platform; IV-1. mobile belt on the southwest margin, IV-2.collision depressed belt on the north margin; 5.Cathaysia oldland; V-1.passive continental margin in the northwest; V-2.inland rift; VI.north margin of Gondwana Continent

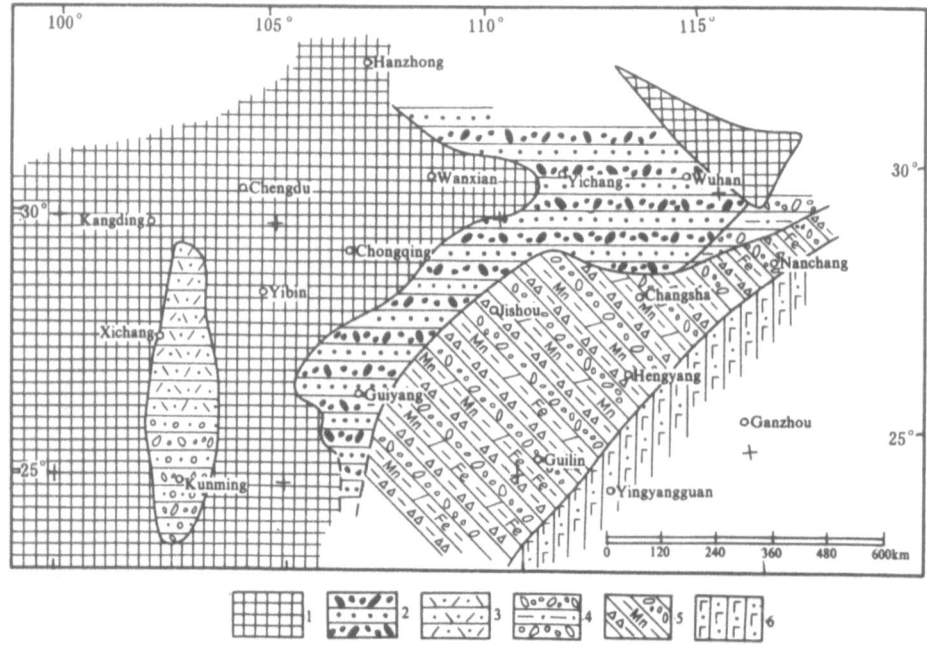

Fig. 5.3. Lithofacies paleogeographic map of Early Sinian of the Yangtze Platform (after Wang Hongzhen 1985): 1. Denudated oldland; 2. terrestrial clasts-till assemblage; 3. terrestrial pyroclasts-bearing fluviaglacial assemblage; 4. littoral clasts-till assemblage; 5. marine clastic gravity flow-till assemblage; 6. marine spilite flysch assemblage

in distribution and lithology, and exhibit large-scale cross-bedding (trough- and wedge-like crossbeds) as well as ripple marks and mud cracks. The thickness of these units is also highly variable, being generally between 300-700 m, but with maximum thicknesses between 2000-3200 m. Furthermore, scattered terrestrial clastics of the Liantuo Formation are found in eastern Guizhou, showing a decrease of grain size, an increase in silt content, a gradual change in rock colour from purplish -red to greyish-green and a reduction in the content of volcanic clasts and lava from west to east.

In early Sinian times, the east part of the Yangtze Platform was covered by marine sediments. The Dongmen Formation in northwestern Jiangxi consists of gravel-bearing quartz sandstone at the bottom overlain by feldspar-quartz sandstone. The sandstone exhibits a well-developed, coarse to fine rhythmic layering and frequent trough- or wedge-like crossbeds, reflecting a littoral, high-energy sedimentary environment. The middle and upper part of the Formation is composed of fine-grained lithoclastic sandstone, siltstone and mudstone, with well developed horizontal stratification and minor lenticular bedding, reflecting a back-barrier low-energy sedimentary environment dominated by tidal action. In comparison the Xiuning Formation in southern Anhui consists of littoral terrigenous and volcanic clastics, consisting of conglomerate and sandstone overlain by sandstone, siltstone and mudstone, commonly intercalated with tuff and other volcanic clastics, containing well-developed primary sedimentary structures such as beach-wash crossbed, tidal bedding and ripple marks. This again reflects a sedimentary environment dominated by tidal action (Xiong Xingwu 1988).

The first episode of Early Sinian sedimentation was terminated by the Chengjiang movement, which caused further uplifting of the entire Yangtze Protoplatform and formed a break in sedimentation with the overlying Nantuo Formation. Towards the end of the Early Sinian the climate abruptly became cold and the platform was covered by mountainous and continental glaciers. In the Yangtze Gorges area, the Nantuo Formation mainly comprises a set of dark-green, greyish green and purplish red continental glacial gompholites. In the west of the platform, the glacier moved towards the topographically lower eastern Yunnan region, a downfaulted basin where moraine conglomerate between ten to more than one hundred metres thick accumulated. Subsequent glacial melting and retreat resulted in the formation of a glacial lake where purple silty shale tens of metres thick with horizontal, banded varved beds were deposited. In the western Sichuan downfaulted basin the Lieguliu Formation was deposited. This is composed chiefly of greyish-green and purple silty shales with underwater laminated banding, locally containing volcanic ash and clasts, totalling 100 to over 300 m in thickness. In comparison, in the eastern part of the platform within the southern Anhui region, there is a sequence called the Leigongwu

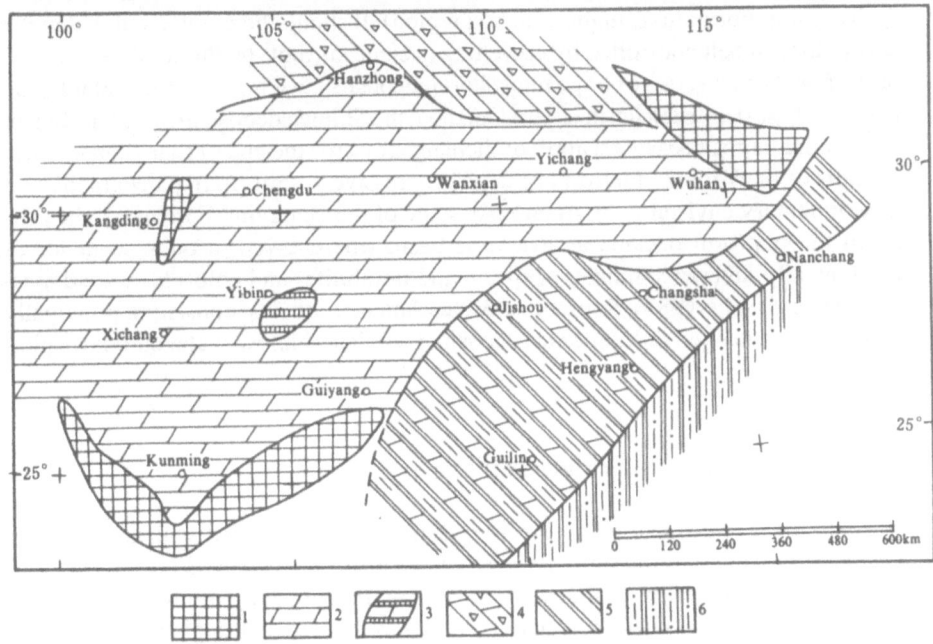

Fig. 5.4. Lithofacies paleogeographic map of Late Sinian of the Yangtze Platform: 1. Denudated oldland; 2. surfacial sea dolomite assemblage; 3. dolomite-gypsum-mirabilite assemblage; 4. gravity flow-carbonate assemblage; 5. carbonaceous siliceous-dolomite assemblage; 6. siliceous-flysch assemblage

Formation, which is equivalent to the Nantuo Formation. It consists of tuffaceous, moraine gompholite intercalated with sandstone, quartz siltstone, silty mudstone and dolomitic limestone, reflecting a marine-glacial sedimentary environment (Wang Xiangfang et al. 1983).

Having undergone a period of prolonged denudation during Late Sinian times, the Yangtze Platform underwent a large-scale transgression from east to west, flooding the fold belt inverted in the Middle and Late Proterozoic on the west margin of the platform. As a result, the platform was largely covered with sea water, leaving only several residual, isolated, island-like oldlands (Fig. 5.3). In the broad, shallow-water, epicontinental sea environment a dolomite-dominated rock assemblage was deposited. The lower part of the Late Sinian Doushantuo Formation in the eastern Yangtze Gorges consists of greyish white siliceous dolomite with disturbed structures overlain by dolomite exhibiting microbedding or slaty bedding and intercalated with carbonaceous, Mn-bearing shale, the products of a closed reducing environment in the central part of the platform. The upper part of the Formation is composed of greyish white massive dolomite intercalated with lenticular chert beds and oolitic silicalite, indicating a progressively more open environment. In the western part of the platform, the sequence equivalent to the Doushantuo Formation is called the Guanyinya Formation. It is made up of sediments of tidal-flat, lagoonal facies. However, the tidal-flat clastic sequence (with carbonate rocks) near the island-like oldlands, is termed the Labagang Formation (in the northwestern Sichuan region) or the Yangshui Formation (in the Guizhou region). It should be pointed out that in the neritic-littoral belt controlled by upwelling oceanic currents on the southeast margin of the platform, a series of large phosphorite deposits have formed: for example, the Kaiyang-type deposit in northeastern Guizhou, the Shimen deposit in northern Hunan, and similar deposits east and north of Yichang. Towards the end of Late Sinian times, all parts of the Yangtze Platform became progressively uniform in paleography and in sedimentary environment. At an early stage of the Dengying Formation, the centre of the platform became an evaporite basin where a 480 m thick sequence of rhythmically layered sulphates (gypsum and mirabilite) and chlorides were deposited (Yang Xianhe 1987). Subsequently a carbonate sequence consisting principally of dolomite and algal dolomite was deposited within a shallow-water carbonate platform environment.

5.2.2 North Margin of the Yangtze Platform

In the Late Proterozoic when the Yangtze Platform coalesced with the Qinling-Dabie massif, the sea water did not retreat from its north margin (Zhou Gaozhi et al. 1989). Therefore no big sedimentary break and unconformity are found in this region. Instead, continual sedimentation is observed between the glaucophane-bearing Doushantuo Formation and the Dengying Formation, which does not contain glaucophane. The glaucophane schist in the Doushantuo Formation is the high pressure metamorphic product formed in the collision zone between the North China Protoplatform and the Yangtze Platform (see chapter 3.4.1). However the carbonate rocks of the Dengying in the middle segment of the north margin are the exclusive products of bathyal or abyssal environments that include gravity-flow sedimentation in continental slope environments, slide and slump type deposits and abyssal basin stagnant-water sediments. These types of deposits are suggestive of the underwater environment of a deep enclosed basin in the collision zone (Xiang Cai, 1988).

5.2.3 Mobile Belts of the Southeast Margin of the Yangtze Platform

In Sinian times the southeast margin of the Yangtze Platform remained as an active continental margin, and as a consequence the Sinian can be divided into two zones on the basis of tectonic activity. The inner zone was developed on a folded basement composed of the Middle Proterozoic Sibao Group or Fanjingshan Group, and Sinian sedimentation began in marine facies environments (Fig. 5.2). In general, the inner zone is an active zone of miogeosyncline character (Fig. 5.4). The lower part of the Lower Sinian (i.e. the Tiesi'ao Formation in eastern Guizhou, the Jiangkou Formation in central Hunan, or the Chang'an Formation in northern Guangxi) is a clastic rock association of complicated lithology and intense facies changes. According to Wang Yangeng et al. (1986), the bottom of the association is an olistostrome composed of rock blocks different in composition and size, irregular in shape and poor in sorting, indicating that the coast was steep at that time. The lower part is a set of massive, polymictic conglomerates with clasts composed of rudaceous sandstone, rudaceous mudstone, and diamictite, whereas the middle to upper part of the sequence is dominated by lithoclastic wacke and sandstone. The peak of the frequency curve of grain size for the matrix material of the polymictic conglomerate appears at 1-2 Phi, comparable to that for a gravity-flow sediment. The polymictic conglomerates are mostly massive and unstratified, but some of them are interbedded with rudaceous sandstone, showing normal and reverse rhythmic changes in grain size, convolute bedding, slump structure, basal scour, parallel and oblique bedding, and gravel orientation.

The chemical composition of the sandy matrix of the conglomerate is also useful in understanding the environment of deposition: the SiO_2 content is relatively stable, ranging from 59.85 to 67.55%, averaging 64.29%; the Al_2O_3 content is higher than that of normal sandstone; $FeO>>Fe_2O_3$; with $K_2O/NaO<<1$. These indicate that it is a gravity-flow sediment formed in a continental margin environment, rather than a tillite or glacial rock (Wang Yangeng et al. 1986). The middle part of the Lower Sinian includes the Datangpo Formation (in Guizhou), the Xiangmen Formation (in central Hunan) and the Fulu Formation (in northern Guangxi), which are composed of marine carbonaceous shale, silty shale and dolomite with rhodochrosite (in Hunan and Guizhou) or hematite (in northern Guangxi), and are widely and persistently distributed. However, this sequence is absent on the Yangtze Platform. Also found in this zone is a till sheet of the Nantuo Formation of the upper part of the Lower Sinian is a set of glaciomarine sediments. The Upper Sinian Doushantuo Formation is made up of dark grey carbonaceous shale with dolomite. The overlying Liuchapo Formation (central Hunan and eastern Guizhou), is equivalent to the Dengying Formation,

Strata	Xiadong[1]	Central Guizhou[2]	Central Yunnan[3]	W. Sichuan[4]	S. Anhui[5]	NW Jiangxi[6]	NE Jiangxi[7]	N. Margin of Yangtze Plt. — N Hubei[8]
Upper Series	Dengying Fm dolomite	Dengying Fm dolomite	Dengying Fm dolomite and shale dolomite	Hongchunping Fm dolomite	Piyuancun Fm silicalite and siliceous shale	Dengying Fm dolomite and limestone	Upper Xifengsi Fm: siliceous rocks	Dengying Fm dolomite
Upper Series	Doushantuo Fm dolomite and black shale	Doushantuo Fm dolomite and phosphorite	Wangjiawan Fm sandstone, shale and carbonate	Guanyinya Fm limestone, marl and sandstone	Lantian Fm carbonate and shale	Doushantuo Fm shale and siltstone	Lower Xifengsi Fm: siliceous shale and dolomite	Doushantuo Fm argillaceous, arenaceous and carbonate rocks
Lower Series	Nantuo Fm till conglomerate	Nantuo Fm till conglomerate	Nantuo Fm shale and till conglomerate	Lieguliu Fm stratified tuff, siltstone, silty mudstone	Leigongwu Fm tillitic conglomerate (tuffic)	Nantuo Fm tillitic conglomerate (pyroclastic)	Leigongwu Fm tillitic conglomerate (tuffic); Menghushan Fm. sandy shale; Langkou Fm. olistostrome	
Lower Series	Liantuo Fm. claystone, siltstone, sandstone, conglomerate	Liantuo Fm. feldspar-quartz sandstone	Chengjiang Fm lithic sandstone		Xiuning Fm siltstone, sandstone and conglomerate	Dongmen Fm shale, siltstone, sandstone and gravel sandstone	Zhitang Fm shale, tuffaceous sandstone conglomerate	
Rock assem. type (Sinian System)	Stable sandstone-shale-carbonate rock assemblage							Collision belt continent slope basin carbonate assemblage

Table 5.1. Correlation of Sinian Rock assemblages in Southern China Sources of data: [1] Zhao Ziqiang et al. (1980); [2] Wang Yangeng et al. (1980); [3] Cao Renguan et al. (1980); [4] Yinjicheng (1980); [5] Geological Survey of Anhui D...

Strata	Southeastern margin of Yangzi Platform					NW margin of Cathysia	Cathysia
	Inner belt		Outer belt				
	E. Guizhou 9	Central Hunan 10	N Guangxi 11	Guangdong-Guangxi-Hunan border 12	Central Jiangxi 13	S Jiangxi 14	SW Fujian 15
Upper Series	Dengying Fm dolomite and limestone	Liuchapo Fm dolomite and silicalite	Laobao Fm siliceous shale carbonaceous shale	Xialong Fm upper part: quartz sericite slate and basalt	Laohutang Fm meta-tuff siltstone and phyllite, with a flyschoid member in the upper part	Baii Fm flint rock and siliceous slate, greywacke, feldspar-quartz sandstone	Huanglian Fm phyllite, meta-quartz sandstone, silicalite and acid volcanics
	Doushantuo Fm carbonaceous shale and dolomite	Doushantuo Fm carbonaceous shale and dolomite	Doushantuo Fm carbonaceous shale	lower part: sericite slate and dolomite lens			Nanyan Fm fine sandstone and phyllite with silicalite
Lower Series	Nantuo Fm tillitic conglomerate	Nantuo Fm tillitic conglomerate	Nantuo Fm tillitic conglomerate	Nantuo Fm tillitic conglomerate	Yangjiaqiao Fm upper part: glacial and sandy conglomerate	Shabahuang Fm upper part: gravel-bearing sandstone	Dingwuling Fm meta-quartz sandstone, siltstone, phyllite with splite-keratophyre and meta-conglomerate at the base
	Datangpo Fm carbonaceous shale bearing carbonate manganese ore	Xiangmeng Fm carbonaceous shale bearing carbonate - Mn ore	Fulu Fm hematite, silicalite and ferriferous ore	Yingyangguan Fm spilite and keratophyre	middle part: Mn-dolomite, tuffic sandstone and magnetite-quartzite	middle part: magnetite (pyrite)	
	Tisiao Fm gravity-flow complex conglomerate	Jiangkou Fm gravity-flow complex conglomerate	Changan Fm gravity-flow complex conglomerate	Changan Fm gravity-flow complex conglomerate	lower part: tholeiitic pyroclastics, olistostrome	lower part: gravel bearing sandstone	
	Liantuo Fm	Liantuo Fm feldspar-quartz sandstone and lithic wacke					
Sinian System	Miosynclinal gravity flow complex conglomerate-abyssal-byssemal silicalite carbonaceous shale assemblage		Island-arc calc-alkalic volcanic-flyschoid assemblage			Passive continental margin terregineous clastics formation	Intracontinental bimodal volcanic-clastic formation

Table 5.1. Continued. Sources of data: [9,10] Wang Yangeng et al. (1986); [11,12] Zhang Taijui et al. (1988); [13,14] Liu Hongyun et al. (1980); [15] Li Genkun (1989).

is dominated by silicallcite with thin siliceous dolomite. The Laobao Formation (northern Guangxi) is formed principally of greyish-black silicalite, with carbonaceous and siliceous shales being present locally. The sedimentary environment gradually became relatively quiet, bathyal-abyssal in character.

The outer zone is more active than the inner one (Fig. 5.3). The lowermost Sinian deposits in the southwest segment of the outer zone are characterised by flysch deposits. These deposits consist of dark-colored polymictic conglomerate, wacke and slate exhibiting massive and graded bedding and very low maturity of composition and texture. The accumulative probability curve of granulometric analysis shows a CM linear distribution. All these factors are characteristic of a turbidity current deposit (Meng Xianghua et al. 1989). These deposits are overlain by the Yingyangguan Formation, which consists chiefly of thick spilite, keratophyric tuff and tuff with less sericite and calcareous phyllites, quartzite, dolomite and phosphatic lenses, and also with banded and podiform silicalites in keratophyre. The formation represents a spilite-keratophyre formation generated in a bathyal environment. Chemical analysis suggests that the Yingyangguan Formation spilite is a slightly intermediate to basic lava characterized by high Fe, high alkalis, low Ca and low Ti. Its Fe_2O_3/FeO ratio (>1) and SiO_2 content is close to that of island-arc andesite. Statistics based on the FeO/MgO versus SiO_2 and the Na_2O+K_2O versus SiO_2 diagram indicate that subalkaline lithologies (made up of approximately equal amounts of tholeiite and calc-alkaline conponents) accounts for 69.9% of the total thickness of the volcanics, whereas alkaline rocks (mainly spilite) constitutes 26.7% of the thickness. The chemistry of these rocks is essentially consistent with the volcanics of island-arc areas and represent late stage products of arc magmatism (Meng Xianghua et al. 1989; Zhang Taigui et al. 1988). The Nantuo tillite sheet is also developed towards the top of the Lower Sinian sequence, but here is notably thin. The Upper Sinian sequence consists of sericite slate containing dolomite lenses, with amygdaloidal dacite occurring towards the top. In the northeast segment (central Jiangxi) of the outer zone, the lower part of the Lower Sinian sequence is made up of volcanic-breccia tuffaceous phyllite and locally tholeiite and pyroclastic rock. The middle horizons of the sequence is made up of metatuff and phyllite with Mn-bearing dolomite and magnetite quartzite; and the upper horizons of approximately 10 m thick greyish black, carbonaceous phyllite with glacial conglomerate and sand-conglomerate. The Laohutang Formation of the Upper Sinian is composed of flyschoid sediments dominated by metatuffaceous silicalite, totalling several hundred to more than one thousand metres thick. Recently a Sinian-Early Cambrian carbosiliceous mudstone-spilite sequence has been found near Nanchang, Jiangxi Province (Di Ruiji et al. 1989), which is taken to be further proof of the island arc nature of the outer zone. In summary, the Sinian of the outer zone is an association of bathyal-abyssal terrigenous flysch and spilite-keratophyre, representing a mobile belt of island-arc character, whereas the corresponding sediments of the inner zone are of miogeosynclinal nature and the products of a back-arc basin. These two zones constitute an active continental margin on the southeast edge of the Sinian Yangtze Platform, with the position shifted southeastward from that in Middle Proterozoic and early Late Proterozoic times.

Strata		Eastern		Southern			Western
		North Sector[1]	South Sector[2]	East Sector — Marine Aulacogen[3]	West Sector — N Qinyu aulacogen[4]	Southern Depression Western Henan Xiaoqinyu Region[5]	Western Depression Yinchuan Region[6]
Lower Cambrian		Jianchang Fm.	Houjiashan Fm.			Xinji Fm.	Suyu Fm.
Sinian System	Upper Series	Langan Group Suxian Group — stable sandstone-shale-carbonate assemblage	Jinxiang Group — stable sandstone-shale-carbonate assemblage	Fozling Group — abyssal flysch assem.	Taouan Group — Miogeo-synclinal coarse clastic-fine clastic-carbonate assemblage	Dongpo Fm Luoquan Fm Dongjia Fm Huanglian Fm — stable sandstone, conglomerate, shale and carbonate series with tillite	Zhengmuguan Fm — Upper member: stable sandstone siltstone assemblage
	Lower Series	Xuhua Group — stable sandstone-shale-carbonate assemblage	Wuxingshan Group — stable sandstone-shale-carbonate assemblage	rift-type bimodel volcanic assem.			Zhengmuguan Fm — Lower member: drift conglomerate
Underlying Series		Xihe Group of Qingbikouan	Bagongshan Group of Qingbikouan	Luanchuan Group		Luoyu Group	Wangquankou Fm of Jixianian

Table 5.2. Correlation of Sinian rock assemblages in North China Platform. [1] Chang Shaoquan 91980); [2] Regional Geological Party of Anhui Province 91985); [3] Xu Guizhong et al. (1988); [4] Geng Shufang (1989); [5] Guan Baode et al. (1980); [6] Yang Zhende et al. (1988).

5. 2. 4 Cathaysia

The Sinian crust of the Cathaysia interior was different from its northwest margin. The sediments at the margin are characterized by abundant coarse clastics. For example, in southern Jiangxi, the lower part of the Lower Sinian is composed of palimpsest feldspar-quartz sandstone and wacke, locally with magnetite beds. This is overlain by black slate and metatuffaceous sandstone; and subsequently by gravel-bearing sandstone, conglomerate and gravel-bearing slate with common slate. In comparison the Bali Formation of the Upper Sinian is several hundred to more than 1000 m in thickness, of which the lower part is feldspar-quartz sandstone with black slate and the upper part is greyish-black, medium- to thick-bedded chert with siliceous slate and metawacke. The abundant terrigenous feldspar, quartz, lithic and other clasts contained in Sinian strata were derived from Cathaysia, and these and other sediments indicate that the area was tectonically a passive continental margin. The Sinian is also developed in the interior of Cathaysia but is only sporadically exposed due to metamorphism, deformation, structural disruption and later covering and is therefore poorly studied.

There are still contrasting views about the Sinian deposited on Cathaysia because of a lack of reliable isotope age data and paleontologic evidence. Available data suggest that the Sinian of southern Fujian is a comparatively representative sequence. It rests unconformably on the Louziba Group of the Middle-Upper Proterozoic (Li Genkun 1989), and lithologically consists of marine volcanics and volcanoclastics, with argilloarenaceous clastic rock and calc-siliceous sediments of normal marine facies (Zhu Yulin et al. 1986). Two phases of volcanism took place, one in the Early Sinian and the other in the Late Sinian to Early Paleozoic times. Early volcanic eruption is dominated by spilite and keratophyre, with quartz keratophyre and subordinate amounts of corresponding pyroclastic lavas and pyroclastics. In contrast later eruption is chemically bimodal, marked by a dacitic (or keratophyric) tuff and rhyolitic tuff with rhyolite. On the K_2O+Na_2O/SiO_2 diagram for discriminating alkaline and nonalkaline rocks (Miyashiro, 1975), the components of the volcanic rocks mostly fall in the alkaline area; the REE distribution of basic volcanic rocks is of a LREE enrichment type, with the LREE/HREE+Y values mostly ranging from 1.2 to 5.2; the REE distribution of acidic to intermediate-acidic volcanic lavas and pyroclastics is also of a LREE enrichment type and has a steep slope. These chemical characteristics are similar to those of continental rift-type volcanics (Zhang Weiquan 1986). In addition, the U-Pb concordia intersection of a group of zircon samples from tuffite yields a crystalline age of 643+73/-78 Ma. Similarly the whole-rock Rb-Sr isochron age of metamorphic silicalite and phyllite is 605.7±57.6 Ma (Zhu Yulin et al. 1986). The micropaleobotanic fossils found in the strata have been determined to be elements of the Doushantuo-Dengying Formation (Li Genkun 1989). The Sinian here is conformable to the overlying Cambrian System and has, together with the lower Paleozoic, undergone a subgreenschist facies metamorphism and tectonic deformation.

5.3 Margin of the North China Platform

After the Jixian epiorogenic uplift, sea water retreated from the North China Platform, which subsequently underwent a long-term denudation, and received sediments only in the marginal depressions (Fig. 5.1). The stratigraphic divisions, correlation and sedimentary characteristics are shown in Table 5.2.

5.3.1 Jiaoliao-Xuhuai Depressed Belt

On the east side of the platform, the Late Proterozoic Jiaoliao-Xuhuai depressed belt continued to develop during the Sinian and to receive sediments which are stratigraphically continuous with the underlying Qingbaikou System. As mentioned in Section 4.2.3, the Sinian of the south and north segments of the belt have similar characteristics too, and can be well compared with the Qingbaikou System (Table 5.3). In the north segment of the southern Liaoning area the Lower Sinian is termed the Wuhangshan Group (Chang Shaoquan 1980; Compilation Group of Regional Stratigraphic Table of Liaoning Province 1978) and in the south segment of the belt the Xuhuai Group (Brigade of Regional Geological Survey, Anhui Bureau of Geology and Mineral Resources 1985). It consists of sandstone at the bottom, overlain by shale and limestone (or argillaceous limestone), and contains Vermes and Pogonophora fossils and algal megafossils in the middle part of the sequence. Carbonate rocks yielding rich stromatolites comprise the upper horizons.

The Upper Sinian is subdivided into the Suixian Group and the overlying Langan Group in the south segment of the belt (Brigade of Regional Geological Survey, Anhui Bureau of Geology and Mineral Resources 1985). In contrast, the Suixian and Langan Groups are referred to as the Jinxian Group in the north segment of the belt (Chang Shaoquan, 1980; Compilation Group of Regional Stratigraphic Table of Liaoning Province 1978). Further north the Jinxian Group is absent from the stratigraphy. On the whole, the lower part of the Upper Sinian is dominated by carbonate rocks containing abundant stromatolites and micropaleobotanic fossils and the upper part is composed of shale, sandstone and carbonate which in turn is overlain by the Lower Cambrian. Therefore sedimentation within the Jiaoliao-Xuhuai belt took place in a tidal environment within a stable epicontinental sea, and resulted in the production of a cover sequence. In the south segment of the belt, at the base of the Lower Sinian Xuhuai Group, there is a glauconite-quartz sandstone which has yielded a glauconite K-Ar age of 738.6 Ma (Brigade of Regional Geological Survey, Anhui Bureau of Geology and Mineral Resources 1985) and 749.8 Ma (Project Co-operation Group 1984). In the north segment of the belt, the shale from the lower part of the Lower Sinian Wuhangshan Group has yielded a whole-rock Rb-Sr isochron age of 723 ± 43 Ma ($^{87}Sr/^{86}Sr=0.722\pm0.0006$), whereas the shale from the upper part of the Jinxian Group has given a similar age of 649.5 ± 20 Ma ($^{87}Sr/^{86}Sr=0.714\pm0.0002$) (Research Group of Upper Precambrian, Shenyang Institute of Geology and Mineral Resources 1986).

Stratigraphic unit			South segment (west of Tancheng-Lujiang fault)[1]	North segment (east of Tancheng-Lujiang fault, southern Liaoning)[2]
Langan Group			Shale, silty shale and dolomite.	Shale, silty shale, glauconite-quartz sandstone with limestone; contains fossils of molluscan metazoans and large acritarchs
			Silty shale, glauconite-quartz sandstone, with limestone, contains molluscan metazoans, large acritarchs and stromatolites	
Suxian Group			Dolomitic limestone	Limestone, contains abundant stromatolites
			Sandstone and shale, with shale, with limestone; contains molluscan metazoans, microfossil plants and stromatolites	
Xuhuai Group			Algal limestone, limestone, argillaceous dolomite and limestone with purplish red calcareous shale; contains abundant fossils of Vermes and stromatolites, large algae, acritarchs, and microfossil plants	Sandy limestone, and limy dolomite; contains abundant stromatolites
				Siltstone and shale with limestone; contains Vermes, Pongonophora, megafossil algae
			base: quartz and calcic sandstone	base: sandstone and feldspar-quartz sandstone
	Langan Group	Suxian Group	Xuhuai Group	Jinxian Group / Wuhangshan Group
	Upper		Lower	
	Sinian System			
	Upper Proterozoic			

Table 5.3. Stratigraphy and sedimentary characteristics of the Sinian in the north and south segments of the Jiaoliao-Xuhuai Depressed Belt. [1]Brigade of the Regional Geological Survey, Anhui Bureau of geology and al Resources, 1985; [2]Compilation Group of Regional Stratigraphic Table of Liaoning Province,1978)

5.3.2 Aulacogens and the Depression Belt on the South Margin of the North China Platform

A number of aulacogens or troughs were formed in the south of the North China Platform which became centres of volcanic and/or sedimentary activity: For example, the Foziling Group occupies a taphrogenic trough on the north side of the Dabie massif, in the east segment of the southern margin of the North China Protoplatform. The main body of the Group is of Sinian age (see Section 5.2.2). The lower part of the Group consists of a set of migmatized gneiss, amphibolite, greenschist, mica-quartz schist and marble, derived from argillo-arenaceous, carbonate and to the largest extent basic and acidic volcanic protoliths. The protoliths of the amphibolite and greenschist are essentially of basalt-spilite affinity, whereas the gneiss is of keratophyric affinity. This indicates that the basic and acidic volcanic associations of the sequence in the lower part of the Foziling Group are similar in character to continental rift type bimodal volcanic rocks. On the Na_2O+K_2O versus SiO_2 diagram (Middlmost 1980), the volcanic rocks almost all fall in the alkaline volcanic field (Fig. 5.5). This suggests that they belong to an alkaline series, and were formed in the taphrogenic environment between the North China Platform and the Dabie massif (Xu Guizhong et al. 1988; Yang Shennan et al., 1983). The middle to upper part of the Foziling Group is composed predominantly of mica quartz schist, mica schist and schistose quartzite. The protolith to these rocks was a very thick (> 8500 m) rhythmic sedimentary assemblage dominated by argillo-arenaceous, argillo-calcareous and siliceous materials. This sedimentary succession was probably a flysch turbidite sequence formed in a bathyal environment (Yang Shennan et al. 1983; Xu

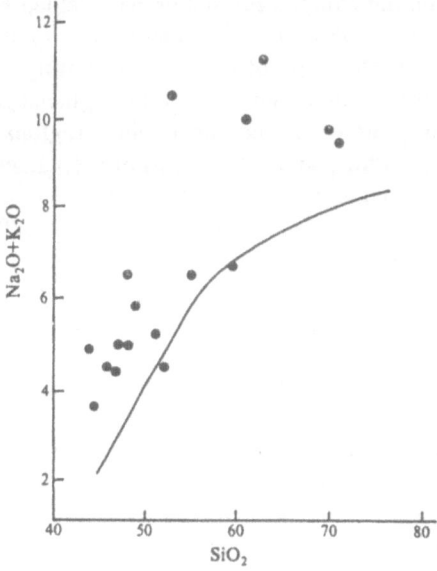

Fig. 5.5. Na_2O+K_2O versus SiO_2 diagram for metamorphic volcanic rocks of the Foziling Group

Guizhong et al. 1988) and its deposition reflects a change from sea trough to marginal sea environment. This taphrogenic process probably continued into Early Paleozoic times. The Foziling Group has experienced an intense folding and a greenschist facies metamorphism, now constituting a synclinorium with an axis which changes in direction eastward from northwest to nearly eastwest. The folding intensifies eastward but weakens northward. Tight second-order folds are well-developed, with the axial planes mostly dipping north. There is a similar change in the metamorphism: the Group has suffered stronger metamorphism and migmatization towards the southeast (Xu Guizhong et al. 1988).

The Taowan Group also includes some strata of Sinian age: the Group is situated in the west segment of the south margin of the platform, along the north side of the Qinling massif (see Section 4.2.2). This Group consists of phyllite schistose conglomerate overlain by schistose marble (Zhang Weiji et al. 1989). It is a parametamorphic sequence containing no volcanic rocks, with a consistent lithology in the E-W direction but a relatively abrupt facies change in the N-S direction, indicating that it is a miogeosyncline-type sedimentary association on a south-dipping continental slope (Geng Shufang. 1989); its related taphrogenesis probably continued into Early Paleozoic time.

North of this zone, on the south side of the protoplatform, a Sinian platform cover sequence is present. It is best developed in western Henan, where the typical section is composed (in ascending order) of the Huanglianduo, Dongjia, Luoquan and Dongpo Formations. In the first two formations, there is a complete marine sedimentary cycle of conglomerate-sandstone-carbonate rocks. Parallel unconformities occur both below and above the cycle; the Huanglianduo Formation overlaps the Luoyukou Formation and the Sanjiaotang Formation of the Luoyu Group is more limited in distribution. Glauconite from the Dongjia Formation has yielded K-Ar ages of 665 and 669 Ma (Guan Baode et al. 1980). The Luoquan Formation is relativly widely distributed throughout the north slope of the eastern Qinling Mountains. It is a sequence of matrix-supported conglomerates and fluvioglacial gravel-bearing argillaceous sandstone (features are described later), which regionally form overlapping strata. An angular unconformity and a disconformity separates the Formation from the underlying Archean Dengfeng Group, and the Sinian Dongjia Formation respectively. The thickness increases from 0.3 m in the north to 288 m in the south. The bottom of the Dongpo Formation is composed of lenticular calcareous quartz sandstone, gravel-bearing shale, and interbedded sandstone and conglomerate, whereas the middle and upper horizons are dominated by shale with glauconite siltstone. The Formation is 11-141 m thick, and conformably overlies the Luoquan Formation. These characteristics indicate that the Sinian in this area is a stable-type sedimentary cover.

The early Late Proterozoic collision and coalescence of the Qinling-Dabie massif with the Yangtze Platform took place in a marine setting. In the Late Sinian, abundant carbonates continued to be deposited, and are mostly products of a bathyal or abyssal environment (for their characteristics see Section 5.2.2). By that time the Qinling-Dabie massif had already become accreted to the north margin of the Yanzi Platform.

5.3.3 Depression Belt on the West Margin of the North China Platform

On the west margin of the North China Platform, some depressed basins received thin Sinian sediments. Due to later structural disruption and uplifting as well as denudation, the sediments only outcrop sporadically. In the mountainous area near Yinchuan, Sinian strata is known as the Zhengmuguan Formation. This Formation is clearly divisible into two members. The lower member consists of variegated, massive, matrix-supported conglomerate, 144 m thick containing the microfossil plants *Leipsophosphara* sp., *L. minor* Schep, *and Trachysphaeridium hyalinum* Sin et Liu, and overlies unconformably the stromatolite-bearing dolomite of the Wangquankou Formation of the Jixian System. The upper member is a magnetite-bearing silty slate with sandstone, containing *Taeniatum* aff. *crassum* Sin et Liu, Sabelliditidae, and trace fossils. It is 108 m thick, and underlies the Lower Cambrian phosphoric conglomerate. The Zhengmuguan Formation is laterally stable in lithology, and ranges in thickness from tens of metres to more than 200 m (Yang Zhende et al. 1988). It is quite similar to the Upper Sinian in the subsidence belt of the south part of the North China Platform: the matrix-supported conglomerate in the lower member is equivalent to the Luoquan Formation, and the upper member to the Dongpo Formation (Table 5.2). The overlying strata are all phosphoric conglomerates of Lower Cambrian age. This succession is therefore interpreted as a stable-type sedimentary cover. It is inferred from this that Upper Sinian and Lower Cambrian sediments jointly formed a northwest-trending belt of subsidence in the southwest part of the North China Platform (II-3 in Fig. 5.2).

In the Beishan depressed belt (II-4 in Fig. 5.2), strata of Sinian age is known as the Hongshankou Group. The Lower Formation is composed of tillites and the Upper Formation consists of neritic carbonate and clastic rocks. They are respectively equivalent to the lower member and upper member of the Zhengmuguan Formation, and belong to a stable platform type of deposit.

5.4 Sinian Crust of the Tarim Platform

In the Sinian, diverse types of sediments were formed in different structural locations on the Tarim Platform: the Sinian sequence in the Tianshan Mountains in the north part of the platform is a sediment of rift origin, whereas the Quanji Group in the Qaidam massif is part of the platform sedimentary cover. Deposits in the Qilian Mountains were formed in active tectonic environments (Table 5.4). The variety of these deposits reflect the different tectonic settings within the Tarim Platform.

During Sinian times, a rift structure was formed at the north margin of the Tarim Platform (III-1 in Fig. 5.2). During taphrogenesis, intense volcanism and multiple glaciations took place, producing a thick glacial-volcanic-flysch accumulation. The Sinian sequence of the Tianshan Mountains, developed in just such a rift, includes sediments of two (or three?) glacial stages, three phases of volcanism, and marine deposits. These extend westward into the northwest margin of the Tarim Platform and reflect a unique tectonic and climatic environment occurring when the rigid crust became active again (Yang Zhende et al. 1988).

The south-central part of the Tianshan Mountains is an area showing a complete development of the Sinian sequence deposited on a folded, metamorphic basement of the Qingbaikou System. The Early Sinian taphrogenesis was accompanied from the very beginning by intense volcanic eruptions. The volcanics of the Beiyixi Formation are most widespread, occurring mainly in the west segment of the Tianshan Mountains. They include acidic volcanic lavas (e.g. rhyolite, quartz keratophyre, felsite) and pyroclastics intercalated with small amounts of banded ferruginous jasper rock as well as minor basic volcanics (e.g. basalt and diabase). In the eastern part of the south-central Tianshan Mountains, there are even slightly alkaline volcanics, and at the northwest margin of the Tarim Platform, pyroclastics of the Beiyixi epoch are also present. Within this epoch, acidic rocks have an average SiO_2 content of 70.32% and basic rocks 45.31% (Gao Zhenjia et al. 1985a). The Beiyixi Formation also contains wacke-conglomerate, gravel-bearing sandstone, siltstone and lamellar slate which were formerly considered to be tillite sheets and correlated with the moraine sheet at Chang'an, South China (Lu Songnian et al. 1983; Gao Zhenjia et al. 1985a). However, judging from their sedimentary characteristics, regional correlation and the fact that the till sheet of the Qiao'enbulake Formation (equivalent to the Beiyixi) on the northwest margin of the Tarim Platform has the characteristics of a typical turbidity current (Gao Zhenjia et al. 1985b), they are most likely to be gravity-flow sediments. Therefore, the question of whether or not the Beiyixi Formation is a tillite sheet needs further study.

Other formations of Early Sinian age are the Zhaobishan, Terui'aiken and You'ermeinake Formations. The Zhaobishan Formation (also termed the Aleitonggou Formation) is a marine terrigenous clastic sequence including feldspar-quartz sandstone, siltstone, sandy slate and lamellar slate that constitute a sedimentary cycle. The Terui'aiken Formation was deposited towards the end of Early Sinian times and is widely distributed in the west segment of the Tianshan Mountains. It constitutes a tillite layer, composed of morainic gompholite, tillite-like conglomerate, mudstone-slate and siltstone, having a lot of clear moraine indicators, dominated by marine-glacial sediments. This is a continental glacial accumulation, equivalent to the You'ermeinake Formation which consists of purplish-red tillite, sandstone and silty slate on the northwest margin of the Tarim Platform (Gao Zhenjia et al. 1985b). The tillite sheet of the Terui'aken Formation is also equivalent to that of the Nantuo Formation in the Yangtze Platform.

In contrast, the Zhamoketi Formation of early Late Sinian age is a sequence of turbidity current deposits (Fig. 5.6), accompanied by volcanics generated during the second-phase volcanism of rift volcanism. The resultant volcanics are distributed in the western part of the south-central Tianshan Mountains and on the northwest margin of the Tarim basin. They are lithologically simple, consisting of basic lavas (mainly basalt and diabase), with volcanic-clastic rocks occurring locally. According to Gao Zhenjia et al. (1985a), their average SiO_2 content is 45.68%. The Late Sinian Yukengou Formation (and the overlying Shuiquan Formation) comprise a sedimentary cycle made up of siltstone, mudstone, dolomitic limestone, sandy limestone and marl. The Shuiquan epoch marks the third-phase of volcanism, and the resultant volcanics are basic volcanic lavas, more restricted in distribution and only found in some areas in the west part of the south-central Tianshan Montains. The Hange'erqiaoke Formation is an upper till sheet, deposited at the end of Late Sinian times, which belongs to a continental glacial accumulation. This till sheet extends discontinuously westward into the USSR, and eastward from the Beishan area to the Qilian Mountains, to southern Shaanxi and Shanxi and finally to western Henan where it is connected with the till sheet of the Luoquan Formation, forming a narrow, long belt more than 3000 km long (Fig. 5.9). It also occurs in the Yinchuan area and the Kunlun Mountains to the west.

Sinian strata is also found in other depressions or rifts on the Tarim Platform: the Sinian Quanji Group forms a cover sequence and lies in the depression in the north of the Qaidam block. The bottom of the Group is marked by a molassic formation. This is overlain by a sandstone-shale-carbonate association formed in ascending order by quartz sandstone, sand-shale and finally by algae- and stromatolite-rich carbonate rocks. In contrast, the Sinian sequence located in the central part of the Qilian Mountains in the Qilianshan aulacogen, termed the Duoruonuo'er Group,

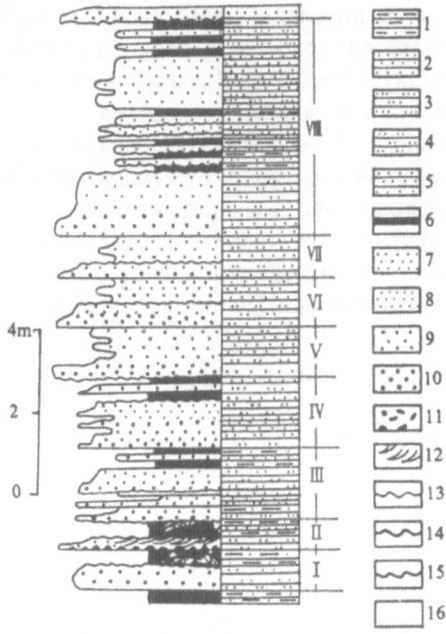

Fig. 5.6. Succession of turbidites of the Zhamoketi Formation in the south-central Tianshan Mountains (after Gao Zhenjia, 1985a). 1. Sandy shale; 3. siltstone; 3. fine-grained sandstone; 4. medium-grained sandstone; 5. coarse-grained sandstone; 6. mud; 7. silt; 8. fine sand; 9. medium-grained sand; 10. coarse sand; 11. lithic fragment; 12. ripple bedding; 13. rugged surface; 14. bottom print; 15. bulged surface; 16. Bouma layer number

was deposited in an active tectonic environment. It consists of chlorite-quartz schist, basic to intermediate-basic volcanics, slate and limestone. The Group reaches a maxium thickness of nearly 5000 m (Institute of Geology, Bureau of Geology and Mineral Resources of Gansu 1986).

Finally, it should be noted that the tillites which commonly occur at the top of the Sinian of the Tarim Platform are similar in characteristics and equivalent in stratigraphic position to the till sheet of the Luoquan Formation of the North China Platform (Table 5.5). This suggests that the two platforms were quite close to each other in Sinian time, but still separated by an ocean as indicated by the tectonically active nature of the sedimentary environment in the Qilian Mountain area. After the end of the Sinian the whole area was uplifted, and the Sinian sequence is disconformably overlain by Lower Cambrian strata.

5.5 High Himalayas Mobile Belt

On the north margin of Gondwanaland, the Sinian sequence is limited in distribution, occurring locally in the high Himalayas area. It has been studied to a very limited degree and therefore only scattered information is available.

The Sinian in the Himalayas area, known as the Bei'ao Formation (Table 5.4), is a greenschist facies metamorphic sequence dominated by black slate. The Bei'ao Formation probably contains sediments formed in the Qingbaikou and Sinian Systems, and has a transitional boundary with the overlying Cambrian Rouqiecun Group. The Group originally belonged to the Precambrian sequence on the Indian plate, and was subducted beneath the Himalayas as the plate moved northward.

Sinian strata also occurs in western Nepal where it has yielded stromatolites. In the west part of Panjab, it is termed the Dogra slate (Xiao Xuchang et al., 1988).

5.6 Paleomagnetic Evidence

The Sinian has been studied both extensively and in detail using palaeomagnetic techniques. Studied Sinian sections include:

(1) the eastern Yangtze Gorges (Zhang Huimin et al. 1982) and other regions within the Yangtze Platform (Li Quan et al. 1987; Zhu Hong 1986);

(2) the upper Proterozoic section of southern Liaoning in the north segment of the Jiaoliao-Xuhuai depressed belt within the eastern North China Platform (Gao Rongfan et al. 1983);

(3) the Sinian Quanji Group sections of the Tarim massif (Li Yong'an et al. 1984, 1989); and Qaidam massif (Li Yanping et al. 1988) within the Tarim Platform.

The results have provided more evidence for further research into the tectonic evolution of the North China, Yangtze and Tarim Platforms. Since their apparent paleomagnetic polar wandering paths in Sinian times have already been discussed in Section 4.6 (Fig. 4-37), the following discussion focuses only on their paleo-positions and relative drifting motions.

Paleomagnetic determinations indicate that during the Sinian the Qaidam massif was near the equator (Li Yanping et al. 1988) whereas the southwest margin of the Tarim massif was at a paleolatitude of 9.7°N and 2.3°N during the deposition of the Kezisuhumu and Hankemakelike Formations respectively (Li Yong'an et al. 1984), hence both massifs were in low-latitude regions. Furthermore, the consistency of

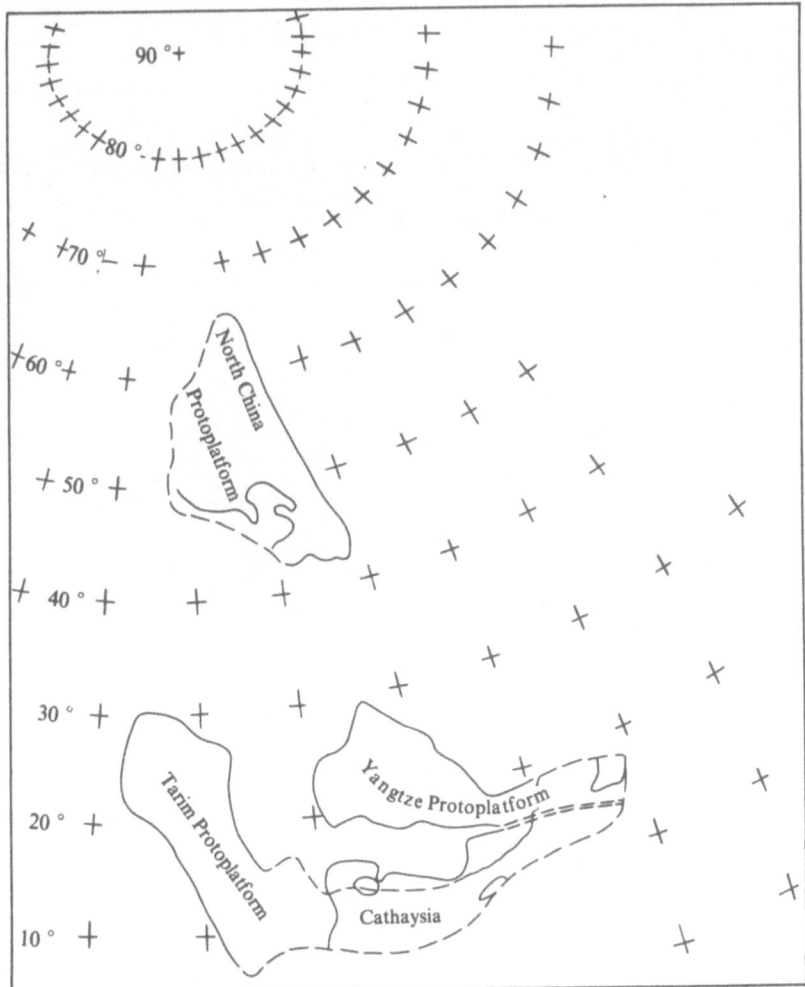

Fig. 5.7. Paleolatitude map of North China, Yangtze, Tarim Platforms and Cathaysia old land in Early Sinian (800-700 Ma) (after Qiao Xiufu et al. 1988)

Upper Carboniferous paleomagnetic inclinations between the two massifs indicates that the Qaidam massif was probably part of the Tarim massif during the Sinian, and in the tectonic evolution, these two acted as an integrated Platform.

During the Early Sinian the North China Protoplatform was located at 40°-60°N (Gao Rongfan et al. 1983), the Yangtze Platform at 10°-30°N (Zhang Huimin et al. 1982; Zhu Hong 1986; Qiao Xiufu et al. 1988) and the Tarim Platform roughly at 0°-30°N (Li Yong'an et al., 1984; Li Yanping et al. 1988). In addition, there are many similarities in the tectonic evolution of the Yangtze and Tarim Platforms. For example, the folded metamorphic sequences which surround the crystalline basement are overlain by Sinian cover sequences of similar sedimentology and paleontology. It is therefore inferred that they might have constituted an integrated oldland. Thus

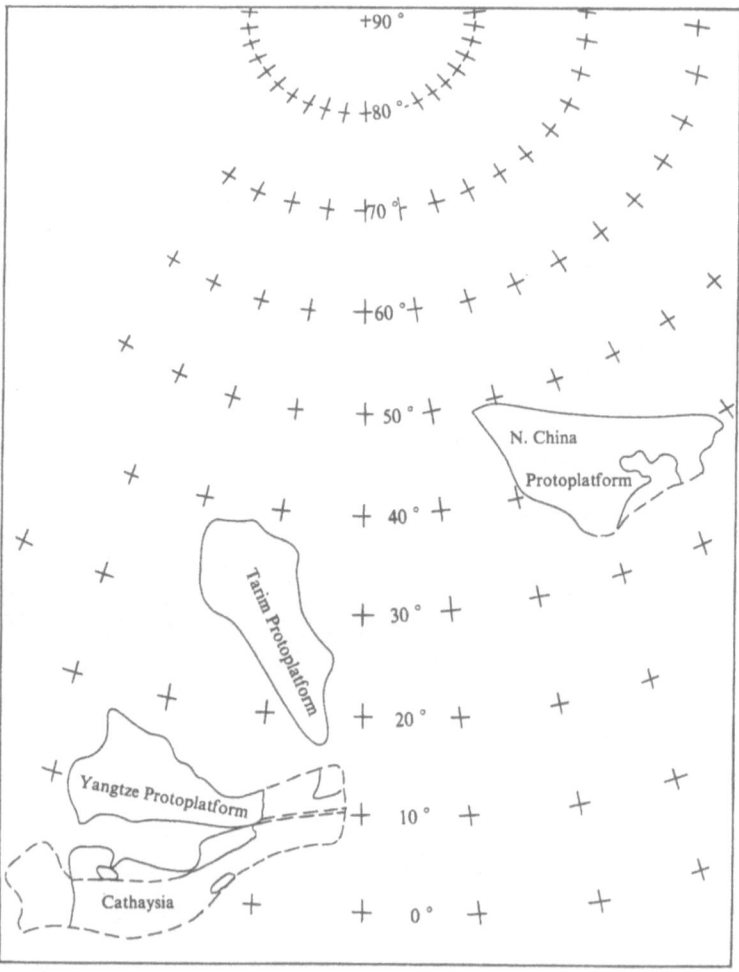

Fig. 5.8. Paleolatitude map of North China, Yangtze, platforms and Cathaysia oldland (after Qiao Xiufu et al. 1988)

from this evidence a paleogeographic framework for the North China massif and Yangtze-Tarim massif was constructed (Fig. 5.7). The Nantuo glacial stage, which forms a major geological event of probably global significance, is consistent with most of the paleolatitude data for the same period world wide. Therefore the distribution of glaciers also indicates the positions of continental plates derived from paleomagnetic data.

During the Late Sinian, the North China Platform was situated at paleolatitude 30°-40°N (Gao Rongfan et al. 1983). It is thus inferred that the North China Platform gradually drifted southward, and at the end of the Sinian its south margin was at approximately 30°N (paleolatitude data from the Luoquan Formation, Li Yong'an et al. 1984). During this time period the Yangtze Platform drifted toward the equator. The Tarim Platform lay at about 30°N, and was therefore separated from the Yangtze Platform but was moving increasingly closer to the North China Platform (Fig. 5.8). At the start of Late Sinian times, glacial ablation and a large-scale transgression took place on the Yangtze Platform. Oceanic currents intruded into its southeast margin, forming a huge phosphorite belt genetically related to currents upwelling along the shoal to the southeast. Towards the end of Late Sinian times, under the hot climate near the equator, an evaporite basin containing gypsum-salt sediments was formed locally within the platform. At the end of the Sinian the North China and Tarim protoplatforms were extensively uplifted and during the Luoquan glacial stage, the till sheet formed chiefly by continental glaciation was widespread across the two platforms.

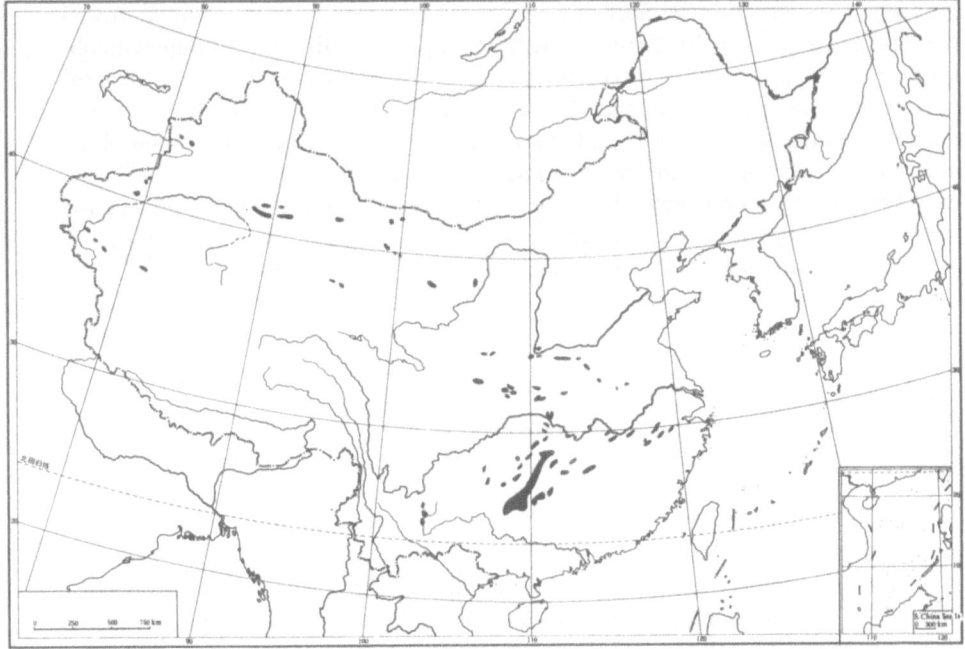

Fig 5.9. Distribution of Sinian glaciation in China

5.7 Characteristics of Sinian Glaciation

The Sinian glaciation had an extensive effect in China. It is now generally accepted that the glaciation is divided mainly into two glacial stages: the Nantuo and the Luoquan, and that these two stages of glaciation occurred widely on the North China, Yangtze and Tarim Platforms (Fig. 5.8), becoming one of the major geological events of Late Proterozoic time. The till sheets associated with this event are listed and discussed below:

Nantuo till sheet: This till sheet was discovered in 1907. Li Siguang (J.S. Lee) defined its stratigraphic position and geological age in 1924 and presented the paper "Sinian Glaciers in China" at the 17th International Geological Congress in 1937 (Lee, 1937). Since the 1950s, large-scale regional geological surveys and mineral reconnaissances were conducted, resulting in the acquisition of plenty of new data, and a more thorough and comprehensive study of the till sheet. It has now been confirmed that the Nantuo till sheet is widely distributed in ten provinces of South China and, as a Nantuo glacial rock, on the north, northwest and southwest margins of the Tarim area in Xinjiang Province. Hence it is the most widely distributed of the Sinian till sheets in China (Lu Songnian et al. 1983).

The Nantuo till sheet in South China can be divided into two types in terms of sedimentary features and environment. The first type is in the interior of the Yangtze Platform (including the Guizhou, southeastern Sichuan, Hubei, western and northern Hunan, eastern Yunnan, northern Jiangxi, western Zhejiang and southern Anhui regions) and is principally a greyish green to purplish red tilly gompholite with obscure or absent bedding. Its gravels are mixed in size, irregularly arranged and without sorting. The gravels make up 20-50% of the content, are generally 3-8 cm in diameter, with a few fragments being up to 1 m or in size. The composition of the gravel is complex, including schist, granite, gneiss, quartzite, diorite and, sandstone indicating varied source areas. They are cemented by sandy-argillaceous matter and in a few places by tuffaceous substances. Several sets of glacial striations and indentations of varying direction occur on the surface of the gravel. This type of till sheet varies greatly in thickness, ranging from several metres to nearly 1000 m, and up to 3000 m in a few places. Sedimentary features suggest that it is a continental glacial sediment.

The second type of till is distributed on the southeast margin of the Yangtze Platform. It is dominated by greyish green gravel-bearing pelitic sandstone and gravel-bearing slate and is generally free of bedding; the gravels are 5-15% of the rock content, and smaller in size than those found in the first till type, and striations are only rarely found on their surfaces. The gompholite commonly contains nodular pyrite and other forms or carbonaceous claystone with star-like pyrite. The thickness of this till type also varies greatly (between 80 and 3174 m), and comparatively large thicknesses occur in Congjiang, Tongdao and Qianyang near the margin of the Yangtze Platform. These glacial deposits clearly show the character of underwater sedimentation and therefore classified as glaciomarine sediments.

The Lower Sinian Aleitonggou and Terui'aiken Formations in the west-central segment of the Tianshan rift of the Tarim Platform, as well as the Youmeinake Formation on the northwest margin, are also products of the Nantuo glacial stage. The Terui'aiken Formation is characterized by as many as five to eight alternations between basal moraine and glaciofluvial facies, and belongs to a sedimentary facies sequence formed at the front of a glacier (Lu Songnian et al. 1983). The Youmeinake Formation consists of purplish red tillite and associated rocks of continental glacial accumulation, and shows conclusive evidence of glaciation: giant glacial boulders, striated pebbles, pressure-pit stones are common and mud-coated stones are occasionally found. Well-preserved striae occur on the bedrock surface of glacial pavement at the bottom (Gao Zhenjia 1985b).

On the southeast margin of the Yangtze Platform, another Early Sinian glacial stage-the Chang'an stage- was defined by many scientists (e.g. Liao Shifan et al. 1979; Wang Yuelun et al. 1980; Liu Hongyun et al. 1980). This unit is also termed the Gucheng glacial stage (Lu Songnian et al. 1983). In recent years, however, some scientists (see Wang Yangeng et al. 1986) suggest on the basis of lithology, rock assemblage and sedimentary structures that this deposit is not a tillite or glacial rock but formed by gravity-flow (see Section 5.2.3). Flow was controlled by the paleotectonic environment and the sediment was distributed along the Yangtze paleocontinent. In the Tianshan rift of the Tarim Platform, the glacial rock deposited at the bottom of the Lower Sinian sequence was also considered to be comparable with the Chang'an (or Gucheng) glacial stage (Lu Songnian et al., 1983; Gao Zhenjia et al. 1985a). The distribution of this unit is quite limited, and its tillite origin of this unit remains to be further verified.

Luoquan till sheet: This till sheet is distributed on the south and west margins of the North China Platform as well as on the Tarim Platform. The Luoquan epoch is marked by many types of glacial sediment and considerable variation in lithofacies between various regions. According to the study by Guan Baode et al. (1986) of the Luoquan Formation in the middle segment (western Henan) of the south margin of the North China Platform, the following facies are generally recognized: (1) massive polymictic clastic rock, which is chiefly derived from lag moraine or meltout moraine; (2) bedded polymictic clastic rock, which is mainly derived from water-area moraine or a glaciomarine/glaciolacustrine sediment containing pebbles dropped from floating ice; (3) deltaic or underwater mass-flow sediments containing dropped pebbles; (4) slaty conglomerate and sandstone, which are the product of glaciofluvial, deltaic or underwater mass-flow sedimentation; (5) lenticular breccia, formed by underwater mass-flow sedimentation; and (6) sand-gravel wedges, formed by infilling of contraction cracks derived from freezing processes. Well-preserved glacial pavements, some with an outcrop area as large as several hundred square meters, are present in strata underlying the Luoquan Formation (Photo 5.1). This phenomenon is rare in the Precambrian geological record. Glacial erosion has produced a large variety of features on the glacial pavement, including glacial striae, grooves, more than two sets of scratches of contrasting orientations, striated cracks and fissures, crescent gouges, plastic model-type (P-type), roche moutonnee, streamlined and glaciated steps. Based on the variation of thickness of the Luoquan Formation and the parallel unconformity between it and its underlying stratum, it is inferred that the basement had a relief of 200 m during glacial sedimentation. From various glacial erosion features of the glacial pavement, it is preliminarily determined that the general direction of movement of glacier was from northwest to southeast (122°-145°) or to the south (180°). All these features suggest that the Luoquan till sheet is characterized by plateau glacial erosion associated with sedimentation under continental glacial and glaciolacustrine environments. The distribution of known outcrops of the Luoquan Formation indicates that it extends westward from southern Anhui, western Henan to southern Shaanxi, then northwestward through the Qilian Mountains to the Tarim Platform, forming a belt of discontinuous outcrop trending nearly northwest and on the whole similar in character.

Available isotope age data show that the whole-rock Rb-Sr isochron age on shale at the bottom of the Datangpo Formation is 728±27 Ma, whereas ages obtained from the Doushantuo Formation overlying it is 700±5 Ma, 693±66 Ma and 691±29 Ma (Lu Songnian et al. 1983). Therefore it is inferred that the overlying Nantuo glacial stage was formed at some time between 730 and 700 Ma ago. Biostratigraphic and geochronologic data show that the upper time bracket for the Luoquan glacial stage is approximately between 620 and 600 Ma ago, near the Precambrian-Cambrian boundary (Guan Baode et al., 1986). Globally, four glacial stages have been defined for Late Proterozoic time, but in China it is mainly the last two of these glaciations which are well developed. The Early Sinian Nantuo glacial stage is roughly comparable with the well-known Lapland Glacial Stage in northwestern Europe, the

Photo 5.1. Glacial erosion surface and whalback rock from the quartz sandstone of Sanjiaotang Formation, with directional striae and grooves, Shimengou Valley, Lushan County, Henan Province (photo by Guan Baode et al. 1983)

Marinoan Glaciation in Australia and the Lower Glacial Rock in western Africa, and represents the most widespread glaciation in the Late Proterozoic time. The terminal Late Proterozoic glaciation equivalent to the Luoquan glacial stage is not as extensive, but has been recorded in Europe, Asia, Africa and South America, and even in Australia.

5.8 Paleontology and Stromatolites

5.8.1 Evolution of Phytosphere and Appearance of Metazoa

During Sinian times, the phytosphere underwent further development. The macroalgae Vendotaenia was widely distributed. The red algae Epiphyton also appeared, multicellular algae such as the brown and red varieties tended to flourish, and aspects of the microflora changed. The microflora of the Early Sinian is dominated by elements of Sphaeromorphide similar in appearance to the varieties found within the Qingbaikou Period. The microflora of the Late Sinian includes, besides abundant elements of Sphaeromorphide, some members of Hystrichosphaerid, some new genera and species of Polygonomorphitae and some elements similar to Cambrian forms, such as *Archaeodiscina, Micrhystridium, Polyedryxium*, and *Razumovskia.*

The first appearance of metazoan and trace fossils represented by coelenterates, Vermes and sponges in the Sinian marks another major event in Precambrian biological evolution. There are mainly two areas that are known to yield metazoan fauna: one is the Jiaoliao-Xuhuai depressed belt in the east of the North China Platform, the other is the Sinian sequence in the eastern Yangtze Gorges within the Yangtze Platform. This zone of subsidence can be divided into lower and upper fossiliferous horizons. The lower horizon was deposited at the begining of Lower Sinian time, which is known as the Wuhangshan Group in the north segment of the belt and the Xuhuai Group in the south segment (Fig. 5.1; Table 5.3). The lower part of the Wuhangshan Group contains fossil worms and medusae. All the fossil worms belong to the family Sabellitidae, dominated by *Sabellidites* and *Paleolina*. The fossil medusae, which occur in a horizon higher in position than that of the fossil worms, include 12 genera and species such as *Cyclomedusa annulata* (Sprigg) Xing et Liu and *C. davidi* Sprigg (Xing Yusheng et al. 1985). The metazoan fauna of the lower part of the Xuhuai Group is similar to that of the lower part of the Wuhangshan Group and, besides *Sabellidites* and *Paleolina*, includes *Huaiyuanella*. The fossiliferous bed in the lower part of the Wuhangshan Group has been dated at 723±43 Ma by Rb-Sr isochron age method (Research Group of the Upper Precambrian, Shenyang Institute of Geology and Mineral Resources 1986). This date is slightly earlier than the appearance of metazoans that is generally internationally accepted. The upper fossiliferous horizon of the Xuhuai-Jiaoliao subsidence belt is located within Upper Sinian strata. In the south segment of the belt, the lower part of the Suixian and Langan Groups have mainly yielded *Huaiyuanella* aff. *jiuliqiaoensis* Yan et Xing (gen. et sp. nov), *Calyptrina striata* Sokolov, and *Paleolina evenkiana* Sokolov (Project Co-operation Group 1984). In recent years, the upper part of the Jinxian Group in the north segment of the belt has also yielded metazoans such as *Medusinites* (Hong Zuomin et al. 1987, 1988). The metazoan fossils discovered in the Sinian sequence of the eastern Yangtze Gorges include coelenterate, vermian and periferan and trace fossils. The coelenterate fossils occur in the Upper Sinian Dengying Formation, represented by *Charnia dengyingensis* Ding et Chen. Vermian fossils occur in the same formation, including *Saarina* sp. and *Micronemaites formosus* Sin et Liu (Xing Yusheng and Liu Guizhi, 1978), and *Sinotubulites beimatuoensis* Chen, Chen et Qian (Chen Meng'e et al. 1981). The periferan fossils occur in the Upper Sinian Doushantuo and Dengying Formations. Those discovered are mainly sponge spicules, including calcareous triactins, siliceous sponge (primitive monoactin) and siliceous ceratosponge, marking the appearance of organisms with an endoskeleton. The trace fossils are also found in the Dengying Formation, including *Skolithes miaoheensis* Chen, Chen et Qian, *Planolites taishanmiaoensis* Ding et Chen, et Chen (Xing Yusheng et al. 1985).

5.8.2 Stromatolites

Stromatolite data from the Sinian System in all provinces of South China suggest that the Sinian is a geological period with poorly developed stromatolites. They are few in type, small in quantity, and limited in distribution, being found only in the Doushantuo and Dengying Formation in a few areas. However, available data from the Yangtze Platform and other areas does suggest that these stromatolites of the Sinian can be grouped into an assemblage (Xing Yusheng et al. 1985; Zhu Shixing et al. 1987). Simple types such as stratified, stratopillar, mound-shaped and knotted forms (e.g. elements of *Stratifera*, *Nucleella* and *Panisgcollenia*) are common. In addition, pillar and even conical types are fairly well-developed locally. The pillar stromatolite type is often closely related with the enrichment of siliceous and phosphatic materials, dominated by elements of *Boxonia*, *Linella* and *Patomia*, with less *Gymnosolen*, *Scopulimorpha*, *Xifengia* and *Baicallia*. The conical *Conophyton* and its related *Jacutophyton* are widely developed in the Dengying Formation and even become a predominant stromatolite type in local areas such as the Zhejiang region. In this assemblage, banded, linear and aggregated microstructures are present and often mixed, and the microtextures are predominantly of the pelletal, lumped and cross filiform types.

In summary, during the Sinian the organic world changed remarkably. Multicellular algae such as the red and brown types flourished and some new types of unicellular algae evolved, but stromatolites were poorly developed. Another distinctive feature of this period is that a metazoan fauna represented by worms, coelenterates and poriferans appeared for the first time and tended to become widespread. In the later part of this period, a few metazoans with an endoskeleton also evolved, which became the predecessors of the Cambrian polytaxonomic fauna. Therefore, based on the Paleontological evolution and the view of Cloud (1983), the period may be assigned to the Paleozoic Era of the Phanerozoic Eon. This analysis has been supported by some geologists in China (e.g. Wang Hongzhen 1986).

Strata	Tarim Platform			N. margin of Gondwanaland
	Tianshan Rift[1] — Central-southern Tianshan Mtns (Xishanbluk Fm)	Northern Qaidam Depression[2] (Xiagaolu Group)	Qilian Mtns aulacogen[3]	High Himalaya mobile belt[4] (Ruqiecun Fm.)
Sinian System — Upper Series	Hangolgok Fm — conglomerate tillite; Shuiquan Fm — siltstone-mudstone-carbonate assemblage; Yukengol Fm — turbidite and basic volcanics; Zhamkti Fm	Quanji Group — stable-type sandstone-shale-carbonate assemblage and molasse	Doro Nor Group — eugeosynclinal thick-bedded sandstone-siltstone-limestone and basic-limestone and basic-intermediate volcanics	Bei'ao Fm — black slate assemblage
Sinian System — Lower Series	Terieken Fm — conglomerate tillite; Altungol Fm — coarse-fine clastics; Zhaobishan — basic-acid volcanics; Beyxi Fm — and greywackes			
Underlying strata	Sanartag Fm	Dakenadaban Group	Gongqa Group	Nyalam Group

Table 5.4: Correlation of Sinian rock assemblages in the Tarim Platform and Northern margin of Gondwanaland. [1]Gao Zhenjia et al (1985a); [2]Wang Yunshan et al. (1980); [3]Institute of Geology, Gansu Bureau of Geology and Mineral Resources (1986); [4]Xiao Xuchang et al. (1986).

Southern depression of North China Platform			Western depression of North China Platform			Tarim Platform	
Huoqlu and Huainan regions of Anhui (Research Team of Upper Cambrian System in northern Jiangsu and Anhui)	West Henan (Guan Baode et al. 1980)	Xiaqinling region of South Shaanxi (Li Qingzhong 1980)	Around Yinchuan (Yang Zhende et al. 1988)	Beishan Mts. Depression (Zhao Xiangsheng et al. 1984)	Qilian Mts Aulacogen (Geochemical Regional Survey and Party Survey of Gansu Province, 1981*)	Northern depression of Qaidam block (Wang Yunshan et al. 1980)	Tianshan Rift (Gao Zhenxi, 1985a)
Houjianshan Fm. grayish black phosphate conglomerate	Xinji Fm. phosphate conglomerate	Xinji Fm. collophane bearing sandy limestone and breccia collophane	Suyukou Fm. phosphate conglomerate	Shuangyingshan Fm. phosphate conglomerate		Xiagaolu Gr. phosphate sandy conglomerate	Blak Fm. in Xishan
Wugangji Fm. purple silty shale	Dongpo Fm. shale with glauconite siltstone — 94 m	Upper: sandy slate, feldspathic quartz sndstone and slate — 64 m	Upper: slate — 108 m	Upper: silty slate, dolomite — 68 m	Thetopnoseen Upper: slate, phyllite and sillicic limestone — 364 m	Zhoujieshan Fm. quartz siltstone and fine-grained sandstone with sandy dolomite in the lower part — 22 m	Upper: Variation of dolomitic sandstone and lenticular dolomite — 71 m
Fengtai Fm. tilly conglomerate	Luoquan Fm. tilly conglomerate	Lower: tilly conglomerate	Lower: tilly conglomerate	Lower: till conglomerate	Lower: tilly conglomerate	Hongtiegou Fm. tilly conglomerate	Lower: tilly conglomerate
	180.8 m	37 m	144 m	247 m	335 m	110 m	377 m
Upper Sinian Jiudingshan Fm.	Sinian System Dongjia Fm.	Middle Proterozoic Jixian System Fengjiawan Fm.	Middle Proterozoic Jixian System Wangquan Fm.	Upper Proterozoic Qingbaikou System Daholuoshan Fm.	Upper Proterozoic Qingbaikou System Daliugou Fm.	Heitupo Fm.	Upper Sinian Shuiquan Fm.

(Underlying strata)

Table 5.5: Comparison of the Luoquan Formation tillite in northern China

6 Concluding Remarks

Bai Jin* and Ma Xingyuan**
*Tianjin Institute of Geology and Mineral Resources, Tianjin, China, 300170.
**Institute of Geology, State Seismological Bureau, Beijing, China, 100029.

6.1 Endogenous and Exogenous Processes in Crustal Evolution

6. 1.1 Magmatism

The numerous plutons in the Archean terranes broadly belong to three rock groups: charnockite, tonalite-trondhjemite-granodiorites (TTG), and the quartz monzonite-granite. The most extensively distributed of these three types are the TTG suite of intrusions: they exhibit a tendency of evolution from early Na-rich TTG to late K-rich granites. Charnockites are associated with granulite grade metamorphism whereas the TTG and quartz-monzonite-granites are mostly associated with metamorphic rocks of amphibolite facies but are products at deeper levels of the crust. This evidence, coupled with the analysis of supracrustal rock associations, regional structural styles and paleogeothermal regime, indicates that they are magmatic arc sequences responsible for the lateral accretion of the continental crust, and may also be a factor in the vertical accretion processes due subduction-related underplating.

Compared with the Archean, the plutonic magmatism of the Proterozoic is generally K-rich, and characterized by the appearance of post tectonic alkaline granite and anorogenic anorthosite-rapakivi granite suite and basic dyke swarms. This reflects the increase in thickening, enlarging and rigidity of the continental crust at this time. Although Proterozoic magmatism continued to display the tendency to evolve from Na-rich to K-rich with time, Na-rich granites such as tonalite might still occur at the active continental margin adjacent to a subduction zone. This indicates the importance of tectonic environment in controlling the composition of magmas.

Eruptive rocks occur mainly at the divergent and convergent margins of plates. In the Archean, the volcanics in greenstone belts are most representative. These belts are generally composed of mafic to felsic volcanics which can be assigned roughly to two series: the calc-alkaline and bimodal series. Available Nd isotope data suggest that the basic magma was derived chiefly from the long-depleted mantle. The two series of volcanics occur in environments similar to modern for- and back-arc basins and island arcs. At the bottom of the greenstone belts ultramafic volcanics that are compositionally equivalent to komatiite commonly occur but are poorly developed: in only a few greenstone belts is spinifex texture is preserved. Greenstone belt formation is related not only to possible structural disturbance and granite trapping but also to the degree of mantle melting and the thickness of primitive oceanic crust (Windley 1984). Proterozoic eruptive rocks belong to a diversified series due to distinct differences in tectonic environment. Besides calc-alkaline and bimodal series rocks, these volcanics may occur as continental tholeiites in intracratonic rift zones as basalt-andesite suite in aulacogens and even as definite ophiolite complexes representative of oceanic crust.

6.1.2 Deposition

The deposition was not only affected by basement compositon, maturity, and atmospheric oxygen concentration, but was also obviously controlled by the differences in tectonic environment. Archean sedimentary rocks are dominated by volcanic and greywackes which give an indication of the tectonic environment of the mobile belt. Stable shelf deposits are also found, and these are similar to Al-rich argillites intercalated with quartzite and carbonate rocks in a khondalite series. In response to the diversification of tectonic environment which began at the beginning of the Proterozoic, depositionary environments have changed greatly. The deposits in mobile belts are principally of the volcanic graywacke-flysch type; those in continental rift zones are associations of immature clastics, argillites and carbonates, whereas those in more stable Middle to Late Proterozoic intracontinental rift zones and subsidence zones are primarily mature quartzite-shale-carbonate associations.

In terms of chemical characteristics, the K_2O/Na_2O ratio is less than 1 for Archean sedimentary rocks but exceeds 5 for their Early Proterozoic equivalents. This indicates not only the presence of abundant low-K mafic volcanics in Archean time, but also a progressively maturing environment (Engel et al. 1974), with the derivation of deposits from more and more fractionated igneous rocks and the gradual thickening of the continental crust. The ratio can also be used as one of the geochemical indicators for distinguishing Archean and Proterozoic sediments.

The Fe^{2+}/Fe^{3+} ratio grows remarkably (from 0.8 to 2.0) from the Archean to the Early Proterozoic, reflecting an abrupt increase in oxygen pressure. This is consistent with the first appearance of red beds in the Early Proterozoic and provides material condition for the flourishing of micro-organisms.

6.1.3 Geothermal Regimes

It is generally considered that the thermal production in the Archean was two to three times higher than the present day. The geothermal regime varied markedly during the evolution of the crust, making it possible to define the paths of thermal diffusion in different times. Furthermore, the spatial distribution of metamorphic facies zones formed by metamorphic types and P-T conditions allows us to understand the geothermal regime during their formation. The evolution of the metamorphic types exhibited a secular tendency from medium-high-temperature to low-temperature dynamic regional metamorphism. The temperature and scope of metamorphism gradually diminished, indicating an irreversible decrease of heat flow in the geological development (Dong Shenbao et al. 1986).

Except for a few Late Archean greenstone belts preserved in synform structures and some progressively metamorphic zones of greenschist to lower amphibolite facies, Archean metamorphism is generally dominated by amphibolite and granulite facies types that are characterized by a temperature of over 700°C, pressures commonly exceeding 0.6 GPa and a geothermal gradient usually of 26-28°C / km. The metamorphic facies are generally "planar" in distribution (Dong Shenbao et al. 1986). The tectonic environments and tectonic activity which caused such a state are unknown due to the complex internal structures in the Early-Middle Archean terrane. However, one point is clear - that such "planar" metamorphic facies are a manifestation of the degree of deep-level or lower crust metamorphism and, after being uplifted and severely denuded, were exposed at the surface before Early Proterozoic depositions. Even if a high- to low-grade metamorphic facies zone were formed at that time, it would not be completely preserved owing to the denudation of the shallow facies zone.

From the beginning of the Proterozoic, crustal structure, tectonic environment, and the geothermal regime changed conspicuously. For example, the metamorphism of the Early Proterozoic is quite different from that of Late Proterozoic times: Early Proterozoic metamorphism is obviously controlled by the extent of mobile belts. Its facies zones have a linear, zonal distribution; and its main types belong to medium- or low-pressure regional dynamothermal metamorphism of low greenschist to low amphibolite facies and show an evidently progressive metamorphic zonation. Another important character of Early Proterozoic metamorphism is that the metamorphic facies zones are parallel to regional tectonic trend and tend to decrease in metamorphic grade from the border adjacent Archean oldland to the centre of the mobile belt, indicating that during the deformation and metamorphism of the mobile belt, the heat flow tended to migrate toward the centre along the compression-shearing direction. These metamorphic zones are found to be consistent with the tectonic regime of crustal shortening and may serve as an indicator of tectonic environment. All the Early Proterozoic tectonic belts - either intracratonic rift, intercratonic collision or cratonic marginal accretion - have undergone metamorphism to various degrees in the process of tectonic evolution. In addition, due to disparities in the nature of the mobile belt, they differ greatly from each other in degree of metamorphism. For instance, the metamorphism in collision zones and continental margin active belts may be generally up to the lower amphibolite facies and in a few belts (e.g. the Jiao-Liao belt) may reach high amphibolite facies, reflecting a high heat flow, thin crust and active environment. However, in intracratonic rift zones (e.g. the Jin-Yu rifted province) the metamorphism is represented principally by lower greenschist facies, reflecting a relatively thick crust and low heat flow. By the middle of Late Proterozoic mobile belts, only regional low-temperature dynamic-metamorphism of lower greenschist facies to greenschist facies is developed.

Although no Phanerozoic granulites have been found in China, deep-level granulites dated at 50 Ma do crop out due to collision of allochthonous terranes at the margins of the Cordillera continent in North America (Hollister 1982). This shows that such granulites are products from deep parts of orogenic belts (Morger et al. 1982; Hutchinson 1982). Young granulites, thrust-nappe tectonics and the root of batholith of Andes-type calc-alkaline series jointly represent a tectonic environment which provides a good modern example for a similar tectonic environment of an Archean granulite-gneiss belt (Windley 1984).

At the north margin of the North China Protoplatform there is a discontinuous granulite belt which is composed of many small, scattered granulite bodies. An analysis of the distribution pattern of small continental nuclei reflected by an isodepth contour of deep-level magnetic interfaces (Fig. 2. 7) suggests that the smaller granulite bodies lie just at the margins of the small nuclei and in the depressions between them. Owing to the intrusion of granitic rocks, structural superimposition and covering by younger rock formations, it is difficult for the time being to get an overall picture of the granulite distribution during the Archean. However according to data available, Archean granulites are not only distributed at the margins of three continental nuclei, i.e. the Dongsheng, the Bohai Sea and the Liaodong (eastern Liaoning) nuclei, but are also found at the west, south and east margins of the Linfen continental nucleus as well as the south margin of the Jining continental nucleus. Judging from the distribution of ophiolite belts and TTG suites, each of the granulite zones implies the existence, at their time of formation, of a subduction plate boundary that is located at deep level today.

Globally, glaucophane schist zones are found mainly in Meso-Cenozoic high-pressure, low-temperature metamorphic belts related to plate subduction, obduction and collision as well as in Paleozoic and Late Precambrian orogenic belts (Windley 1984). In the welded area between the North China and Yangtze Protoplatforms there occurs a glaucophane schist-eclogite belt. Although no direct age evidence has been obtained for the belt, its stratigraphic position indicates that a Late Precambrian age is suitable. This suggests that the modern-type plate subduction and collision zones occurred as early as the Late Precambrian.

Due to subsequent tectonic superimposition and reworking, active belts (e.g., the Early Proterozoic belts at the north and south margins of the North China Protoplatform and the Early Proterozoic Liao-Ji belt) can not be completely reconstructed. In particular, owing to the intensity of metamorphism and denudation, even the original high-pressure metamorphism zones cannot easily be mapped. Therefore, the absence of glaucophane schist in Precambrian mobile belts in China does not seem to be a decisive indicator of the existence of plate subduction boundaries: - even magnificent modern orogenic belts, such as the Himalayan orogen over 3000 km long, contain only a few glaucophane schist zones less than several km long each (Windley 1984). It is therefore not suprising that let alone Precambrian medium- to deep-level orogens exposed at the surface after being strongly eroded do not show such features.

6.1.4 Development of Life Forms Related to Atmospheric Evolution

The development of first simple and then complex life forms with time indicates the secular rise of oxygen content, as the increase in oxygen concentration is favourable to the evolution of life. As the Earth's crust developed, the life form changed from lowly to highly complex levels in a series of evolutionary leaps.

Organic photosynthesis is the main way to produce free O_2. As seen in the eastern section of Liaoning Province on the North China craton, the banded iron formations (BIF) in the Archean terranes contain bacterial fossils, suggesting that oxygen production by organic photosynthesis occurred as early as 3.0-2.5 Ga ago and met the demand for oxygen during the formation of the BIF. However, since the beginning of Early Proterozoic time, there have appeared large amounts of stromatolites which · contain globular cells without sheaths, indicating that blue algae, formed by the life activities of primitive prokaryotes, reached an unprecedented stage of development. Due to the photosynthesis of such algae, the content of free O_2 in the atmosphere may have broken through the Pasteur level, i.e. 1% higher than that in today's atmo-

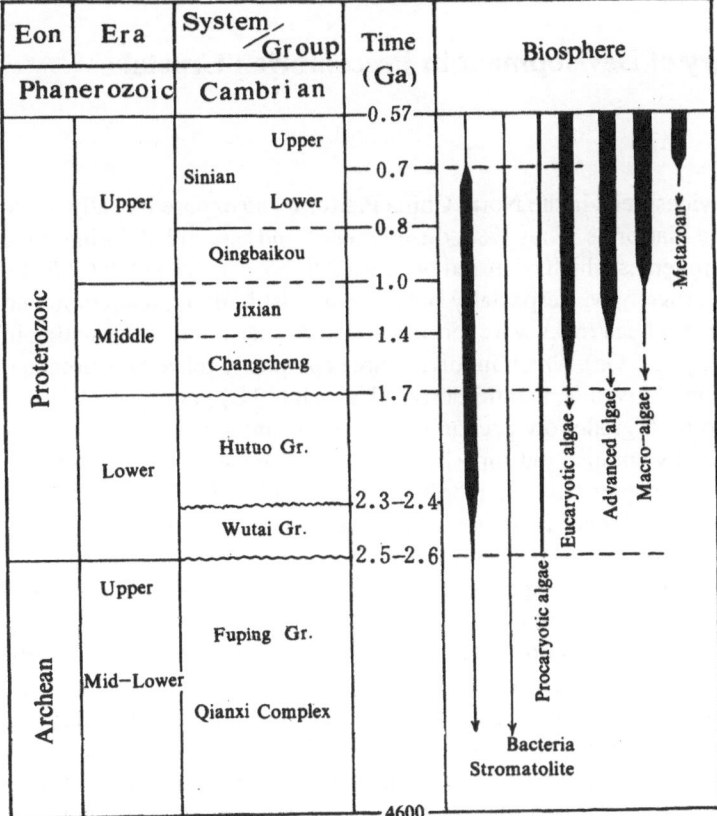

Fig. 6.1. Biological evolution of Precambrian in China

sphere. Furthermore, the presence of numerous red beds also implies the existence of large amounts of free O_2. All these suggest that an evolution from an essentially reducing atmospheric environment to an oxidizing one occurred, opening a new chapter in crustal history.

The Middle to Late Proterozoic is the most important time for stromatolites. More importantly, eukaryotic algae occurred since the Middle Proterozoic Changchengian, although procaryotes dominated by blue algae and bacteria were predominant at this time. During the Jixianian, eukaryotic algae appeared in large quantities. During the Late Proterozoic Qingbaikouian period, the microflora was characterized not only by a remarkable increase in the number of genera and species and in the size of individuals, but also by more complicated ornamentation; and macroscopic algae (e.g. brown algae) occurred in large amounts, marking a very successful period for higher algae. The Sinian shows the first appearance of metazoans represented by worms, coelenterates and sponges. At a later stage of the Sinian a few metazoans with internal skeleton came into existence and became the ancestors of the Cambrian polytaxon fauna (Fig. 6. 1).

6.2 A Summary of Development in Precambrian Crustal Evolution

Archean crust is widespread in the North China Platform and exposed locally in the Tarim and Yangtze Platforms. They were differentiated and assembled during three main periods of orogenies which occurred between 3.8-3.5, 3.1-2.9 and 2.7-2.5 Ga. However the last orogeny was especially potent as a crust-building operation, and thus the bulk of the Archean rocks were formed in later Archean time, culminating in the Fuping Orogeny (2.6 Ga), which involved three complexly related phenomena: 1) major production of juvenile, mantle-derived Na-rich (TTG) granitoids; 2) widespread medium to high grade low pressure metamorphism; and 3) pervasive tectonism charaterized by compressed superimposed folding patterns and thrusting.

As mentioned in Chapter One the North China Protoplatform in Archean time contains NE-trending and locally east-trending alternating superbelts composed respectively of low to medium metamorphic grade granite-greenstone belts and medium to high metamorphic grade gneiss terranes. There was extensive formation of granulite, charnockite and amphibolite in the period of 3.0-2.6 Ga inside or along the margins of the North China Protoplatform. The metamorphic rocks are distributed in isolated small patches or in a zonal pattern. The mineralogy and geochemistry of these granulites indicate that they formed in the middle or upper middle parts of the lower crust. The lower part of the lower crust has not yet been recognized (Shen Qihan, 1992). The major period of greenstone development, with concurrent Na-granitoid plutonism, was 2.9-2.5 Ga. However, the Archean greenstone belts are of generally higher metamorphic grade as compared with those in the type localities of other countries, which may imply higher heat flow and uniform metamorphism.

A striking feature in the Archean crustal development in China is that in contrast to the global mega-BIF accumulation which occurred mostly in the period of Early Proterozoic time (2.5-2.0 Ga), in response to rising oxygen production globally, BIF deposits in China were developed mainly in Late Archean time, and are especially rich in Anshan-Benxi of Liaoning Province and Luanxian districts in eastern Hebei Province. This implies that sufficient oxygen has produced during the period of 2.9-2.5 Ga. However BIF was not well developed in Early Proterozoic times, for instance those of the Wutai Group and Lüliang Group of Shanxi Province are of much smaller size and limited distribution.

The stratigraphic, tectonic and petrochemical evidence for crustal thickening at the end of the Archean time provide a clue to the marked increase in lithospheric stability in Early Proterozoic time. This process of progressive stabilization has generally been regarded as cratonization. The extensive thick and rigid Early Proterozoic plates were beginning to respond to deposition, deformation and intrusion in a more modern style of tectonic regime. However, the Archean-Proterozoic boundary in China is transitional: granulite gneiss and granite-greenstone belts continue to develop through the Proterozoic, as is apparent from recent isotopic age determinations and geological reevaluation. A large amount of Early Proterozoic granulites or granulite-amphibolite belts has been found: of particular note are the Tongbai-Dabie, the Altun, and the Dulan-Golmud granulite-amphibolite belts of Early Proterozoic age. The greenstone belts also continued to develop in Early Proterozoic time, for instance the Qinglong greenstone belt and the Wutai granite greenstone belt. In the latter granitoid rocks were emplaced during different evolutionary stages, with the syntectonic diapiric granite bodies emplaced at 2.4-2.3 Ga ago (Tian Yongqing et al. 1991). The presence within such belts of occasional ophiolites or ophiolitic melange suggests the possible operation of subduction-related plate tectonic process.

Early Proterozoic crust typically adjoins and partly surrounds Archean nuclei. The metamorphic strata are highly variable in pattern and composition, thereby representing a major watershed in Earth's continental evolution. The Early Proterozoic rocks of China generally belong to the basement complex and are thereby separated in a fundamental way from widespread Mid- to Late Proterozoic crust. The Early Proterozoic sequence is characterized by shelf sedimentation in the craton and other tectonic elements, ranging from intracratonic rifts, troughs, and substantial major multizoned intercratonic mobile belts are collectively products of cratonic break up, enlargement and eventual reconsolidation. For example, the Wutai'an aulacogenic trough in between the Hengshan and Fuping Archean nuclei was filled with volcanic-wacke-BIF assemblages. Commonly these comparatively local basement structures are themselves unconformably overlain by broader, more extensive arenite-pilite-carbonate synclinora, such as the Hutuo Group. Long intercratonic or craton-enclosing mobile belts up to several hundred km wide and attributable to protocontinental rifting and drifting are also salient features of the Early Proterozoic crust.

The major part of the craton was consolidated through the Lüliangian Orogeny around 1.8 Ga. Subsequently, the stabilization of the Protoplatform was completed at about 1.7 Ga, especially along the southern and eastern border parts of the North China Protoplatform (Ma Xingyuan et al. 1981; Qiao Xiufu and Ma Lifang, 1982). For instance, following the close of the Lüliang-Zhongyue Orogenies by 1.8-1.7 Ga the North China Protoplatform and major part of the Tarim Protoplatform and the Qaidam Massif were fully consolidated. Together they formed the North China continent, which is characterized by Mid Proterozoic aulacogen-anorogenic taphrogenesis, and basin subsidence with bimodal anorogenic suite mantle-derived anorthosite and crust derived rapakivi granites at around 1.5 Ga, which are products of long-sustained craton stability and linear subcrustal thermal regimes. The whole region of the North China Protoplatform underwent uplift (termed the Qinyu Uplift) at the end of the Mid Proterozoic 1.0 Ga ago with the development in the Qingbaikou Period of genuine cover sequences. In this sense the North China Platform was also finally established at the end of the Jinningian Stage.

During Mid- to Late- Proterozoic time, the south margin of the North China Protoplatform had twice undergone northward subduction from the Qinling marine realm to the south, while the margins on the other sides were mainly passive. The unconformable contact between the upper Mid-Proterozoic Jixianian Pengjiawen Formation and Late Proterozoic Qingbaikouan strata, together with the emplacement of 1.0 Ga granitic rocks in the Xiao Qinling area of western Henan Province, are indicative of northward subduction from the Qinling marine realm. This subduction was also probably responsible for the general uplift of the North China Protoplatform in the beginning of the Late Proterozoic time. The gentle northward tilt of the protoplatform southern margin may act as the triggering force for the gravity gliding tectonic in the Wufoshan area (Ma Xingyuan et al. 1981b).

The Yangtze Protoplatform differs from the North China Protoplatform in its younger basement, which is mainly consolidated in Late Proterozoic time. The Archean-Early Proterozoic nuclear basement encompasses only the central Sichuan, eastern Yunnan Provinces and the northern border part of the craton. The boundary nature of the Yangtze Protoplatform also differs remarkably from that of the North China Protoplatform. The northern margin of the protoplatform was mainly passive during Mid-Late Protreozoic time. Extensive and complicate continental margin terranes had developed along the western margin and to the southeast of the craton on the Jiangnan Uplift. The formation of island arcs with marginal seas and subduction zones caused successive accretions of the continental crust during the Sibao'an and Jinningian Orogenies at about 1.0 Ga and 800 Ma respectively. For example, the 2000 km long Qinling mobile belt marks the E-W trending junction between the North China craton to the north and the Yangtze craton to the south. The discovery of Late Proterozoic (about 1.1-0.7 Ga) glaucophane schists along southern flank of the Qinling-Dabie orogenic belt has suggested to many authors (Dong Shenbao 1989; J.G.Liou et al. 1989) that the Yangtze plate was subducted to a more profound depth during Late Proterozoic convergence, thereby developing high-pressure phase assemblages. A second example is the Northeast Jiangxi Fault on the Jiangnan Uplift. This marks a Late Proterozoic collision, amalgamation and suture zone between the

Huaiyu terrane and the Jiuling terrane, as evidenced by 866±14 Ma glancophane schists. These two Late Proterozoic tectono-stratigraphic terranes are of contrasting rocks: the Huaiyu terrane is oceanic and contains an 1024-1154 Ma ophiolite whereas the Jiuling terrane is made up of continental crust.

After the culminating Jinningian thermo-tectonic event at ~ 1.0 Ga, the Yangtze and northern part of the Tarim Platform and central Tianshan were consolidated. The pronounced unconformity developed at that time provided the subsequent basement for cover rocks of the Sinian System (800-570 Ma). The Yangtze Platform subsided as a whole with extensive sedimentation resulting in well developed Sinian sequences. The main part of North China Platform underwent epeirogenic uplift at the end of Qingbaikou'an time (850 Ma). Only the eastern, southern and western margins subsiding to receive significant Sinian sequences.

The internal differentiation of South China is manifested by the NE-trending sedimentary provinces, from the Yangtze Province in the northwest to the southeast Maritime Provinces. There are more complex basement configurations in the Yangtze Platform involving basement highs with adjoining thicker sub-depocentres, - a legacy of their taphrogenic heritage. These include the Early Sinian Xiaoxiangling-Suxiong aulacogen in southern Sichuan and eastern Yuanan formed on the basement, which was filled with very thick medium to acidic volcanic eruptives and clastics and terminated at the end of Early Sinian time, being covered by Upper Sinian deposits formed in a stable sedimentary environment (Wang Hongzhen et al. 1984).

Additional Sinian strata occur on the Tarim Platform and are widely distributed in other parts of China, and include the cover sequences in the Qaidam Massif and central-south Tianshan. There are mobile type Sinian sequences in the Qilian mobile belt, the High Himalayas and northern Heilongjiang Province. The twin association of global diamicites, including at least two periods of glaciogenic tillites and a soft-bodied metazoan fauna (Ediacara) at 600 Ma were also well developed in China.

From the above discussion one can see that Precambrian crustal development in China is the result of progressive Archean-especially Late Archean-cratonization, subsequent Early Proterozoic protoplatformal consolidation and Mid-Late Proterozoic aulacogen-anorogenic taphrogenesis and basin subsidence. The evolution concluded with the generation of fully consolidated, well defined Sinian continental platforms under the influence of plate tectonic processes.

References

Allegre CJ, Li TD (1984) Structure and evolution of the Himalaya-Tibet orogenic belt Nature 307:17-22

An SY, Zhang WJ, Yang JX, Lin XJ (1990) Rock associations and metamorphism of the Qinling Group, Taibai, Shaanxi. In: Liu GH Zhang SG (eds) Geological Memoirs of the Qinling-Daba Mountains (1) Metamorphic Geology, Beijing Scientific and Technical Publishing House, Beijing 25-39 (in Chinese with English abstract)

Bai WJ, Gan QG, Yang JS, Xing FM, Xu X (1986) Discovery of well-preserved ophiolite and its basic characters in the southeastern margin of the Jiangnan ancient continent. Acta Petrologica et Mineralogica 5:289-299 (in Chinese with English abstract)

Bai J, Zhang XQ (1981) The structure of Dahongshan ore district and the Dahongshan Group. Bulletin of the Tianjin Institute of Geology and Mineral Resources GAGS, 3:103-116 (in Chinese with English abstract)

Bai J (1985) On the features of the early Precambrian Qinglong rift. Abstracts of the International Symposium on Deep Internal Processes and Continental Rifting 1985, China. p 2.

Bai J (ed) (1986) The early Precambrian geology of Wutaishan. Tianjin Science and Technology Press, Tianjin. (in Chinese with detailed English summary)

Bai J (ed) (1991) The Precambrian geology and its control of Pb-Zn mineralisation in the northern margin of the North China platform.

Bai J, Dai FY (1990) Judgement of the tectonic settings of the early Precambrian mobile belts. Selected writings of the International Symposium on Tectonic Evolution and Dynamics of the Continental Lithosphere. Third All China Conference on Tectonics. Scientific Publishing House, Beijing. (In Chinese).

Bai J, Li H, Wang R (1991) U-Pb single zircon isotope ages for the Wutai granite greenstone and their possible geological significance. Bulletin of the Tianjin Institute of Geology and Mineral Resources, No.2. pp12. (in Chinese with English abstract)

Bai J, Li HM, Wang RZ (1990) U-Pb single zircon isotopic ages for the Wutai granite-greenstone terrain and their possible geological significance. Bulletin of The Tianjin Institute of Geology and Mineral Resources, CAGS, No.24. (in Chinese with English abstract)

Bai J, Yang CL (1984a) Early Proterozoic crustal evolution of Eastern Hebei. In: Sun DZ (ed). The Early Precambrian geology of eastern Hebei, Tianjin Science and Technology Press 219-221 (in Chinese with English Summary)

Bai J, Yang CL (1984b) The lower Proterozoic low grade stratigraphy of eastern Hebei. In: Sun DZ (ed) The Early Precambrian geology of eastern Hebei, Tianjin Science and Technology Press, 14-24 (in Chinese with English abstracts).

Bai YL, Sou ST, Liu RQ, Tang JF, Ma XY (1984) Fold interference patterns in Precambrian rocks from three key areas of eastern China. In: A. Kroner and R Greiling, E.Schweizerbart'sche (eds) Precambrian Tectonics Illustrated, Verlagsbuchhandlung. Stuttgart, 335-352.

Baode BD, Pan ZC, Geng UC, Rong ZQ, Du HY (1980) Sinian Suberathen in the northern slope of eastern Qinling Ranges. Research on Precambrian Geology: Sinian Suberathem in China, Tianjin Science and Technology Press, Tianjin. 288-313 (in Chinese with English abstract)

Ben FG (1986a) The essential character of volcanic-sedimentary rocks of Dahongyu Formation in Jixian. Bulletin of the Tianjin insitute of geology and Mineral Resources, CAGS, No.16. Geological Publishing House, Beijing, 91-106 (in Chinese with English Abstract).

Ben FG (1986b) On the volcanic-intrusive magma activity during Changcheng Period at Jixian. Bulletin of the Tianjin institute of Geology and mineral Resources, CAGS, No.16 Geological Publishing House, Beijing, 109-119 (in Chinese with English abstract)

Brigade of Regional Geological Survey, Anhui Bureau of Geology and Mineral Resources (1985) Stratigraphy of Anhui Province, Precambrian subdivision. Anhui Science and Technology Press, Hefei, p 174 (in Chinese)

Bureau of Geology and Mineral Resources of Anhui Province (1987) Regional geology of Anhui province. Geological Publshing House, Beijing, p 721   (in Chinese with English abstract)

Bureau of Geology and Mineral Resources of Fujian Province (1985) Regional Geology of Fujian Province. Geological Memoirs, MGMR, PHC, Series 1, Vol No.4 Geological Publishing House, Beijing, 1-67 (in Chinese with English abstracts)

Bureau of Geology and Mineral Resources of Gansu Province (1990), Regional Geology of Gansu Province. Geological Memoirs, MGMR, PHC, Series 1, Vol. 19. Geological Publishing House, Beijing, 1-67 (in Chinese with English abstracts)

Bureau of Geology and Mineral Resources of Hubei Province (1986) Regional Geology of Hubei Province. Geological Memoirs, MGMR, PHC, 705. Geological Publishing House, Beijing, 1-67 (in Chinese with English abstracts)

Bureau of Geology and Mineral Resources of Xinjiang Uygur Autonomous Region (1985) Geological Map of Xinjiang Uygur Autonomous Region, China (1:2,000,000). Geological Publishing House, Beijing. (with a Chinese explanation)

Bureau of Geology and Mineral Resources of Xinjiang Uygur Autonomous Region (1993) Regional Geology of Xinjiang Uygur Autonomous Region. Geological Publishing House, Beijing, p 841.

Chang SQ (1980) Subdivision and correlation of late Precambrian in southern Liaodong peninsula. In: Sinian Suberthern in China, Tianjin Science and Technology Press, Tianjin, 266-281 (in Chinese with English ahstract)

Chen HS, Zhou XQ (1987) Lithosphere structure characteristics and geological significance in the Yangtze area. International symposium on Tectonic Evolution and Dynamics of Continental Lithosphere, Abstracts (1), 314-315

Chen JB, Zhang HM, Zhu SX, Zhao Z, Wang ZG (1980) Research on Sinian suberathem of Jixian, Tianjin. In: Research on Precambrian Geology Sinian Suberathem in China, Tianjin Science and Technology Press. 56-112 (in Chinese with English abstract)

Chen Yifei (1979) Prilininary study on BO value of muscovite in the Precambrian meta-pelite, Qinglong County, Eastern Hebei Province. Kexue Tongbao, 38-43. (Chinese Science Bulletin).

Chen ZL, Chen SY (1987) On the tectonic evolution of the west margin of the Yangtze block. Chongqing Publishing House, China. p 172 (in Chinese)

Chung FT, Compston W, Foster J, Bai J, Sun DC (1979) Age of the Chiensi (Qianxi) Group, North China. Annual Report 1979, The Australian School of Earth Sciences, 145-151

Commission on stratigraphy of China (1983) Summary of the conference on stratigraphic classification and nomenclature of the Upper Precambrian. Journal of Stratigraphy Vol 7. No.1 78-80.

Condie KC (1982) Plate tectonics and crustal evolution. 2nd edition, Pergamon Press, New York, p310

Cong BL, Li JL, Zhang RY (1982) Primary rocks of the early Archean metamorphic rocks in the Qianxi-Qianan regions of Eastern Hebei Province and their geoiogical significance. Scientia Geologica Sinica 2: 125-133.

Cong Y, Liu G (1990) Metamorphism and Evolution of the Kuaping Group in the Shan-Luo Region, East Qinling Mountains. In: Geological Memoirs of the Qinling-Dabie Mountains (I) Metamorphlic Geology. Edited by Liu Gohui and Zhang Shouguang, Beijing Scientific and Technical Publishing House, p 72-88.

Deng YH (1980) Discovery of plenty of metamorphosed volcanic rocks in Yexian County, Shadong Province, and a preliminary search of the Fenishan greenstone belt. Abstract collection of the 1980 symposium, Geological Society of Shandong Province, 51 (in Chinese)

Dong SB (1986) Metamorphism in China and its relation with crustal evolution. Geological Memoirs, Series 3, No.4, Geological Publishing House p 233

Dong SB (1989) The general features and distribution of the glaucophane schist belts of China. Acta Geologica Sinica 3. No.1

Dong YJ, Xu HF (1989) A study on the Archaean stratigraphy and its mineralization in Xintai area, western Shandong.   No.1 Team of Geology and Mineral Resources, Bureau of Shangdong. (Domestic research report, typescript in Chinese)

Du RL, Tian LF, Li HB (1986) Discovery of megafosslls in the Gaoyuzhuang Formation of the Changchengian System, Jixian. Acta Geoilogica Sinica 60: 115-120 (in Chinese with English abstract)

Ganshu Bureau of Geology and Mineral Resources 1990 Regional Geology of Ganshu Province. Geological Memoirs, MGMR PRC, Series 1, Vol.19. Geological Publishing House, Beijing. p 692 (in Chinese with English Abstract).

Gao HX, Cheng XT, Li SL (1988) Excursion guide to Proterozoic geology of the Qinling range. International symposium on geochemistry and mineralisation of Proterozoic mobile belts.

Gao LD, Liu ZG (1988) New discovery of microfossils from the Nanwan Formation of the Xinyang Group, Hena, and its geological significance. Geological Review 34: 421.

Gao ZJ, Chen JB, Lu SN, Pong CN, Qin ZY (1989) Research report on the Precambrian and its mineralizing property in the northern Xinjiang. (In press) p 396 (in Chinese)

Gao ZJ, Wang WY, Peng CW, Li YA, Xiao B (1985b) The Sinian system of Aksu-Wushi region, Xinjiang, China. Xinjing People's Publishing House, p 184. (in Chinese with English abstract).

Goldich SS (1973) Ages of Precambrian banded iron formations. Economic Geology 68: 1126-1134

Goodwin AM, Ridler RH (1970) The Abitibi orogenic belt. Geological Survey of Canada. Paper 70-74, 1-30

Guan BD, Geng WC, Rong ZQ, Du HY (1983) On the age of Luoquan formation in Henan province. Precambrian Geology. In: The Collected works of late Precambrian Glacigenous Rocks in China, Vol.1. Geological Publishing House, Beijing, 183-206. (in Chinese with English abstract)

Guan ZN, An YL, Wu CJ (1987) Inversion of magnetic boundaries and investigation into the deep-seated geological structure of north China. In: Wang MJ, Cheng JY (eds), Contributions to the Exploration Geophysics and Geochemistry, Vol.6. Regional Geophysical Features of Eastern China, Geological Publishing House, Beijing 80-10 (in Chinese with English abstract)

Guo JJ, Wang RZ (1989) Geological features on Bantukou Formation in the east of Wutai Mt. and its attribution. Precambrian Geology. Geological Publishing House, Beijing, Vol 4. 89-98.

He TX, Jin W (1988) The geological features of the Chencai Group, Zhejiang Province, and its significance in stratigraphy.Geology of Jiangxi Vol 2. 146-150 (in Chinese with English Abstract).

Hong ZM, Huang ZF, Yang XD, Lan J, Xian BC, Yang YJ, Liu XL (1988) Medusoid fossils from the Sinian Xingmincun Formation of southern Liaoning. Acta Geologica Sinica, 62:200-209 (in Chinese with English abstract)

Hong ZM, Liu XL (1987) The discovery of metazoan fossils in the Sinian Xingmincum and Getum Formations in the Fuzhou-Dalian depression, southern Liaoning. Regional Geology of China No . 2, 45- 147 (in Chinese with English abstract).

Hsu KJ,, Sun S, Li JL (1988) Huanan Alps, Not South China platform. Scientica Sinica (Series B), 29:109-119

Hu SL,Wang SS (1988) The ^{40}Ar/^{39}Ar age spectrum of Damiao anorthosites and its geological implication. International Symposium on Geochemistry and Mineralization of Proterozoic Mobile Belts Abstracts, p 40

Hu SX Lin QL et al. (1988) The geology and metallogeny of the amalgmation zone between the ancient North China plate and the South China plate (taking Qinling-Tongbai as an example). Nanjing University Publishing House, 1-558 (in Chinese)

Huang JQ (1978) An outline of the geotectonic characteristics of China. Eclog Geol. Helv. 71, 3.

Huang JQ (director), Ren JS, Jiang CF, Zhang ZK, Qin DY (1980) The tectonic evolution of China. Science Press, BeiJing. p 124 (in Chinese)

Huang JQ, Jiang CF, Wang ZX (1990) On the opening-closing tectonics and accordion movement of plates in the Xinjiang and adjacent regions. Geoscience of Xinjiang, Vol 1. 3-16. (in Chinese with English abstract)

Huang XG (1985) The evolutionary characteristic of llthofacies-palaeogeography of the Sangshuan Period of the Gaoyuzhuang Formation in middle Yanshan range. Bulletin of the Tianjin Institute of Geology and Mineral Resources, Vol.13. 1-30 (in Chinese with English abstract)

Jahn BM Zhang ZQ (1983) Archaean granulite gneisses from eastern Hebei Province, North China; Part 2: Rare earth geochemistry and petrogenesis, In: G.Hanson et al., (eds); Archaean Geochemistry. Springer-Verlag.

Jahn BM Zhang ZQ (1984) Radiometric ages (Rb-Sr, Sm-Nd, U-Pb) and REE geochemistry of Archaean granulite gneisses from eastern Hebei province, China. In: A Kroner, GH Hanson AM Goodwin (eds) Archaean Geochemistry. Springer-Verlag 204-234

Jakes P. White A.J.R. (1972) Major and trace elements abundances in volcanic rock of orogenic areas. Geological Society of America Bulletin Vol 83. 29-40

Jia BW (1986) Petrological features and genesis of Damiao-Heishan complex in Hebei Province. Shanxi Geology. Vol.1. No.1 51-65. (in Chinese with English abstract)

Jiang CC (1986) The basic outline of the Precambrian crustal evolution of the Sino-Korean paraplatform. Proceedings of the International Symposium on Precambrian Crustal Evolution, No.1, pp75-86 (in Chinese with English abstract)

Jin WS, Sun DZ, Wu CH, Han G, Li BY (1992) Characteristics of Metamorphism of the early Precambrian, south belt of Yinshan, North China platform. Bulletin of the Tianjin Institute of Geology and Mineral Resources, CAGS. 2 (in Chinese with English abstract)

Kroner A (1983a) Archaean to early Proterozoic tectonics and crustal evolution: A review. Review of Brazilia Geocience, 12:15-31

Kroner A (1983b) Proterozoic Mobile belts compatible with the plate tectonic concept. In: Medaris, Jr. G.L. (ed), Proterozoic geology, Geological Society of America Memoir 161, 59-74

Kroner A (1989) Plate motion, crustal accretion and supercontinental assemblage since the early Archaean. Abstracts of 28th International Geological Congress, Vol 2. 230-231

Lan YQ, Shi XM (1984) Research on Precambrian volcanic sedimentary cycle, Qian-an, eastern Hebei. Journal of Changchun College of Geology, Vol.3. 14-25. (in Chinese with English abstract).

Lee JS, Chao YT (1924) Geology of the Gorges District of the Yangtze from Yichang to Tsekuei, with special reference to the development of the Gorges Bulletin of the Geological Society of China, Vol . 3. 350-392

Li BH (1980) Petrological characteristics of keratophyrein the Dahongshan copper-iron ore district, Yunnan Province. Geological science and technology information of Yunnan. Vol. 1. 1-9 (in Chinese)

Li GK, Lin HC (1988) Isotopic ages and their tectonic significance in Fujian. Geology of Fujian, Vol.7. No.2 80-118 (in Chinese with English abstract)

Li J, Hu F (1981) The serpentinitic distostrome in Bayanobo group, Scientia Geologica Sinica, Vol. 3. p269-272 (in Chinese with English abstract)

Li JL (1984) Eugeosyncline rock association of Yanbian Group in western Sichuan, China. Bulletin of the Chinese Academy of Geological Sciences, Vol. 8. 21-38 (in Chinese with English abstract)

Li Q, Leng J (eds) (1987) The upper Precambrian in the Shennongjia region. Tianjin Science and Technology Press, Tianjin, China. p 503 (in Chinese with English Summary)

Li SG (1986) Greenstone belt tholeiite and modern island arc tholeiite in Cr and Ni abundances. In: Proceedings of the international symposium on Precambrian crustal evolution, No.2. Geological Publishing House, Beijing, 80-94

Li Y (1988) A dating model of illite grain Rb-Sr isochron and its application to Precambrian stratigraphy. Acta Gelogica Sinica, 62:268-275 (in Chinese with English Abstract).

Li YA, Gao ZJ, Wang JH (1984) Preliminary palaeomagnetic study of the Tarim late Precambrian paleo-block. Xinjiang Geology, Vol. 2 81-95 (in Chinese with English abstract)

Li YA, Li YP, Gao ZJ, Li Q, Zhai YJ, Zhang ZK, Sharps R, McWilliams M, Cox A (1989) Sinian palaeomagnetic research of Aksu-Keping region. Xinjian Geology, 7: 78-88 (In Chinese with English abstract)

Li YM, Zhao ZF (1988) The primary study of the geochemical characterics of the metavolcanic series from the Bikou Group in southern Gansu Province. International Symposium on Geochemistry and Mineralisation of Proterozoic Mobile Belts Abstracts, 54

Li YP, McWilliams M, Tan CZ (1988) The paleomagnetic study of the Quarlji Group (Sinian) of the Chaidam terrane. Xinjiang Geology, 6:35-40 (in Chinese with English abstract)

Li ZH Gao YD (1988) Geochemical character and environment of occurrence of the Xiong'er Group in the south margin of the North China Platform. International Symposium on Geochemcastry and Mineralization of Proterozoic Mobile Belts Abstracts, 5

Liang WT, Wang HJ, Jing YR, Zhang LT, Xia ML. (1989) The petrological and mineralogy characteristics of a high pressure metamorphic belt in Zhangbaling, Anhui. Journal of Changchun University of Earth Science, special issue on the blueschist belt in Hubei and Anrlui Provinces, 97-122 (in Chinese with English abstract)

Liang YF, Qi YR (1988) Metanorphism of the Pre-changcheng period in Shanxi. Shanxi Geology, 3: 335-356 (In Chinese with English Abstract)

Liao SF (1979) Sinian glacial deposites in China. Scientific Papers on Geology for International Exchange-prepared for ttle 26th International Geological congress, 2. Geological Publishing House, Beijing, 36-40 (in Chinese)

Liou JG, Wang XM, Coleman RG, Zhang ZM, Maruyama S (1989) Blueschists in major suture zones of China. Tectonics, 8:609-620.

Liu DY, Page RW, Compston W, Wu JS (1985) U-Pb zircon geochronology of late Archaean metamorphic rocks in the Taihangshan-Wutaishan area, North China. Precambrian Research 27: 85-109

Liu DY, Shen QH, Zhang ZQ, BM Jahn, B Auvray (1990) Archean crustal evolution in China: U-Pb geochronology of the Qianxi Complex. Precambrian Research 48: 223-244.

Liu GD, Zhao JH (1986) A brief discussion. Science Journal: 38. p75.

Liu GL (1987) New progress in the study of the age of the Kongling Group. Regional Geology of China, No. 1. p 93 (in Chinese).

Liu HY (1991) The Sinian Systen in China. Science Press, Beijing. p388.

Liu RG, Zhang JQ, Zhang BH, Cai YT, Ma WN (1984) Tectonic styles of the Archaean rock groups in the north-central Liaoning Province, China. Journal of Geodynamics 1:279-300.

Lu GY, Huang JH (1987) Recent results on Rb-Sr date of epirocks in Eastern Hebei Province and their geological significance. China Regional Geology 3: 219-224 (in Chinese wlth English abstract).

Lu JM (1986) Early Proterozoic metamorphic rock series, west Sichuan and east Yunnan, China: metamorphism and crustal evolution. Journal of Changchun College of Geology, Vol.3. 12-22 (in Chinese with English abstract)

Lu XM, Li CG (1988) Discussion on spitites of the Huashan Group in Dahongshan area, Hubei Province and its origin. Hubei Geology 2:12-22 (in Chinese with English abstract)

Luo YN (1983) The evolution of paleoplates in the Kang-Dian tectonic zone. Earth Science 21:93-102 (in Chinese with English abstract)

Ma XY (1989) Geological observations along the northern and southern parts of the geoscience transect from Xiangshui, jiangsu to Manduls, Inner Mongolia. Earth Science-Journal of China University of Geosciences 14: 1-7 (in Chinese with English abstract)

Ma XY He GQ (1989) Precambrian crustal evolution of eastern Asia. Journal of South East Asian earth Sciences 13: 9-15

Ma XY, Suo XT, You ZD, Liu RQ (1981) Tectonic deformation of the Songshan area, henan province, China - Gravitational tectonics, structural analysis. Geological Publishing House, Beijing, p 256.

Ma XY, Wu ZW (1981) Early tectonic evolution of China. Precambrian Research 14: 185-202.

Mao JW, Zhou KZ, Zhu Z (1988) Preliminary study on middle Proterozoic komatiites and Related copper-nickel deposits in Jiuwandashan Area, northern Guangxi. Bulletin of the Institute of mineral deposits, CAGS. No. 1. 147-160 (in Chinese with English abstract)

McLennan SM, Fryer BJ, Young GM (1979) Rare earth element in Huronian (Lower Proterozoic) sedimentary rocks: composition and evolution of the post-Kenoran upper crust. Geochimica et Cosmochimica Acta 43: 375-388.

Meng X, Ge M, He Z (1989) Sedimentary suites (SDS) of the early Sinian in China. Geoscience - Journal of Graduate school, China University of Geosciences 3: 1-16 (in Chinese with English abstract)

Monger JWH, Price RA, Tempelman-Kluit DJ (1982) Tectonic accretion and the origin of the two major metamorphic and plutonic belts in the Canadian Cordillera. Geology 10: 70-75.

O'Connor JT (1965) A classification for quartz-rich igneous rocks based on feldspar ratios. U.S. Geological Survey Paper. Vol. 525B. 79-84

Ouyang JP, Zhang BR (1988) Geochemical Characteristcs and tectonic setting of the Volcanic Rock Suite of the Xiong'er Group in the northern Qinling. International Symposium on Geochemistry and Mineralisation of Proterozoic Mobile Belts Abstracts, 67

Pearce TH, Gorman BE, Birkett TC (1975) the TiO_2-K_2O-P_2O_5 diagram: a method of discriminating between oceanic and non-oceanic basalt. Earth and Planetry Science Letters 24: 419-426.

Project Co-operation Group (1984) Research on the upper Precambrian of northern Jiangsu and Anhui Provinces. Anhui Science and Technology Press p 209. (in Chinese with English abstract

Qi CM, Yuan C, Jia KS, Zhou YC (1988) Geology and geochelnistry of the Damiao anorthosite-norite intrusion, Hebei, China. International Symposium on Geochemistry and Mineralization of Proterozoic Mobile Belts Abstracts, 7

Qiao J, Shen Y (1990) Ancient volcanic iron-copper deposit in Dahongshan district, Yunnan province. Geological Publishing House, Beijing, 205.

Qiao XF, Ma LF, Zhang HM (1988) The terminal Precambrian palaeogeographical framework in China. Acta Geologica Sinica 62: 290-300

Shen BF, Luo H, Han GG, Peng XL, Li JJ (1989) The Archaean geology of the Qingyuan region in Liaoning Province. Precambrian Geology, No.4, 9-18 (in Chinese, with English abstract)

Shen QH (1992) Early Precambrian granulites in China. Advancement in Earth Sciences, Vol.7. 1, 95-96.

Shen QH, Wu JS, Gen YS, Liu DY, Li WL, Zhao DM, Song B, Zhang ZQ, Ye XJ (1990) Early Precambrian major geological events and their age determinations in North China Platform.(in press).

Shu LS, Li YJ (1987) On Terrance Tectonics of North Jiangxi. Geology of Jiangxi 1: 31-37 (in Chinese with English abstract).

Shui T (1987) Tectonic framework of the continental basement of southeast China. Scientia Sinica (series B) Vol. 31. 885-896.

Spear FS (1986) PT path: A Fortran Program to caculate pressure-temperature paths from zoned metamorphic garnets. Computers and Geoscience 2: 1-12.

Sun DZ (ed) (1984) The early Precambrian geology of eastern Hebei Tianjin Science and Technology Press. p 273 (in Chinese with English Summary).

Sun DZ, Zhang HM, Hu WX, Liu WX, Tang M, Zhao FQ, Mei HL (1988) Proterozoic geology, geochemistry and mineral deposits of the Zhongtiao mountains (excursion guide). International Symposium on Geochemistry and Mineralization of Proterozoic Mobile Belts, Tianjin, p 89

Sun S, Zhang GW, Chen ZM (1985) Geological evolution of the southern part of North China Block, Metallurgical Industry Press, Beijing, p 216 (in Chinese)

Suo ST, You ZD, Han YQ, Zhong ZQ (1987) Tectonic features of Precambrian metamorphic rocks in China. Earth Science - Journal of Wuhan College of Geology 12. No.5. (in Chinese with English Abstract)

Tan Y, Li S, Zhao W (1983) On the characteristics of tectonics and its evolution of Zunhua Group in the East Hebei Province, Earth Science - Journal of Wuhan College. vol? 103-116. (in Chinese with English abstract)

Tang M, Sun DZ (1988) Geochemical comparisons of the early and middle Proterozoic volcanics in Zhongtiaoshan mountains, Shanxi Province. International Symposium on Geochemcastry and Mineralization of Proterozoic Moblle Belts Abstracts 95.

Taylor SR (1977) Island arc models and the composition of the continental crust. In M. Talwani and W.C.Pitmall (Eds), Island Arcs, Deep Sea Trenches and Back Arc Basins. American Geophysical Union Maurice. Ewing Service 1, 325-376.

Taylor SR (1979) Chemical composition and evolution of the continental crust: the rare earth elament evidence. In M.W.McElhinny (ed) The Earth: its Origin, Structure and Evolution, Academic Press, London, 353-376.

Taylor SR, Mclennan SM (l981) The rare earth element evidence in Precambrian sedimentary rocks: implications for crustal evolution. In: A. Kroner (ed), Precambrian plate tectonics, Elsevier, Amsterdam, 527-548.

Tian YQ (ed) (1991) Geology and gold mineralization of the Wutaishan-Henshan greenstone belt. Shanxi Science and Technology Press, p244.

Upper Precambrian Research Group Shenyang Institute of Geology and Mineral Resources (1986) Study on the upper Proterozoic in southern Liaodong peninsula. Precambrian Geology. Geological Publishing House, No.3. 129-140.(in Chinese with English abstract)

Walter MR, Bujck R, Dunlop JSR (1980) 3400-3500 Myr. old Stromatolites from the North Pole area, Western Australia. Nature 208: 443-445.

Wan YS, Cong YX, Zhao ZR, Liu GH, Zhang SG (1988) Geochemical characteristics and origin of metamorphic basalts of Kuanping Group. International Symposium on Geochemistry and Mineralization of Proterozoic Mobile Belts Abstracts, 103

Wang HZ (1978) On the subdivision of the stratigraphic province of China, Acta Stratigraphica Sinica, Vol.2. No.2.

Wang HZ (1981) Geotectonic units of China from the view-point of mobilism. Earth Science, Journal of the Wuhan College of Geology, 1, Vol.14 65-66 (in Chinese with English abstract)

Wang HZ (1982) Tectonic devolopement of the continental margins on both sides of palaeo-Qinling marine realm. Acta Geologica Sinica, Vol.56. No.3. (in Chinese with English abstract)

Wang HZ (Chief compiler) (1985) Atlas of the palaeogeography of China. Cartographic Publishing House, Beijing, p281

Wang HZ, Qiao XF (1984) Proterozoic stratigraphy and tectonic framework of China. Geological Magazine 121: 599-614

Wang J, Li SQ (1987) The Langshan-Bayunobo Rifting and Its Metallogenesis. Contributions to the project of Plate Tertonics in North China, No. 2. Geological Publishing House, Beijing, 59-72 (in Chinese with English abstract)

Wang KH (l983) Primary rock restoring for metamorphic rocks of the late Archaean Jiaodong Group and its geological significance. Bulletin of the Shenyang Institute of Geology and Mineral Resources, CAGS, No.6. 11-31 (in Chinese with English abstract)

Wang KY, Yan YH, Yang RY, Chen YF (1985) REE geochemistry of early Precambrian charnockites and tonalites-granodioritic gneisses of the Qinan Region, eastern Hebei, North China. Precambrian Research 27: 63-84

Wang QC, Sun S, Li JL, Zhou D, Hsu KJ, Zhang GW (1989) The tectonic evolution of the Qinling Mountain belt. Scientia Geologica Sinica, No.2, 129-142 (in Chinese with English abstract)

Wang RM, He SY, Chen ZZ, Li PF, Dai FY (1985) Geochemical evolutlon and metamorphic development of the early Precambrlan in eastern Hebei, China. Precambrian Research 27: 111-129

Wang XF, Bi ZG (1983) The Sinian tillite of south Anhui Province. Precambrian Geology, No.1. Geological Publishing House, Beijing. 245-260 (in Chinese with English abstract)

Wang YG, Xie ZQ, Wang LX, Chen DC, Zhu SC (1986) The stratigraphic sequence and origin of the depositional environments of the Tiesiao Formation in eastern Guizhou and its adjacent areas. Regional Geology of China, No.4 341-384 (in Chinese with English abstract)

Wang YL, Lu ZB, Xing YS, Gao ZJ, Lin WX, Ma GG, Zhang LY, Lu SN (1980) Subdivision and correlation of the upper Precambrian in China. In: Sinian Subarethem in China,. Tianjin Science and Technology Press -30 (in Chinese with English abstract).

Wang YS, Chen JN (1989) Synthetic research on the metamorphic series of the northern Qinghai-Xizang (Tibet) plateau. p 345. (in Chinese)

Wang Z, Tang R, Yen K, Uang F (1989) The discovery of Early Cambrian fossils in the "Taowan Group" in Shaanxi and their geological significance. Regional Geology of China, No.2 113-122. (in Chinese with English abstract)

Wen QY, Yang TQ, Li GL, Jiao FC (1988) Sedimentary features ad palaeogeography of the rift belt basin in Huili Group. Panxi Collected Papers of Lithofacies and Palaeogeography, No.4. Geological Publishing House, Beijing. 58-79 (in Chinese with English abstract).

Wen XD (1987) Turbidites or the Xiajiang Group in Taijiang-Jinping region of eastern Guizhou and its geological significance. Precambrian Geology, No. 3 257-272 (in Chinese with English abstract)

Williams HR (1977) African Archean mobile belts and granite-greenstone terranes. Nature 266: 163-164.

Wills B, Blackwelder E, Sargent RH (1904) Research in China Vol. 1 part 1. p 98

Windley BF (1984) The evolving continents (2nd Edition). John Wiley and Sons, Chichester, p 399.

Working group on the geological time scale of China, MGMR (1987). A geological time scale of China. Geological Memoirs, Series 2. No. 8. Geological Publishing House, Beijing, p 148. (in Chinese with detailed English Abstract).

Wu JS, Chen T, Liu DY (1988) The character of geochemistry and isotopic age Gantaohe Group basic volcanic rocks, Hebei. Abstract of International symposium on geochemistry and mineralization of Proterozic mobile belts, Tianjin, China.

Wu JS, Liu DY, Jin LG (1986) The zircon U-Pb age of metamorphosed basic volcanic lavas from the Hutuo Group in the Wutai Mountain area, Shanxi Province. Geological Review 32: 178-184 (in Chinese with English abstract)

Wu TS, Lu HR, Li ZH (1984) Precambrian granites of Shanxi Province. (internal publication in Chinese) p 227

Xiao XC,Li TD, Li GC, Gao YL, Xu ZQ (1990) Tectonic evolution of the Qingbai-Xizang (Tibet) plateau. Bulletin of Chinese Academy of Geological Sciences 20: 123-125.

Xie DK, Guo KY (1984) A study on the paleosuture in the Dahie Mountain. Bulletin of the Chinese Academy of Geological Sciences 10: 167-178

Xie GH, Wang JW (1988) A Preliminary Study on the emplacement age of the Damiao anorthosite complex. Geochimica 1: 13-17

Xing XY, Suo ST, You ZD, Liu RQ (1981) Tectonic deformation of the Songstan Area, Henan province, China - Gravitational tectonics, structural analysis. Geological Publishing House, Beijing, p 841

Xing YS (1979) The Sinian of China. Scienltific Papers on geology for International Exchange - prepared for the 27th International Geological Congress (2). Geological Publishing House, Beijing. 1-12 (in Chinese)

Xing YS, Duan C L YZ, Cao RG (1985) Late Precambrian Gntology of China. Geological Memoirs, MGMR, Series 2, No.2 Geological Publisiling House, Beijing p288.(in Chiniese with Erlglish abstract)

Xu B, Qiao GS (1989) Sm-Nd isotopic age and tectonic setting of the late Proterozoic ophiolites in northeastern Jiangxi Province. Journal of Nanjing University (Earth Sciences) 3: 108-114 (in Chinese with English abstract)

Xu Bei (1990) The late Proterozoic trench-basin-arc system in northeastern Jiangxi-southern Anhui provinces. Acta Geologica Sinica 64: 33-42 (in Chinese with English abstract).

Xu GZ, Hao J (1988) The characteristics of Foziling Group and the geotectonic environment of its formation in the northern foot of the Dabie Mountain. Scientia Geologica Sinica 2: 97-109. (in Chinese with English abstract)

Xu H, Dong Y, Shi Y, Jin R, Shen K, Li S (1992) Granite-greenstone belt in western Shandong Province. Geological Publishing House, Beijing. p 84.

Yan E, Li XQ, Han GG (1981) The geological character of the greenstone belt in Qingyuan area, Liaoning Province. Bulletin of the Geological Society of Liaoning Province, China 1: 158-177

Yan E, Liang YH (1986) The division of the strata of the Archean greenstone belt in northern Liaoning, based on the cyclicity of the metamorphic volcano-sedimentary rock series. In: proceedings of International Symposium on Precambrian crustal evolution, No.2.geological Publishing House 45-52 (in Chinese with English abstract)

Yang LZ (1990) Middle Proterozoic komatiitte in northern Guangxi. Regional Geology Of China 1: 14-24. (in Chinese with English abstract)

Yang ZS, Yu BX (1984) Polydefomation of the Archean greenstone belt in the Hongtoushan area, Northern Liaoning Province. Journal of Changchun College of Geology 1: 20-35

Ye BD, Zhu JP (1990) The time of the Baoban Group and gold ore at Erjia Dongfang, Hainan Province, China. Contributions to Geology and Mineral Resources Research 5: 17. (in Chinese with English abstract)

Yin BX, Ma J, Chen XS, Zhao ZC (1988) The characteristics of the mineralizing geology for the vermiculite deposit of Qiegan bulak in Yuli county, Xinjiang Uygur Autonomous region. 129.(unpublished paper)

You ZD, Han YQ (1986) The geochemistry and paleotectonic setting of the Archaean trondhjemite near Dengfeng, Henan Province. In: Proceedings of the International symposium on Precambrian crustal evolution No.2. Geological Publishing House, Beijing, 153-162

You ZD, Suo ST, Chen NS, Zhong ZQ, Hao Y (1987) The metamorphic petrography and early crustal evolution of Qinling Group in western Henan. earth Science - Journal of Wuhan College of Geology 12: 32-328. (in Chinese with English abstract)

You ZD, Suo ST, Han YQ, Zhong ZO, Chen NS (1990) Metamorphic and deformational history of the Qinling-Daba Mountains (1) Metamorphic Geology, Beijing Scientific and Technical Publishing House, Beijing. 1-10 (in Chinese with English abstract)

YU JH (1990) Geochemistry of a rapakivi granite- suite in a Proterozoic rift in Beijing and its vicinity. Acta Geologica Sinica 64: 322-336. (in Chinese with English abstract).

Yuan HH (1985) Archean age information in Dukau District, Sichuan, China. Journal of Chengdu College of Geology 12: 79-844 (in Chinese with English abstract).

Zhai MG, Yang RY (1986) Early Precambrian gneiss basement in the Panxi area, Southwest China. Acta Petrologica Sinica 2: 22-37 (in Chinese with English abstract)

Zhai MG, Yang RY, Lu WJ, Shao JB (1984) Major and trace element geochemistry of the Archaean Qingyuan granite-greenstone terrane. Geological Review, China 30: 523-535

Zhai MG, Yang RY, Lu WJ, Zhou JB (1985) Geochemistry and evolution of the Qingyuan Archaean granite-greenstone terrain, N. China. Pecambrian Research 27: 37-62

Zhang GW, Bai YB Sun Y, Guo AL, Zhou DW, Li TH (1985) Composition and Evolution of the Archaean Crust in central Henan, China, Precambrian Research 27: 7-35

Zhang GW, Zhou DW (1990) The Qinling complex and the Qinling orogenic belt. In: Liu GH and Zhang SG (eds.) Geological Memoirs of the Qinling-Daba Mountains (1) Metamorphic Geology. Beijing Scientific and Technical Publishing House, Beijing. 11-24 (in Chinese with English abstract)

Zhang HM, Zhang WZ (1989) Continental drift for Late Precambrian. Bulletin of the Tianjin Institute of Geology and Mineral Resources, CAGS, 59-73 (in Chinese with English abstract)

Zhang HM, Zhang WZ, Li P (1982) Paleomagnetism of the Sinian System of eastern Yangzi Gorges in China.Bulletin of the Tianjin Institute of Geology and Mineral Resources, CAGS, 57-68 (in Chinese with English abstract)

Zhang K, Deng HG (1981) The development features of Ordos block in the early stage of late Proterozoic and its oil and gas prospect. Petroleum Exploration and Development, No. 5. 18-26 (in Chinese).

Zhang PY, Yan XL (1984) Microfossils from the Gaoyuzhuang Formation in Laishui, Hebei, China. Acta Geologica Sinica 58: 196-204 (in Chinese with English abstract).

Zhang PY, Zhu M, Song W (1989) Middle Proterozoic (1200-1400 Ma) microfossils from the Western Hills near Beijing, China. Canadian Journal of Earth Science 26: 322-328

Zhang QS (1988) Early crust and mineral deposits of Liaodong peninsula. Geological Publishing House, Beijing. p 574 (in Chinese with English summary)

Zhang QS (1984) Geology and metallogeny of the Early Precambrian in China. Jilin Press, p 536

Zhang RY, Cong BL (1982) Mineralogy and P-T conditions of crystallization of early Archaean granulites from Qianxi county, N.China. Scientica Sinica 25: 96-112.

Zhang SG (1989) Discussion on some features of metamorphism in China. Progress in Geosciences of China (1985-1988), Papers to 28th IGC, Vol.III 83-86

Zhang TG, Zhang JY (1988) Proterozoic strata and characteristics of the evolution of the sedimentary basin In Guangxi. Regional Geology of China 2: 109-116 (in Chinese with English abstract)

Zhang WJ, Li YJ (1989) The sequence and the age of the Taowan Group. Journal of Xi'an College of Geology 11: 1-10 (in Chinese with English abstract)

Zhang WQ (1986) Geochemical properties of metamorphic volcanites in northwest Fujian. Geology of Fujian 5: 36-50

Zhang WZ, Li P (1980) Paleomagnetism of the Sinian Suberatem In: Jiixian, China. Bulletin Acad Geol. Sci. Series Vl, 1: 111-122.

Zhang Y (1981) Proterozoic stromatolite microfloras of the Gaoyuzhuang formation (early Sinian: Riphean, Hebei, China. Journal of Palaeontology 55: 485-506.

Zhang YX, Luo YN, Yang CX (1988) Panzhihua-Xichang rift in China. Geological Memoirs Series 5, No.5, Geological Publishing House, Beijing, p 324

Zhang YX, Ye TS, Yan HO, Zheng SY, Duan CQ, Chen MR, Zhao WH (1986) The Archaean geology and banded iron formation of Jidong, Hebei Province. Geological memoirs, Series 1, No. 6 Geological Publishing House, Beijing (in Chinese with English summary)

Zhang ZY, Li ZH (1985) Microflora from the Gaoyuzhuang Formation (Changchengian System) of western Yanshan range, North China. Acta Micropalaeontologica Sinica 2: 219-233.(in Chinese with English abstract)

Zhao YS, Li JH (1987) Micropalaeofiora of Dagushi Group in northern Hubei and its stratigraphic division and correlation. Hubei Geology 1: 11-21 (in Chinese with English abstract)

Zhao ZR, Zhang SG (1990) Research on the Jiaoangou ductile shear zone in Shangzhou, Shaanxi province. In: Liu GH and Zhang SG (eds), Geological memoirs of the Qinling-Daba mountains, Beijing Scientific and Technical Publishing House, Beijing, 60-70. (in Chinese with English abstract).

Zhu Xinren (1988) Geochemical characters of the volcanic rock of Wudangshan Group and the tectonic environment. International Symposium on Geochemistry and Mineralisation of Proterozoic Mobile Belts. Abstracts, 139.

Zhu YL, Ge RG (1986) Isotopic dating and chronology of the Sinian System in the region of Qinliu-Ninghua, Fujian Province. Geology of Fujin 5: 1-14.

Subject Index

A

acritarchs 182
Agulugou Formation 182, 183
Alai Mountains 140
Alax
– block 87
– greenstone belt 60, 82-85
– Group 59
– region 59
Alehuduge Formation 185
Aleitonggou Formation 282, 289
algae 159, 226, 247, 254, 257, 258, 277, 283, 291, 299, 300
Altun
– granulite-amphibolite belt 301
– Group 142
– lithospheric left-lateral shear fault 132
– Mountains 11, 140, 141, 143, 195, 196, 253
An'shan Group 29
Anhui
– region 225, 269
– Shanghai profile 146
Anning River fault 81
Anninghe-Yimen fault 226
Anshan
– complex 39, 41, 50
– Group 19, 21, 50
– iron ore field 151
– orogeny 51
Aolaishan body 27
apparent polar wandering path 241, 243
Ardeng Formation 185
Arksu Group 196

aulacogen 112, 116, 117, 152, 161, 163, 183, 217, 244, 246, 279, 296

B

Bacteriophyta 254,
Badaohe Group 21, 56
Badu Group 13
Bagongshan Group 181
Baihu Group 188
Baiyigou Group 211, 244
Bajiazhai rock body 250
Bali Formation 276
banded iron Formation (BIF) 7, 21, 22, 28, 33-36, 50, 51, 53, 57, 67, 72, 75, 150, 151, 299, 301
Banxi
– flysch 225
– Group 144, 214, 219, 220, 225, 246
Banyukou Formation 67, 71, 73
Baoban Group 226
Barberton greenstone belt 48
Bashikurgan Group 195
Bayan Obo
– aulacogen 187
– Group 182, 185
Bedong granodiorite 218
Bei'ao Formation 284
Beidahe Group 137-139
Beidajian Formation 173, 174
Beidashan Group 129, 131
Beishan
– depression 188, 281
– Group 136
– Mountains 132, 136
Beiyixi Formation 282
Bikou
– Group 209-211